About Island Press

Since 1984, the nonprofit Island Press has been stimulating, shaping, and communicating the ideas that are essential for solving environmental problems worldwide. With more than 800 titles in print and some 40 new releases each year, we are the nation's leading publisher on environmental issues. We identify innovative thinkers and emerging trends in the environmental field. We work with world-renowned experts and authors to develop cross-disciplinary solutions to environmental challenges.

Island Press designs and implements coordinated book publication campaigns in order to communicate our critical messages in print, in person, and online using the latest technologies, programs, and the media. Our goal: to reach targeted audiences—scientists, policymakers, environmental advocates, the media, and concerned citizens—who can and will take action to protect the plants and animals that enrich our world, the ecosystems we need to survive, the water we drink, and the air we breathe.

Island Press gratefully acknowledges the support of its work by the Agua Fund, Inc., The Margaret A. Cargill Foundation, Betsy and Jesse Fink Foundation, The William and Flora Hewlett Foundation, The Kresge Foundation, The Forrest and Frances Lattner Foundation, The Andrew W. Mellon Foundation, The Curtis and Edith Munson Foundation, The Overbrook Foundation, The David and Lucile Packard Foundation, The Summit Foundation, Trust for Architectural Easements, The Winslow Foundation, and other generous donors.

Hope Is an Imperative

PREVIOUS PUBLICATIONS BY DAVID W. ORR

Down to the Wire (Oxford, 2009)

Design on the Edge (MIT Press, 2006)

The Last Refuge (Island Press, 2004)

The Nature of Design (Oxford, 2002)

Earth in Mind (Island Press, 1994/2004)

Ecological Literacy (State University of New York Press, 1992)

Sage Handbook of Environment and Society (Sage Publications, 2007)
edited with Jules Pretty et al.

The Campus and Environmental Responsibility (Jossey-Bass, 1992)
edited with David Eagan

The Global Predicament: Ecological Perspectives on World Order
(University of North Carolina Press, 1979)
edited with Marvin Soroos

Hope Is an Imperative

The Essential David Orr

David W. Orr

Foreword by Fritjof Capra

ISLANDPRESS
Washington | Covelo | London

Hope Is an Imperative: The Essential David Orr

Grateful acknowledgment is made to all those who granted permission to reprint the selections in *Hope Is an Imperative*. A complete list of permissions can be found on page 353.

Library of Congress Cataloging-in-Publication Data
Orr, David W., 1944-
Hope is an imperative : the essential David Orr / David W. Orr ;
foreword by Fritjof Capra.
p. cm.
Includes bibliographical references and index.
ISBN-13: 978-1-59726-699-4 (cloth : alk. paper)
ISBN-10: 1-59726-699-X (cloth : alk. paper)
ISBN-13: 978-1-59726-700-7 (pbk. : alk. paper)
ISBN-10: 1-59726-700-7 (pbk. : alk. paper)
1. Environmental quality. 2. Sustainability. 3. Environmental education.
4. Human ecology. 5. Sustainable design. 6. Environmental engineering.
7. Global environmental change. 8. Climatic changes—
Environmental aspects. I. Title.
GE140.O77 2011
333.72—dc22

Printed on recycled, acid-free paper

Keywords: *Ecological design, sustainability, education, energy and climate, climate change, ecological literacy, environmental essays, environmental movement, environmental studies, green campus movement, energy policy, architectural design*

Manufactured in the United States of America

10 9 8 7 6 5 4 3 2 1

To David Ehrenfeld

Contents

PART 3: On Ecological Design

PART 4: On Education

PART 5: On Energy and Climate

Foreword

As THE TWENTY-FIRST CENTURY unfolds, it is becoming increasingly evident that the state of our natural environment is no longer one of many "single issues." It is the context of everything else—of our lives, our work, our politics.

The great challenge of our time is to build and nurture communities that are ecologically sustainable. As David Orr explains in this fine collection of essays, a sustainable community is designed in such a manner that its ways of life, businesses, economy, physical structures, and technologies respect, honor, and cooperate with nature's inherent ability to sustain life.

The first step in this endeavor must be to understand the basic principles of organization that the Earth's ecosystems have evolved over billions of years to sustain the web of life. We need to understand the language of nature, as it were—its flows and cycles, its networks and feedback loops, and its fluctuating patterns of growth and development. Almost 20 years ago, David Orr coined the term *ecological literacy* for this basic ecological knowledge and chose it as the title of his first book. Since then, ecological literacy, or *ecoliteracy*, has become a widely used concept within the environmental movement. Being ecologically literate means understanding the basic principles of ecology and living accordingly.

Ecoliteracy is the first step on the road to sustainability; the second step is ecodesign. We must apply our ecological knowledge to the fundamental redesign of our technologies, physical structures, and social institutions so as to bridge the current gap between human design and the ecologically sustainable systems of nature.

Design, in David Orr's memorable phrase, consists in "shaping flows of energy and materials for human purposes." Ecodesign, he writes, is "the careful meshing of human purposes with the larger patterns and flows of the natural world." Thus, ecodesign principles reflect the principles of organization that nature has evolved to sustain the web of life.

Once we become ecologically literate, once we understand the processes and patterns of relationships that enable ecosystems to sustain life, we will also understand the many ways in which our human civilization, especially since the Industrial Revolution, has ignored or interfered with these ecological patterns and processes. And we will realize that these interferences are the fundamental causes of many of our current world problems.

Because of the fundamental interconnectedness of the entire biosphere, the problems caused by our harmful interferences are also fundamentally interconnected. None of the major problems of our time can be understood in isolation. They are systemic problems—all interdependent and mutually reinforcing—and they require corresponding systemic solutions. Thinking systemically means thinking in terms of relationships, patterns, and context. To use a popular phrase, it means being able to "connect the dots."

The interconnectedness of world problems, the need to become ecologically literate, and the principles of ecodesign are three major strands that weave through the essays in this book. David Orr is a systemic thinker par excellence and a longtime friend and colleague whose thoughts and writings have influenced and inspired my own work for many years. With impeccable clarity, he demonstrates again and again that the current obsession of economists and politicians with unending growth is a fatal illusion; that our persistent failure to formulate sound energy policies has resulted in terrorism, oil wars, economic vulnerability, and climate change; that climate change is a challenge not only to consumerism and the growth economy but also to our political institutions, worldviews, and philosophies.

While all these systemic links are lucidly analyzed, David's writing is also deeply moving, thoughtful, and poetic. His intention is always to foster and expand our awareness of "the connections that bind us to each other, to all life, and to all life to come." I am really glad that Island Press is now publishing David's essential writings in one volume. They include many of his classic essays, which I have savored again and again over the years—for example, Place and Pedagogy, What Is Education

For?, Loving Children: A Design Problem, and my personal favorite, Slow Knowledge.

This book is both wake-up call and inspiration. Its trenchant analysis of the dire state of our world, combined with its passionate call to action, remind me of the famous maxim by the Italian political theorist Antonio Gramsci that we need both the pessimism of the intellect and the optimism of the will. Or, as David himself puts it in his introduction, "*Hope* is a verb with its sleeves rolled up."

These essays are eloquent and full of great wisdom. But what shines through them most of all is their author's deep passion for humanity and for the living Earth.

Fritjof Capra, founding director
Center for Ecoliteracy
Berkeley, California

Introduction

For bad electrical wiring, the sensible response is to call an electrician sooner than later. When sparks fly, the sensible thing to do is pull the breaker and reach for the fire extinguisher. When the house is on fire, the sensible thing to do is to call the fire department. But it would not be sensible to call the fire department when the problem is bad wiring or to call an electrician when the house is on fire. The word *sensible*, in other words, is relative to the gravity of the situation.

In the past quarter century, something analogous has happened to us as a nation and to the entire planet. Faced with the overwhelming evidence of environmental stresses, it would have been sensible decades ago to assemble the expertise necessary to redesign energy, food, materials, and manufacturing systems in order to eliminate waste and to coincide with laws of physics and ecology. As things worsened, it would have been sensible to develop global responses by aggressively implementing Agenda 21, the Rio Accords, the Kyoto Protocol, and more. Now, in the second decade of the twenty-first century, it would be sensible to recognize that we have squandered any margin of safety we once had and are in a planetary emergency and need to act accordingly. But there is no global equivalent of a 911 call and no intergalactic emergency squad to come to our rescue. It's up to us.

Meanwhile, as the years tick by, we are nearing (some say we have passed) irreversible and irrevocable changes in the oceans, atmosphere, soils, forests, and entire ecosystems. Now the sensible things we must do everywhere are merely extraordinary, unprecedented, and heroic at a scale sufficient to avert global catastrophe.

We are in the process of evicting ourselves from the only paradise humankind has ever known—what geologists call the Holocene. This 12,000-year age has been abnormally benign with a relatively stable and warm climate, more or less perfect for the emergence of *Homo sapiens*. But CO_2 levels are now higher than they've been in hundreds of thousands of years and rising still higher each year. We are creating a different and more capricious and hostile planet than the one we've known for thousands of years—what writer and activist Bill McKibben denotes as "Eaarth" (McKibben 2010). The challenge of living on this emerging planet is the challenge of our time, exempting no one, no organization, no nation, and no generation from here on as far as one can imagine.

The essays that follow, now the chapters of this book, were written between 1985 and 2010 as human civilization entered the historical equivalent of rapids on a white-water river. No one knows whether the frail craft of civilization will capsize because of climate destabilization, terrorism, economic collapse, technology run amuck, governmental ineptitude, or any number of other threats or whether it will somehow survive, chastened and hopefully improved. It is clear, however, that our previous unwillingness to do what was sensible, obvious, and necessary has now rendered our situation far more difficult and dangerous than it otherwise might have been.

Every writer works with the refracted influences of other people, places, and experiences. My interest in things environmental was enhanced by the great landscape architect Ian McHarg in the early 1970s when I was a doctoral student at the University of Pennsylvania. I was inspired to read everything I could find on the subject and discovered that the study of the "environment" came with an imperative to roam intellectually in order to connect things otherwise isolated by department, discipline, and narrow perspectives. But despite the great range and diversity of disciplines and perspectives necessary to an informed ecological worldview, the subject comes down to the one big question of how we fairly, durably, and quickly remake the human presence on Earth to fit the limits of the biosphere while preserving hard-won gains in the arts, sciences, law, the open society, and governance, which is to say civilization.

The urgency and excitement of that time was palpable. Some of the best thinking and writing ever about the human place in nature occurred in that decade. New nongovernmental organizations formed to defend particular places, ecologies, and the larger environment. The U.S. political system responded by creating the Environmental Protection Agency and the Council on Environmental Quality. Republicans and Democrats worked together to establish a National Environmental Policy and pass legislation to protect air, water, rivers, wilderness, open space, and endangered species. There were surely differences between Democrats and Republicans, but not paralysis, because there were still enough people in government with a sufficient regard for the issues that bind us together, to the web of life, and to all life to come to justify rethinking crusty old ideas and crossing party lines from time to time in order to protect the common good.

No road map existed then to define the path ahead, but by the late 1970s a global conversation about the sustainability of humankind was gathering steam. Many of us were optimistic that with enough science, better technology, and rational policy reforms, monumental problems could be solved. In hindsight it is obvious that things were not so simple, and neither are they today. Many factors come between what we should do and what we actually do, beginning with the daunting complexity of the problems and potential responses, whether market based or led by government or by cultural change, or all of the above. As well, we have to contend with competing political and economic interests that have become rigid ideologies rooted in tattered beliefs that humans can do as they please with nature without consequences. The stranglehold of bad ideas is deeply rooted often in the inability or unwillingness to see what's right before our eyes. And always the gap between what we should do and what we actually do is widened by ignorance, garden-variety stupidity, and the tendency to put off to tomorrow what should have been done yesterday. And lurking in the shadows there is the darker side of human nature that can't be wished away. But the fact remains that we know enough to act much better than we do. More science and better technology won't be nearly enough without a larger and more rational rationality. And even that won't suffice without summoning help from what Abraham Lincoln once called "the better angels of our nature."

The perplexities of human nature aside, we navigate between two rapidly flowing currents. Nothing in nature is static, but we have accelerated the pace of ecological change to a rate that rivals or exceeds that of the great extinction events of the distant past. The other current is the quickening pace of technological, demographic, social, and economic changes. In such unpredictable circumstances, no one can say for sure what it means for humankind to come to terms with nature, but we know that the road ahead will not be easy or smooth. Along the way, we will be tempted to do things that in less vexing times we would recognize as foolish or risky. We will be urged to deploy magic bullet technologies with vast implications without dealing with underlying problems or larger systemic issues. As long as civilization lasts, however, we will have to monitor and manage our demands and impacts on the planet and find widely acceptable and effective ways to limit what we do, whether by law, regulation, cultural norms, or religion or by some other means. We will also have to muster the wisdom to confront old and contentious issues having to do with the fair distribution of wealth and the balance of rights between generations and between humans and other life-forms. That in turn will require robust and competent governments and an ecologically literate and competent citizenry. However such things play out, we've long since passed the time when we could change atmospheric chemistry or the acidity of oceans, or unravel ecologies, or even procreate with little thought for the morrow or the health of the larger whole.

I have organized this collection of essays in five parts that reflect issues and subjects that caught my attention over the past 25 years. Most all of the essays were initially written as aids in solving one practical problem or another. Running through the entire book is the question of how humankind can fit harmoniously in the ecosphere—which invites controversy, multiple opinions, and lots of conjecture. I have only lightly edited the chapters to take out redundancies and update where necessary, so they are mostly as initially published but with fewer references and without footnotes.

The first part deals with fundamental principles. I'll let those essays speak for themselves without further comment and without any presumption that they are exhaustive or scriptural, including the one presumptuously titled "Orr's Laws." The second part, on the challenges of sustainability, is a bit like a brush-clearing operation that aims to get the lay of the land. However conceived, described, or analyzed, sustainability is *the* issue of our time, all others being subordinate to the global conversation now under way about whether, how, and under what terms the human experiment will continue.

The third part deals with possible responses to the challenges of sustainability. Most, if not all, of our environmental problems result from poor design—factories that produce more waste than product; buildings that squander energy; farms that bleed soil, excess nitrogen, and pollution; cities designed to sprawl; and so forth. The logical response, then, is better design or what is coming to be known as ecological design. It includes the design professions such as architecture and engineering but is a much bigger enterprise. It is quite literally about what McHarg described as "design with nature" in order to remake the human place on Earth. But the change toward ecological design in the fields of urban planning, agriculture, manufacturing, and energy systems as well as architecture will require a major change in how we think and so changes in education at all levels.

The fourth part, then, deals with education and specifically with the problem *of* education, not problems *in* education. Tinkering at the edge of the status quo characteristic of most educational reforms is a kind of nickel solution to a dollar-sized problem. But in the not-too-distant future, I can imagine schools, colleges, and universities designed ecologically, becoming models for the transition ahead and leaders toward a better future than the one now on the horizon. But that future is now clouded by the largest challenge humankind has ever faced, which is the onset of rapid climate destabilization.

The final part of the book is the most troubling of all and requires more explanation. We are, indeed, evicting ourselves from the very conditions in which we emerged as a species. Everything we've done—all of our accomplishments and failures, our arts, literatures, cultures, history, and organizations—occurred under, and partially because of, conditions that we are now changing for the worse. The increasing temperature of Earth, rising seas, extinction of species, changing hydrology, and shifting ecologies are effectively permanent changes that will render the future progressively more difficult for our descendants. That fact runs against the grain of the American tendency to regard problems as always solvable with enough technology or money. But the climate destabilization now under way is not solvable in that sense. We hope that the worst can be contained, but as geophysicist David Archer and others point out, we have already set planet-changing forces in motion that cannot be stabilized for centuries. If there was ever an issue that required clarity of mind, steadiness of purpose, and wisdom, this is it. I close with thoughts on the nature of hope in a progressively hotter and less stable ecosphere. But *hope* is a verb with its sleeves rolled up. In contrast to optimism or despair, hope requires that one actually do something to improve the world. Authentic hope comes with an imperative to act. There is no such thing as passive hope.

My thinking and writing have been much influenced by some of the great minds and personalities of our time. This book is gratefully dedicated to one such person, David Ehrenfeld, who is a physician, a biologist, a teacher, a renaissance man, and a friend and a teacher to me and many others. As the founding editor of *Conservation Biology*, David invited me to write many of the essays included here and helped to improve the results.

I also gratefully acknowledge Gary Meffe, who, like David Ehrenfeld, served with great distinction as editor of *Conservation Biology* and in that capacity improved the column I wrote for 20 years. Whatever clarity and felicity are evident in those essays included here owe much to David and Gary's skill, judgment, and, not the least, friendship.

For many years Wes Jackson has been a friend, provocateur, teacher, and a source of some of the best humor I've ever heard. His life has been one long seminar on soils, farming, civilization, philosophy, religion, ecology, literature, and more and how all of this is related. In long telephone conversations and visits to the Land Institute in Salina, I have been privileged to be a part of many of those mostly impromptu and brilliant sessions which I count mostly as a blessing, occasionally as an irritant, but always as convivial and often profound stimulation.

It is not possible to acknowledge Wes Jackson without saying an appreciative word about Wendell Berry. One of the most interesting and important dialogues of our time is that between Jackson and Berry, who over three decades have mutually influenced each other in a synergy of science, literature, good stories, friendship, inspiration, and devotion to land, agriculture, and rural people. Wendell Berry is described variously as a prophetic voice, one of the great writers of our time, and the wisest among us, all of which I believe to be true. For more than 40 years he has eloquently probed and defined our connections to land and community without ever being repetitive or tiresome. Above all, he has taught us the importance of words faithfully spoken and lived and our connectedness to places and real communities.

Finally, I thank Barbara Dean, Todd Baldwin, and Chuck Savitt at Island Press for their editorial help, advice, and friendship. And I gratefully confess to having been improved, instructed, inspired, sometimes chastised, but always nurtured by many others too numerous to list. But to all, my thanks for much that is beyond the saying.

Hope Is an Imperative

The Essential David Orr

PART I

The Fundamentals

⁖ AUTHOR'S NOTE 2010 ⁖

Having nearly boxed ourselves into a corner, we hear a growing chorus of folks proposing that we search for "breakthrough" solutions that consist mostly of expensive, untested, and risky technologies to do what, in saner times and with a bit of reflection, we'd prefer not to do. Talk about "geoengineering" the Earth is mounting, along with new proposals for perpetually bad, old, and highly subsidized ideas such as more nuclear power and clean coal—an oxymoronic, people-killing, and land-destroying absurdity. Absent in most of this is any thought that we might consider changing course or question why we should want a perpetually growing, consumer-driven economy in the first place, even if that were a biophysical possibility. We might do a great deal better with less stuff, less energy, less hassle, less frenzy, and more conviviality, more leisure, better poetry, more silence, slower food, more bike trails, and more face-to-face friends.

Rather than technological breakthroughs, what we need, I think, is more like a homecoming that requires fewer highly paid experts and consultants, fewer conferences in exotic and expensive places, and more local knowledge, a more competent and empowered citizenry, and more reflection on what's important and what's not. To some, that will sound quaint and old-fashioned, and I suppose that in some ways it is. So be it. I propose a return to fundamentals and offer a starting point for a few—having to do with words and language, how we know and what we presume to know, our fetish with speed, our capacity for love, the miracle of water, gratitude, and a summing-up that I've ostentatiously called "Orr's laws," which is just a distillation of lots of other people's wisdom—the most elusive and important fundamental of all.

Chapter 1

Verbicide

(1999)

In the beginning was the word. . . .

HE ENTERED MY OFFICE for advice as a freshman advisee sporting nearly perfect SAT scores and an impeccable academic record—by all accounts a young man of considerable promise. During a 20-minute conversation about his academic future, however, he displayed a vocabulary that consisted mostly of two words: *cool* and *really*. Almost 800 SAT points hitched to each word. To be fair, he could use them interchangeably, as in "really cool" or "cool . . . really!" He could also use them singly, presumably for emphasis. When he was a student in a subsequent class, I later confirmed that my first impression of the young scholar was largely accurate and that his vocabulary, and presumably his mind, consisted predominantly of words and images derived from overexposure to television and the new jargon of "computer-speak." He is no aberration, but an example of a larger problem, not of illiteracy but of diminished literacy in a culture that often sees little reason to use words carefully, however abundantly. Increasingly, student papers, from otherwise very good students, have whole paragraphs that sound like advertising copy. Whether students are talking or writing, a growing number of them have a tenuous grasp on a declining vocabulary. Excise "uh . . . like . . . uh" from virtually any teenage conversation, and the effect is like sticking a pin into a balloon.

This article was originally published in 1999.

In the past 50 years, by one reckoning, the working vocabulary of the average 14-year-old has declined from some 25,000 words to 10,000 words (Harper's Index 2000). This is a decline in not merely numbers of words but in the capacity to think. It is also a steep decline in the number of things that an adolescent needs to know and to name in order to get by in an increasingly homogenized and urbanized consumer society. This is a national tragedy virtually unnoticed in the media. It is no mere coincidence that in roughly the same half century, by one estimate, the average person has learned to recognize over 1000 corporate logos, but can now recognize fewer than 10 plants and animals native to their locality (Hawken 1994, 214). That fact says a great deal about why the decline in working vocabulary has gone unnoticed—few are paying attention. The decline is surely not consistent across the full range of language but concentrates in those areas having to do with large issues such as philosophy, religion, public policy, and nature. On the other hand, vocabulary has probably increased in areas having to do with sex, violence, recreation, consumption, and technology. Words like *twitter* and *google* have been appropriated or invented to describe entirely new ways to be illiterate and incoherent. As a result we are losing the capacity to say what we really mean and ultimately to think about what we mean. We are losing the capacity for articulate intelligence about the things that matter most. "That sucks," for example, is a common way for budding young scholars to announce their displeasure about any number of things that range across the spectrum of human experience. But it can also be used to indicate a general displeasure with the entire cosmos. Whatever the target, it is the linguistic equivalent of duct tape, useful for holding disparate thoughts in rough proximity to some vague emotion of dislike.

The problem is not confined to teenagers or young adults. It is part of a national epidemic of incoherence evident in our public discourse, street talk, movies, television, and music. We have all heard popular music that consisted mostly of pre-Neanderthal grunts. We have witnessed "conversation" on TV talk shows that would have embarrassed retarded chimpanzees. We have listened to many politicians of national reputation proudly and heatedly mangle logic and language in less than a paragraph, although they can do it on a larger scale as well. However manifested, it is aided and abetted by academics, including whole departments specializing in various forms of postmodernism and the deconstruction of one thing or another. Not so long ago they propounded ideas that everything was relative, hence largely inconsequential, and that the use of language was an exercise in power, hence to be devalued. They taught, in other words, a pseudointellectual contempt for clarity, careful argument, and felicitous expression. Being scholars of their word, they also wrote without clarity, argument, and felicity. Remove half a dozen arcane words from any number of academic papers written in the past 10 years, and the argument—whatever it was—evaporates. But the situation is not much better elsewhere in the academy, where thought is often fenced in by disciplinary jargon. The fact is that educators have all too often been indifferent trustees of language. This explains, I think, why the academy has been a lame critic of what ails the world, from the preoccupation with self to technology run amuck. We have been unable to speak out against the barbarism engulfing the larger culture because we are part of the process of barbarization that begins with the devaluation of language.

The decline of language, long lamented by commentators such as H. L. Mencken, George Orwell, William Safire, and Edwin Newman, is nothing new. Language is always coming undone. Why? For one thing it is always under assault by those who intend to control others by first seizing the words and metaphors by which people describe their world. The goal is to give partisan aims the appearance of inevitability by diminishing the sense of larger possibilities. In our time, language is under assault by those whose purpose it is to sell one kind of quackery or another: economic, political, religious, or technological. It is under attack because the clarity and felicity of language—as distinct from its quantity—is devalued in an industrial-technological society. The clear and artful use of language is, in fact, threatening to that society. As a result we have highly distorted and atrophied conversations about ultimate meanings, ethics, public purposes, or the means by which we live. Since we cannot solve problems that we cannot name, one result of our misuse of language is a growing agenda of unsolved problems that cannot be adequately described in words and metaphors derived from our own creations such as machines and computers.

Second, language is in decline because it is being balkanized around the specialized vocabularies characteristic of an increasingly specialized society. The highly technical language of the expert is, of course, both bane and blessing. It is useful for describing fragments of the world, but not for describing how these fit into a coherent whole. But things work as whole systems whether we can say it or not or whether we perceive it or not. And more than anything else, it is coherence our culture lacks, not specialized knowledge. Genetic engineering, for example, can be described as a technical thing in the language of molecular biology. But saying what the act of rearranging the genetic fabric of Earth means requires an altogether different language and a mind-set that seeks to discover larger patterns. Similarly, the specialized language of economics does not begin to describe the state of our well-being, whatever it reveals about how much stuff we may or may not possess. Regardless, over and over the language of the specialist trumps that of the generalist—the specialist in whole things. The result is that the capacity to think carefully about ends, as distinct from means, has all but disappeared from our public and private conversations.

Third, language reflects the range and depth of our experience. But our experience of the world is being impoverished to the extent that it is rendered artificial and prepackaged. Most of us no longer have the experience of skilled physical work on farms or in forests. Consequently words and metaphors based on intimate knowledge of soils, plants, trees, animals, landscapes, and rivers have declined. "Cut off from this source," Wendell Berry writes, "language becomes a paltry work of conscious purpose, at the service and the mercy of expedient aims" (Berry 1983, 33). Our experience of an increasingly uniform and ugly world is being engineered and shrink-wrapped by recreation and software industries and pedaled back to us as "fun" or "information." We've become a nation of television watchers, googlers, face bookers, text messengers, and twitterers, and it shows in the way we talk and what we talk about. More and more we speak as if we are voyeurs furtively peeking in on life, not active participants, moral agents, neighbors, friends, or engaged citizens.

Fourth, we are no longer held together, as we once were, by the reading of a common literature or by listening to great stories and so cannot draw on a common set of metaphors and images as we once did. Allusions to the Bible and great works of literature no longer resonate because they are simply unfamiliar to a growing number of people. This is so in part because the consensus about what is worth reading has come undone. But the debate about a worthy canon is hardly the whole story. The ability to read serious things with seriousness is diminished by overexposure to television and computers that overdevelop the visual sense. The desire to read is jeopardized by the same forces that would make us a violent, shallow, hedonistic, and materialistic people. As a nation we risk coming undone because our language is coming undone, and our language is coming undone because one by one we are being undone.

The problem of language, however, is a global problem. Of the roughly 5000 languages now spoken on Earth, only 150 or so are expected to survive to the year 2100. Language everywhere is being whittled down to the dimensions of the global economy and homogenized to accord with the imperatives of the "information age." This represents a huge loss of cultural information and a blurring of our capacity to understand the world and our place in it. And it represents a losing bet that a few people armed with the words, metaphors, and mindset characteristic of industry and technology that flourished destructively for a few decades can, in fact, manage the Earth, a different, more complex, and longer-lived thing altogether.

Because we cannot think clearly about what we cannot say clearly, the first casualty of linguistic incoherence is our ability to think well about many things. This is a reciprocal process. Language, George Orwell once wrote, "becomes ugly and inaccurate because our thoughts are foolish, but the slovenliness of our language makes it easier for us to have foolish thoughts" (Orwell 1981, 157). In our time the words and metaphors of the consumer economy are often a product of foolish thoughts as well as evidence of bad language. Under the onslaught of commercialization and technology, we are losing the sense of wholeness and time that is essential to a decent civilization. We are losing, in short, the capacity to articulate what is most important to us. And the new class of corporate chiefs, global managers, genetic engineers, and money speculators has no words with which to describe the fullness and beauty of life or to announce their role in the larger moral ecology. They have no metaphors by which they can say how we fit together in the community of life and so little idea beyond that of self-interest about why we ought to protect it. They have, in short, no language that will help humankind navigate through the most dangerous epoch in its history. On the contrary, they will do all in their power to reduce language to the level of utility, function, management, self-interest, and the short term. Evil begins not only with words used with malice; it can begin with words that merely diminish people, land, and life to some fragment that is less than whole and less than holy. The prospects for evil, I believe, will grow as those for language decline.

We have an affinity for language, and that capacity makes us human. When language is devalued, misused, or corrupted, so too are those who speak it and those who hear it. On the other hand, we are never better than when we use words clearly, eloquently, and civilly. Language does not merely reflect the relative clarity of mind; it can elevate thought and ennoble our behavior. Abraham Lincoln's words at Gettysburg in 1863, for example, gave meaning to the terrible sacrifices of the Civil War. Similarly, Winston Churchill's words moved an entire nation to do its duty in the dark hours of 1940. If we intend to protect and enhance our humanity, we must first decide to protect and enhance language and fight everything that undermines and cheapens it.

What does this mean in practical terms? How do we design language facility back into the culture? My first suggestion is to restore the habit of talking directly to each other—whatever the loss in economic efficiency. To that end I propose that we begin by smashing every device used to communicate in place of a real person, beginning with automated answering machines. Messages like "your call is important to us . . ." or "for more options, please press 5, or if you would like to talk to a real person, please stay on the line" are the death rattle of a coherent culture. Hell, yes, I want to talk to a real person, and preferably one who is competent and courteous!

My second suggestion is to restore the habit of public reading. One of my very distinctive childhood memories was attending a public reading of Shakespeare by the British actor Charles Laughton. With no prop other than a book, he read with energy and passion for 2 hours and kept a large audience enthralled, including at least one 8-year-old boy. No movie was ever as memorable to me. Further, I propose that adults should turn off the television, disconnect the cable, undo the computer, and once again read good books aloud to their children. I know of no better or more pleasurable way to stimulate thinking, encourage a love of language, and facilitate the child's ability to form images.

My third suggestion is that those who corrupt language ought to be held accountable for what they do—beginning with the advertising industry. Advertisers spend hundreds of billions to sell us an unconscionable amount of stuff, much of it useless, environmentally destructive, and deleterious to our health. They fuel the fires of consumerism that are consuming the Earth and our children's future. They regard the public with utter contempt—as little more than a herd of sheep to be manipulated to buy whatever at the highest possible cost and at any consequence. Dante would have consigned them to the lowest level of Hell, only because there was no worse place to put them. We should too. Barring that excellent idea, we should insist that they abide by community standards of truthfulness in selling what they peddle, including full disclosure of what the products do to the environment and to those who buy them.

AUTHOR'S NOTE 2010: *Upon reflection I would move peddlers of misinformation up to the next level and reserve the very basement of hell for those using the public airwaves to incite hate, intolerance, fear, and violence. I would condemn them to listen to their own broadcasts 24/7 for eternity.*

Fifth, language, I believe, grows from the outside in, from the periphery to center. It is renewed in the vernacular where human intentions intersect particular places and circumstances and by the everyday acts of authentic living and speaking. It is, by the same logic, corrupted by contrivance, pretense, and fakery. The center where power and wealth work by contrivance, pretense, and fakery does not create language so much as exploit it. In order to facilitate control, it would make our language as uniform and dull as the interstate highway system. Given its way, we would have only one newspaper, a super *USA Today*. Our thoughts and words would mirror those popular in Washington, New York, Boston, or Los Angeles. From the perspective of the center, the merger of entertainment companies with communications companies is OK because it can see no difference between entertainment and news. In order to preserve the vernacular places where language grows, we need to protect the independence of local newspapers and local radio stations. We need to protect local culture in all of its forms from the domination by national media, markets, and power. Understanding that cultural diversity and biological diversity are different faces of the same coin, we must protect those parts of our culture where memory, tradition, and devotion to place still exist.

Finally, since language is the only currency wherever men and women pursue truth, there should be no higher priority for schools, colleges, and universities than to defend the integrity and clarity of language in every way possible. We must instill in our students an appreciation for language, literature, and words well crafted and used to good ends. As teachers we should insist on good writing. We should assign books and readings that are well written. We should restore rhetoric—the ability to speak clearly and well—to the liberal arts curriculum. Our own speaking and writing ought to demonstrate clarity and truthfulness. And we, too, should be held accountable for what we say.

In terms of sheer volume of words, factoids, and data of all kinds, this is surely an information age. But in terms of understanding, wisdom, spiritual clarity, and civility, we have entered a darker age. We are drowning in a sea of words with nary a drop to drink. We are in the process of committing what C. S. Lewis once called "verbicide." The volume of words in our time is inversely related to our capacity to use them well and to think clearly about what they mean. It is no wonder that following a dreary century of gulags, genocide, global wars, and horrible weapons, our use of language has come to be dominated by propaganda and advertising, and controlled by language technicians. "We have a sense of evil," Susan Sontag has said, but we no longer have "the religious or philosophical language to talk intelligently about evil" (quoted in Miller 1998, 55). That being so for the twentieth century, what will be said at the end of the twenty-first century when the stark realities of climatic destabilization and biotic impoverishment will become fully apparent? Can we summon the clarity of mind to speak the words necessary to cause us to do what in hindsight will merely appear to have been obvious?

Chapter 2

Slow Knowledge

(1996)

There is no hurry, no hurry whatsoever.
ERWIN CHARGAFF

It takes all the running you can do, to keep in the same place.
LEWIS CARROLL

BETWEEN 1978 AND 1984, the Asian Development Bank spent $24 million to improve agriculture on the island of Bali. The target for improvement was an ancient agricultural system organized around 173 village cooperatives linked by a network of temples operated by "water priests" working in service to the water goddess, Dewi Danu, a deity seldom included in the heavenly pantheon of development economists. Not surprisingly, the new plan called for large capital investment to build dams and canals, and to purchase pesticides and fertilizers. The plan also included efforts to make idle resources, both the Balinese and their land, productive year-round. Old practices of fallowing were ended, along with community celebrations and rituals. The results were remarkable but inconvenient: yields declined, pests proliferated, and the village society began to unravel. On later examination (Lansing 1991), it turns out that the priests' role in the religion of Agama Tirtha was that of ecological master planners whose task it was to keep a finely tuned system operating productively. Western development experts dismantled a system that had worked well for more than a millennium and replaced

This article was originally published in 1996.

it with something that did not work at all. The priests have reportedly resumed control.

The story is a parable for much of modern history in which increasingly homogenized knowledge is acquired and used more rapidly and on a larger scale than ever before and often with disastrous and unforeseeable consequences. The twentieth century is the age of fast knowledge driven by rapid technological change and the rise of the global economy. This has undermined communities, cultures, and religions that once slowed the rate of change and filtered appropriate knowledge from the cacophony of new information.

The culture of fast knowledge rests on assumptions that

- only that which can be measured is true knowledge;
- the more of it we have the better;
- knowledge that lends itself to use is superior to that which is merely contemplative;
- the scale of effects of applied knowledge is unimportant;
- there are no significant distinctions between information and knowledge;
- wisdom is indefinable, hence unimportant;
- there are no limits to our ability to assimilate growing mountains of information, and none to our ability to separate essential knowledge from that which is trivial or even dangerous;
- we will be able to retrieve the right bit of knowledge at the right time and fit it into its proper social, ecological, ethical, and economic context;
- we will not forget old knowledge, but if we do, the new will be better than the old;
- whatever mistakes and blunders occur along the way can be fixed by yet more knowledge;
- the level of human ingenuity will remain high;
- the acquisition of knowledge carries with it no obligation to see that it is responsibly used;
- the generation of knowledge can be separated from its application;
- all knowledge is general in nature, not specific to or limited by particular places, times, and circumstances.

Fast knowledge is now widely believed to represent the very essence of human progress. While many acknowledge the problems caused by the accumulation of knowledge, most believe that we have little choice but to keep on. After all, it's just human nature to be inquisitive. Moreover, research on new weapons and that undertaken for new corporate products is justified on the grounds that if we don't do it, someone else will and so we must. Others, of course, operate on identical assumptions. And increasingly, fast knowledge is justified on purportedly humanitarian grounds that we must hurry the pace of research in order to meet the needs of a growing population.

Fast knowledge has a lot going for it. Because it is effective and powerful, it is reshaping education, communications, communities, cultures, lifestyles, transportation, economies, weapons development, and politics. For those at the top of the information society, it is also exhilarating, perhaps intoxicating, and for the few at the very top, it is highly profitable.

The increasing velocity of knowledge is widely accepted as sure evidence of human mastery and progress. But many, if not most, of the ecological, economic, social, and psychological ailments that beset contemporary society can be attributed directly or indirectly to knowledge acquired and applied before we had time to think it through carefully. We rushed into the fossil fuel age only to discover the giant problem of climate destabilization. We rushed to develop nuclear energy without the faintest idea of what to do with the radioactive wastes. Nuclear weapons were created before we had time to ponder their full implications. Knowledge of how to kill more efficiently is rushed from research to application without much question about its effects on the perceptions and behavior of others, about its effects on our own behavior, or about better and cheaper ways to achieve real security. CFCs and a host of carcinogenic, mutagenic, and hormone-disrupting chemicals, too, are products of fast knowledge. High-input, energy-intensive agriculture is also a product of knowledge applied before much consideration has been given to its full ecological and social costs. Economic growth, in large measure, is driven by fast knowledge, with results everywhere evident in mounting environmental problems, social disintegration, unnecessary costs, and injustice.

Fast knowledge undermines long-term sustainability for two fundamental reasons. First, for all of the hype about the information age and the speed at which humans are purported to learn, the facts say that our collective learning rate is about what it has always been: rather slow. Many decades after their deaths we have scarcely begun to fathom the full meaning of Gandhi's ideas about nonviolence or that of Aldo Leopold's "land ethic." A century and a half after *The Origin of Species*, apparently over half of the American public still doesn't believe in evolution and the rest are still struggling to comprehend its full implications. And several millennia after Moses, Jesus, and Buddha we are about as spiritually inept as ever. The problem is that the rate at which we collectively learn and assimilate new ideas has little to do with the speed of our communications technology or with the volume of information available to us, but it has everything to do with human limitations and those of our social, economic, and political institutions. Indeed, the slowness of our learning—or at least of our willingness to change—may itself be an evolved adaptation; short-circuiting this limitation reduces our fitness.

Even were humans able to learn more rapidly, the application of fast knowledge generates complicated problems much faster than we can identify them and respond. We simply cannot foresee all of the ways complex natural systems will react to human-initiated changes at their present scale, scope, and velocity. The organization of knowledge by a minute division of labor further limits our capacity to comprehend whole systems effects, especially when the creation of fast knowledge in one area creates problems elsewhere at a later time. Consequently, we are playing catch-up but falling farther and farther behind. Finally, for reasons once described by the historian of science Thomas Kuhn, fast knowledge creates power structures whose function it is to hold at bay alternative paradigms and worldviews that might slow the clockspeed of change to manageable rates. The result is that the system of fast knowledge creates social traps in which the benefits occur in the near term while the costs are deferred to others at a later time.

The fact is that the only knowledge we've ever been able to count on for consistently good effect over the long run is knowledge that has been acquired slowly through cultural maturation. Slow knowledge is knowledge shaped and calibrated to fit a particular ecological and cultural context. It does not imply lethargy, but rather thoroughness and patience. The aim of slow knowledge is resilience, harmony, and the preservation of what Gregory Bateson once called the "patterns that connect." Evolution is the archetypal example of slow knowledge. Except for rare episodes of punctuated equilibrium, evolution seems to work by the slow trial-and-error testing of small changes. Nature seldom, if ever, bets it all on a single throw of the dice. Similarly every human culture that has artfully adapted itself to the challenges and opportunities of a particular landscape has done so by the patient and painstaking accumulation of knowledge over many generations: what the English writer George Sturt once described as the "age-long effort to fit close and ever closer" into a particular place. Unlike fast knowledge generated in universities, think tanks, and corporations, slow knowledge occurs incrementally through the process of community learning motivated more by affection than by idle curiosity, greed, or ambition.

The worldview inherent in slow knowledge rests on beliefs that

- wisdom, not cleverness, is the proper aim of all true learning;
- the velocity of knowledge is inversely related to the acquisition of wisdom;
- the careless application of knowledge can destroy the conditions that permit knowledge of any kind to flourish (a nuclear war, for example, made possible by the study of physics, would be detrimental to the further study of physics);
- what ails us has less to do with the lack of knowledge than with too much irrelevant knowledge and the difficulty of assimilation, retrieval, and application, as well as the lack of compassion and good judgment;
- the rising volume of knowledge cannot compensate for a rising volume of errors caused by malfeasance and stupidity generated in large part by inappropriate knowledge;
- the good character of knowledge creators is not irrelevant to the truth they intend to advance and its wider effects;
- human ignorance is not an entirely solvable problem; it is, rather, an inescapable part of the human condition.

The differences between fast knowledge and slow knowledge could not be more striking. Fast knowledge is focused on solving problems, usually by one technological fix or another; slow knowledge has to do with avoiding problems in the first place. Fast knowledge deals with discrete things, while slow knowledge deals with context, patterns, and connections. Fast knowledge arises from hierarchy and competition; slow knowledge is freely shared within a community. Fast knowledge is about know-how; slow knowledge is about know-how and know-why. Fast knowledge is about "competitive edges" and individual and organizational profit; slow knowledge is about community prosperity. Fast knowledge is mostly linear; slow knowledge is complex and ecological. Fast knowledge is characterized by power and instability; slow knowledge is known by its elegance, complexity, and resilience. Fast knowledge is often regarded as private property; slow knowledge is owned by no one. In the culture of fast knowledge, "man is the measure of all things." Slow knowledge, in contrast, occurs as a coevolutionary process among humans, other species, and a shared habitat. Fast knowledge is often abstract and theoretical, engaging only a portion of the mind. Slow knowledge, in contrast, engages all of the senses and the full range of our mental and spiritual powers. Fast knowledge is always new; slow knowledge often is very old. The besetting sin inherent in fast knowledge is hubris, the belief in human omnipotence now evident on a global scale. That of slow knowledge can be parochialism and resistance to needed change.

Are there occasions when we need fast knowledge? Yes, but with the caveat that a significant percentage of the problems we now attempt to solve quickly through complex and increasingly expensive means have their origins in the prior applications of fast knowledge. Solutions to such problems often resemble a kind of Rube Goldberg contraption that produces complicated, expensive, and often temporary cures for otherwise unnecessary problems. The point, as every accountant knows, is that there is a difference between gross and net. And after all of the costs of fast knowledge are subtracted, the net gains in many fields have been considerably less than we have been led to believe.

What can be done? Until the sources of power that fuel fast knowledge run dry, perhaps not a thing. Then again, maybe we are not quite so helpless. The problem is clear: we need no more fast knowledge cut off from its ecological and social context—"ignorant knowledge," in Erwin Chargaff's words (Chargaff 1980). In principle, the solution is equally clear: we need to discover and sometimes rediscover the knowledge of things like how the Earth works, how to build sustainable and sustaining communities that fit their regions, how to raise and educate children to be decent people, and how to provision ourselves justly and within ecological limits. We need to remember all of those things necessary to re-member a world fractured by competition, fear, greed, and shortsightedness. If there is no quick cure, neither are we without the wherewithal to create a better balance between the real needs of society and the pace and kind of knowledge generated. For colleges and universities, in particular, I propose the following steps aimed to improve the quality of knowledge by slowing its acquisition to a more manageable rate.

First, scholars ought to be encouraged to include practitioners and those affected in setting priorities and standards for the acquisition of knowledge. Professionalized knowledge is increasingly isolated from the needs of real people and, to that extent, dangerous to our larger prospects. It makes no sense to rail about participation in the political and social affairs of the community and nation while allowing the purveyors of fast knowledge to determine the actual conditions in which we live without so much as a whimper. Knowledge has social, economic, political, and ecological consequences as surely as any act of Congress, and we ought to demand representation in the setting of research agendas for the same reason that we demand it in matters of taxation. Inclusiveness would slow research to more manageable rates while improving its quality. And there are good examples of participatory research involving practitioners in agriculture, forestry (Banuri and Marglin 1993), land use (Appalachian Land Ownership Task Force 1983), and many others.

Second, faculty ought to be encouraged in every way possible to take the time necessary to broaden their research and scholarship to include its ecological, ethical, and social context. They ought to be encouraged to rediscover old and true knowledge and to respect old and hard-won wisdom. And colleges and universities could do much more to encourage and reward efforts by their faculty to teach well and to apply existing knowledge to solve real problems in their communities.

Third, colleges and universities ought to foster a genuine and ongoing debate about the velocity of knowledge and its effects on our larger prospects. We bought in to the ideology of faster is better without taking the time to think it through. Increasingly, we communicate electronically by e-mail, iPhone, and BlackBerry. As a consequence, one can detect a decline in the salience of our communication, and perhaps in its civility as well, in direct proportion to its velocity and volume. It is certainly possible to detect a growing frustration with the rising deluge of electronic messages that makes it increasingly difficult to separate chaff from grain. For comparison, consider the collected correspondence of, say, Thomas Jefferson and James Madison, letters written slowly by quill pen, perhaps by candlelight, delivered by horse, and still full of magic and power two centuries later. Would that magic and power be present still had Jefferson and Madison twittered, blogged, or e-mailed? I doubt it.

Fast knowledge has played havoc in the world because *Homo sapiens* is just not smart enough to manage everything that it is possible for the human mind to discover and create. In Wendell Berry's words, there is a kind of idiocy inherent in the belief "that we can first set demons at large, and then, somehow, become smart enough to control them" (Berry 1983, 65).

Slow knowledge really isn't slow at all. It is knowledge acquired and applied as rapidly as humans can comprehend it and put it to consistently good use. Given the complexity of the world and the depth of our human frailties, this takes time and it always will. Mere information can be transmitted and used quickly, but new knowledge is something else. Often it requires rearranging worldviews and paradigms, which we can only do slowly. Instead of increasing the speed of our chatter, we need to learn to listen more attentively. Instead of increasing the volume of our communication, we ought to improve its content. Instead of communicating more extensively, we should converse more intensively with our neighbors without the help of any technology whatsoever. "There is no hurry, no hurry whatsoever."

Chapter 3

Speed

(1998)

But is the nature of civilization "speed"? Or is it "consideration"? Any animal can rush around a corral four times a day. Only a human being can consciously oblige himself to go slowly in order to consider whether he is doing the right thing, doing it the right way, or ought in fact to be doing something else. . . . Speed and efficiency are not in themselves signs of intelligence or capability or correctness.

JOHN RALSTON SAUL 1993, 259

Water

PLUM CREEK BEGINS in drainage from farms on the west side of the city of Oberlin, Ohio, and flows eastward through a city golf course, a college arboretum, and the downtown area. East of the city, the stream receives the effluent from the city sewer facility before it joins with the Black River, which flows north through two rust-belt cities, Elyria and Lorain, before emptying into Lake Erie 25 miles west of Cleveland. Plum Creek shows all of the signs of 150 years of human use and abuse. As late as 1850 the stream ran clear even in times of flood, but now it is murky brown year-round. Because of pollution, sediments, and the lack of aquatic life, the Environmental Protection Agency considers it to be a "non-attainment" stream. Yet it survives, more or less. To most residents of Oberlin, Plum Creek is little more than a drain and sewer useful for moving water off the land as rapidly as possible. Few regard it as an esthetic asset or ecological resource.

This article was originally published in 1998.

The character of Plum Creek changes quickly as it flows eastward into the downtown. Runoff from city streets enters the stream where the creek runs under the intersection of Morgan and Professor streets. One block to the east, a larger volume of runoff polluted by oil and grease from city streets enters the creek as it flows under Main Street, past a Midas Muffler shop, a NAPA AUTO PARTS store, and city hall, located in the floodplain. Where Plum Creek flows under Main Street, an increased volume of storm water and consequently increased stream velocity have widened the banks and cut the channel from several feet to a depth of 10 feet or more. The city has attempted to stabilize the stream by lining the banks with concrete or by riprapping with large chunks of broken concrete. The aquatic life that exists upstream mostly disappears as Plum Creek flows through the downtown. Bending to the northeast, the creek passes through suburban backyards, past the municipal wastewater plant and a landfill, and on toward the west fork of the Black River and Lake Erie.

Whatever Plum Creek once was, it is now fundamentally shaped by the fact that European settlers cut the forests and drained marshes which once absorbed rainfall and released water slowly throughout the year. The wetlands and forests that once made up the floodplain are now mostly gone, replaced by roads, lawns, buildings, and parking lots. Rainfall is quickly channeled from lawns, streets, and parking lots into storm drains and culverts and diverted into the creek. The result is a landscape that sheds water quickly, contributing to floods, reducing water quality, and degrading aquatic habitats. Mathematics tells the story: doubling the speed of water increases the size of soil particles transported by 64 times.

The history of the Plum Creek watershed is not unusual. Over 90 percent of Ohio wetlands have been drained. As a nation, we have lost over 50 percent of the wetlands that existed before European settlement, and despite federal laws we continue to lose wetlands at a rate of thousands of acres each year. The total paved area in the lower 48 states is equivalent to a land area larger than Kentucky. As a result water moves more quickly across our landscapes than it once did, so flooding, particularly downstream from urban areas, is more common and more severe than ever. Measured in constant dollars, floodplain damage has risen by 50 percent in the 15 years between 1975 and 1990. We labor in vain to control flooding and prevent flood damage by the heroic engineering of dams, levees, and diversion channels while continuing to clear forests, drain wetlands, and pave. The results shown in floods on the Mississippi, Missouri, and

Ohio rivers in recent years are now part of the escalating price we pay for engineering, as if the velocity of water moving through the landscape did not matter.

Money

The city of Oberlin is a fairly typical Midwestern college town with a square around which are arrayed college buildings, a historic church, an art museum, a hotel, and downtown businesses, including three banks, two bookstores, a bakery, a five-and-dime store, an Army Navy Store, an assortment of restaurants, a gourmet coffee shop and pizza parlors and one hardware store. Most struggle to survive, and in the past 6 years, the downtown lost, among other businesses, a car dealership, a drugstore, a bicycle repair shop, a travel office, and an appliance store. Going back even further, the economic changes are more striking. Older residents remember the six grocery stores that would deliver to your home, local dairies that delivered milk in glass bottles, trolleys, and a train station. All that changed after World War II. A large mall with the standard assortment of national merchants located 10 miles away now drains off the largest part of what had once been mostly local business. Going south out of town, new development in Oberlin begins unsurprisingly with a McDonald's and a chain drug store. Further on, a Pizza Hut newly relocated from the downtown has opened beside a large discount store and a Wal-Mart, with more strip development on the way. If this sounds familiar, it should. It is the American pattern of automobile-driven development by which capital moves from older downtowns to the periphery where land is cheaper and zoning regulations are lax.

Despite the fact that the city includes a well-endowed college, a vocational school, an FAA air traffic control center, and several growing businesses located in an industrial park, 38 percent of the residents of Oberlin are estimated to fall below the poverty line. Money does not stay in the local economy for long. Most of the salary and wages paid in Oberlin quickly exits the city economy. Hence the multiplier effect, or the number of times a dollar is spent and re-spent locally before being used to purchase something outside, is low.

By contrast, 55 miles to the south in the Amish economy of Holmes County, the economic multiplier would be very high and unemployment and poverty virtually nonexistent. The Amish buy and sell from each other.

They make their own tools, farm implements, and furniture. They grow a large percentage of their food, much of which they process themselves so that the value is added locally. Their expenditures for fuel, health care, consumer goods, luxury items, and expensive things like cars or retirement is low to zero. They have their own insurance system, which to a great extent consists of the applied arts of neighborliness toward those in need. They accept neither welfare nor social security. The contrast between the Oberlin economy and that of the Amish could hardly be greater.

An Amish friend of mine recently told me that "the horse is the salvation of the Amish society." The Amish culture operates at the clock-speed of the horse and the sun. Because they farm with horses, they aren't tempted to farm large amounts of land. Farming with horses, in other words, serves as a brake to the temptation to take over a neighbor's land. And because the effective radius of a horse-drawn buggy is about 8 miles and its hauling capacity is low, the Amish are not much tempted by consumerism at the local mall. But horse speed does more. It slows the velocity of work to a pace that allows close observation of soils, wildlife, and plants. He often uses only a walking plow, which he believes preserves soil biota and prevents erosion. The natural pace of the horse, in other words, allows the Amish to pay attention to the "minute particulars" of their farm and how they farm. By a similar logic, he waits to cut hay until the bobolinks in the field have fledged. The loss in protein content in the hay, he believes, is more than compensated by the health of the place and the pleasure derived from having birds on the farm.

The capital tied up in an Amish farm is mostly in land and buildings, not in equipment. Their cash flow seldom goes to banks or vendors of petrochemicals and fossil fuels. It is not bundled as collateralized debt obligations and peddled in the global market. It is small wonder that Amish farms typically thrive while nearly 5 million non-Amish farms have disappeared in the past 60 years.

Information

Most of us are now indentured to e-mail, Facebook, twittering, and BlackBerries, all said to improve our efficiency and the ties that bind. But the results, at best, are mixed. The most obvious is a large increase in the sheer volume of stuff communicated, much of it utterly trivial. There is also, as noted above, a manifest decline in the grammar, literary style, and civility of communication. I believe that we often talk to each other less than before. Students remain transfixed before computer screens for hours, often doing no more than playing computer games. Our conversations, thoughts, and institutional routines are increasingly shaped to fit the imperatives of technology. Not surprisingly, more and more people feel overloaded by incessant "communication." But to say so publicly is to run afoul of the technological fundamentalism now dominant virtually everywhere.

By default and without much thought, it has been decided (or decided for us) that communication ought to be cheap, easy, and quick. Accordingly, more and more of us are instantly wired to the global nervous system with cell phones, beepers, pagers, fax machines, BlackBerries, iPhones, and e-mail. If useful in real emergencies, the overall result is to confuse the important with the trivial, making everything an emergency and an already frenetic civilization even more frenetic. As a result we are drowning in unassimilated information, most of which fits no meaningful picture of the world. In our public affairs and in our private lives we are, I think, increasingly muddleheaded because we have mistaken volume and speed of information for substance and clarity.

It is time to consider the possibility that for the most part, communication ought to be somewhat slower, more difficult, and more expensive than it is now. Beyond some relatively low threshold, the rapid movement of information works against the emergence of knowledge, which requires the time to mull things over, to test results, and, when warranted, to change perceptions and behavior. The clockspeed of genuine wisdom, which requires the integration of many different levels of knowledge, is slower still. Only over generations through a process of trial and error can knowledge eventually congeal into cultural wisdom about the art of living well within the resources, assets, and limits of a place.

Synthesis

Water moving too quickly through a landscape does not recharge underground aquifers. The results are floods in wet weather and droughts in the summer. Money moving too quickly through an economy does not recharge the local wellsprings of prosperity, whatever else it does for the global economy. The result is an economy polarized between those few who do well in a high-velocity economy and the majority left behind. Information moving too quickly to become knowledge and grow into wisdom does not recharge moral aquifers on which families, communities, and entire nations depend. The results are moral atrophy and public confusion. The common thread between all three is velocity. And they are tied together in a complex system of cause and effect, mostly overlooked.

There is an appropriate velocity for water set by geology, soils, vegetation, and ecological relationships in a given landscape. There is an appropriate velocity for money that corresponds to long-term needs of whole communities rooted in particular places and the necessity of preserving ecological capital. There is an appropriate velocity for information, set by the assimilative capacity of the mind and by the collective learning rate of communities and entire societies. Having exceeded the speed limits, we are vulnerable to ecological degradation, economic arrangements that are unjust and unsustainable, and, in the face of great and complex problems, to befuddlement that comes with information overload.

The ecological impacts of increased velocity of water are easy to comprehend. We can see floods, and with a bit of effort, we can discern how human actions can amplify droughts. But it is harder to comprehend the social, political, economic, and ecological effects of increasing velocity of money and information, which are often indirect and hidden. Increasing velocity of commerce, information, and transport, however, requires more administration and regulation of human affairs to ameliorate congestion and other problems. More administration means that there are fewer productive people, higher overhead, and higher taxes to pay for more infrastructure necessitated by the speed of people and things and problems of congestion. Increasing velocity and scale tends to increase the complexity of social and ecological arrangements and reduce the time available to recognize and avoid problems. Cures for problems caused by increasing velocity often set in motion a cascading series of other problems. As a result we stumble through a succession of escalating crises with diminishing capacity to foresee and forestall. Other examples fit the same pattern, such as the velocity of transportation, material flows, extraction of nonrenewable resources, introduction of new chemicals, and human reproduction. At the local scale, the effect is widening circles of disintegration and social disorder. At the global scale, the rate of change caused by increasing velocity disrupts biological evolution and the biogeochemical cycles of the Earth.

The increasing velocity of the global culture is no accident. It is the foundation of the corporate-dominated global economy that requires quick returns on investment and the obsession with rapid economic growth. It is the soul of the consumer economy that feeds on impulse, obsession, and instant gratification. The velocity of water in our landscape is a direct result of too many automobiles, too much paving, sprawling development, deforestation, and a food system that cannot be sustained in any decent or safe manner. The speed of information is driven by something that more and more resembles addiction. But above all, increasing speed is driven by minds seemingly unaware that the race is not, and has never been, to the swift.

Upshot

We are now engaged in a great global debate about how we might lengthen our tenure on the Earth. The discussion is mostly confined to options having to do with better technology, more accurate resource prices, and smarter public policies—all of which are eminently sensible, but hardly sufficient. The problem is simply how a species pleased to call itself *Homo sapiens* fits on a planet with a biosphere. This is a design problem and requires a design philosophy that takes time, velocity, scale, evolution, and ecology seriously. We will neither conserve biotic resources nor build a sustainable civilization that operates at our present velocity. But here's the rub:

The same forces that have combined to give us a high-velocity economy and society reform themselves at a glacial pace. The ideas that we need to build a sustainable civilization need to be widely disseminated and put into practice quickly. But results from our efforts to stabilize the climate, preserve biological diversity, reform agriculture policy and practices, deploy renewable energy technologies, reestablish natural systems agriculture, and reform our democracy occur ever so slowly if at all. What can be done?

First, we need a relentless analytical clarity to discern the huge inefficiencies of high-speed "efficiency." We have contrived a high-technology, high-speed economy that is neither sustainable nor capable of sustaining what is best in human cultures. On close examination, many of the alleged benefits of ever-rising affluence are fraudulent. Thoughtful analysis reveals that our economy often works like a kind of Rube Goldberg device, doing with great expense, complication, and waste what could be done more simply, elegantly, and harmoniously or, in some cases, what should not be done at all. Most of our mistakes were a result of hurry in the name of economic competition or national security or "progress." Now many must be expensively undone or written off as permanent losses. The clockspeed of the industrial economy must be reset to take account of evolution, natural rhythms, and genuine human needs. That means recalibrating public policies and taxation to promote a more durable prosperity.

Next, we need a more robust idea of time and scale that takes the health of people and communities seriously. In Leopold Kohr's words,

> the only way that can induce us to reduce our speed of movements is a return to a spatially more contracted, leisurely, and largely pedestrian mode of life that makes high speeds not only unnecessary but as uneconomic as a Concord would be for crossing the English Channel. . . . In other words, slow is beautiful in an appropriately contracted small social environment of beehive density and animation not only from a political and economic but, in the most literal sense, also from an aesthetic point of view, releasing an abundance of long abandoned energy not by patriotically making us drive slowly, but by depriving us materially of the need for driving fast. (Kohr 1980, 58)

Our assumptions about time are crystallized in community design and architecture. Sprawling cities, economic dependency, and long-distance transport of food and materials require high-velocity transport and high-speed communication and result in higher costs, community disintegration, and ecological deformation. Rethinking velocity and time will require rethinking our relationship to the land as well. Here too we have options for increasing density through "open space development" and smarter planning that creates proximity between housing, employment, shopping, culture, public spaces, recreation, and health care—what is now being called the "new urbanism."

Finally, in a society in which people sometimes talk about "killing time," we must learn rather to take time. We must learn to take time to study nature as the standard for much of what we need to do. We must take time and make the effort to preserve both cultural and biological diversity. We must take time to calculate the full costs of what we do. We must take time to make things durable, repairable, useful, and beautiful. We must take the time, not just to recycle, but rather to eliminate the very concept of waste. In most things, timeliness and regularity, not speed, are important. Genuine charity, good parenting, true neighborliness, good lives, decent communities, conviviality, democratic deliberation, real prosperity, mental health, and the exercise of true intelligence have a certain pace and rhythm that can only be harmed by acceleration. The means to control velocity can be designed into daily life like speed bumps designed to slow auto traffic. Holidays, festivals, celebrations, sabbaticals, Sabbaths, prayer, good conversation, storytelling, music making, the practice of fallowing, shared meals, a high degree of self-reliance, craftwork, walking, and shared physical work are speed-control devices used by every healthy culture.

Chapter 4

Love

(1992)

We cannot win this battle to save species and environments without forging an emotional bond between ourselves and nature as well—for we will not fight to save what we do not love.
STEPHEN J. GOULD

STEPHEN JAY GOULD's is one view of the issue and, in most academic disciplines, a decidedly minority view. Mainstream scholars who trouble themselves to think about disappearing species and shattered environments appear to believe that cold rationality, fearless objectivity, and a bit of technology will get the job done. If that were the whole of it, however, the job would have been done decades ago. Except as pejoratives, words such as *emotional bonds*, *fight*, and *love* are not typical of polite discourse in the sciences or social sciences. To the contrary, excessive emotion about the object of one's study is in some institutions a sufficient reason to banish the miscreant to the black hole of committee duty or administration, on the grounds that good science and emotion of any sort are incompatible, a kind of Presbyterian view of science.

Gould's view raises a number of questions. For example, how is it possible to reconcile the procedural requirements of science with those of emotional bonding and love for the creation, a process that Gould believes to be necessary to save species and environments? Do these work at cross-

This article was originally published in 1992.

purposes? Do they entail different curricula and different kinds of education from that now offered almost everywhere? For survival purposes, must curiosity be tempered by affection? Might the same be equally true for physics and economics? Assuming that we could define love satisfactorily, would it set limits on knowledge or on the way in which knowledge is acquired? In any conflict between the requirements of love and those of knowledge, which should have priority? Inherent in Gould's view is the paradox that we must learn to act selflessly, as sages and religious prophets have said all along, but for reasons that are increasingly indistinguishable from our self-interest in survival.

These are important and difficult issues on which reasonable people can disagree, but the stakes are high enough to warrant serious and sustained discussion about love in relation to education and knowledge. But hardly a whisper of that debate is evident. In a cursory survey of indexes to biology texts, for example, I found no mention of the word *love*. Neither did I find any entry under *emotional bonding* or *fight*, for that matter. The same was true of other textbooks in physics, chemistry, political science, and economics. Why is it so hard to talk about love, the most powerful of human emotions, in relation to science, the most powerful and far-reaching of human activities? And why is this so for textbooks written to introduce the young to the disciplined study of life and life processes? An introduction would appear to be a good point at which to say a few words about love, awe, and mystery and perhaps a caution or two about the responsibilities that go with knowledge. This might even be a good place to discuss emotions in relation to intellect and how best to join the two, because they are joined in one way or another. It is as if there were a conspiracy of silence to hide what drives the effort to acquire knowledge. Perhaps it is only embarrassment about what does or does not move us personally.

There are other reasons, as well, why we have failed to think much about the issue of love in relation to science and education. Undoubtedly, the loudest objection to any such discussion will be made by the more rigorous than thou, the academic equivalent of the fundamentalist, who will argue that science works inversely to passion, their own passion for pure science notwithstanding. But in psychologist Abraham Maslow's words,

> science and everything scientific can be and often is used as a tool in the service of a distorted, narrowed, humorless, de-eroticized, de-emotionalized,

> desacralized and desanctified Weltanschauung. This desacralization can be used as a defense against being flooded by emotion, especially the emotions of humility, reverence, mystery, wonder, and awe. (Maslow 1966, 139)

Fundamentalists have mistaken the relation among passion, emotion, and good science. These are not antithetical, but complexly interdependent. Science, at its best, is driven by passion and emotion. We have emotions for the same reason we have arms and legs: they have proved to be useful over evolutionary time. The point in either case is not to cut off various appendages and qualities, but rather to learn to coordinate and discipline them to good use. The problem with scientific fundamentalism is that it is not scientific enough. It is, rather, a narrow-gauge view of things that is ironically unskeptical, which is to say, unscientific, about science itself and the larger social, political, economic, and ecological conditions that permit science to flourish in the first place.

For all of our information and communications prowess, we talk too little about our motives and feelings in relation to our occupations and professions. I recall, for example, a conversation in which a group of distinguished ecologists and environmentalists was asked to describe the sources of their beliefs. In trying to describe their deepest emotions as if they were the result of carefully considered career plans, these otherwise eloquent people descended into a pit of muddled incoherence. But as the conversation continued, deeply moving stories about experiences of the most personal kind began to emerge. Most of us have had similar experiences. But we tend to talk about "career decisions" as if our lives were rationally calculated and not the result of likes, fascinations, serendipity, accidental associations, inspirations, and sensory experiences stitched into our childhood or early adult memories. I believe that most of us do what we do as environmentalists and profess what we do as professors because of an early, deep, and vivid resonance between the natural world and ourselves. We need to be more candid with ourselves and our students who have chosen to study biology or the human place in the environment because of a similar resonance.

Third, I think the power of denial in a time of cataclysmic changes undermines our willingness to talk about important things. There is a scene, for example, in the movie *The Day After* in which a woman, knowing that an H-bomb is about to hit, scurries about to tidy things up. A good bit of what goes on in the modern university likewise seems to me like a kind of tidying up before all hell breaks loose. Of course, it has already broken loose, and more is on the way. The twentieth century was the age of world wars, atomic bombs, gulags, totalitarianism, death squads, and ethnic cleansing. The twenty-first century is a time of terrorism, climate destabilization, loss of species, and economic uncertainty. In the light of such prospects, it is understandable that many find it easier and safer to tidy things up rather than roll up their sleeves to turn these trends around.

Fourth, love is difficult to talk about because we do not know much about how it is built into our deepest emotions and exhibited in our various behaviors. We have good reason, however, to believe that what Edward O. Wilson calls "biophilia," or "the connections that humans subconsciously seek with the rest of life," is innate (Wilson 1992, 350). It would be surprising indeed if several million years of evolution had resulted in no such affinity. But even if we could find no trace of subconscious biophilia, our concern for survival would cause us to invent it for the same reason Gould (1991) described above: We are not likely to fight to save what we do not love. This means that biophilia must become a conscious part of what we do and how we think, including how we do science and how we educate people to think in all fields. What does this mean for teachers and scholars?

For one thing, it means that biophilia, conscious and subconscious, deserves to be a legitimate subject of inquiry and practice (Kellert and Wilson 1993; Kellert et al. 2008). We need to become students of biophilia in order to understand more fully how it comes to be, how it prospers, and what it requires of us. For another, it requires a greater consciousness about how language, models, theories, and curricula can sometimes alienate us from our subject matter. Words that render nature into abstractions of board feet, barrels, sustainable yields, and resources drive out such feelings and the affinities we have at a deeper level. We need better tools, models, and theories, calibrated to our innate loyalties—ones that create less dissonance between what we do for a living, how we think, and what we feel as creatures evolved over several million years.

Finally, and most difficult, from other realms we know that love sets limits to what we do and how we do it. It is, as Erich Fromm once wrote, an art requiring "discipline, concentration, and patience" (Fromm 1956, 100). What does the art of love have to do with the discipline of science? On one side of the question, love is not a substitute for careful thought. On the other side, when the mind becomes, in Abraham Heschel's words, "a mercenary of our will to power . . . trained to assail in order to plunder rather than to commune in order to love" (Heschel 1951, 38), ruin is the logical result. In either case it is evident that personal motives matter, and different motives lead to very different kinds of knowledge and very different ecological results. At a recent meeting of conservation biologists, some of the participants wondered out loud why so few of their colleagues had joined the effort to conserve biological diversity. No good answer was given. On reflection, however, I think the reason lies in the difficulty we have in joining professional science with our love of life and those things that probably attracted most biologists to study science in the first place and continue to attract our students.

Chapter 5

Reflections on Water and Oil

(1990)

WATER MIGHT BE BEST understood in comparison with that other liquid to which we in the twentieth century are beholden: oil. Water as rain, ice, lakes, rivers, and seas has shaped our landscape. But oil has shaped the modern mindscape, with its fascination and addiction to speed and accumulation. The modern world is in some ways a dialogue between oil and water. Water makes life possible, while oil is toxic to most life. Water in its pure state is clear; oil is dark. Water dissolves; oil congeals. Water has inspired great poetry and literature. Our language is full of allusions to springs, depths, currents, rivers, seas, rain, mist, dew, and snowfall. To a great extent our language is about water and people in relation to water. We think of time flowing like a river. We cry oceans of tears. We ponder the wellsprings of thought. Oil, on the contrary, has had no such effect on our language. To my knowledge, it has given rise to no poetry, hymns, or great literature and probably to no flights of imagination other than those of pecuniary accumulation.

Our relation to water is fundamentally somatic, which is to say it is experienced bodily. The brain literally floats on a cushion of water. The body consists mostly of water. We play in water, fish in it, bathe in it, and drink it. Some of us were baptized in it. We like the feel of salt spray in our faces and the smell of rain that ends a dry summer heat wave. The sound

This article was originally published in 1990.

of mountain water heals what hurts. We are mostly water and have an affinity for it that transcends our ability to describe it in mere words.

Oil and water have had contrary effects on our minds. Water, I think, lies at the origin of language. It is certainly a large part of the beauty of language. Water has also given rise to some of our most elegant technologies: water clocks, sailing ships, and waterwheels. The wise use of water is quite possibly the truest indicator of human intelligence, measurable by what we are smart enough to keep out of it, including oil, soil, toxics, and old tires. The most intelligent thing we could have done with oil was to have left it in the ground or to have used it very slowly over many centuries. Oil came to Western civilization as a great temptation to binge, devil take the hindmost. Our resistance had already been lowered by the intellectual viruses introduced by the likes of Galileo, Bacon, and Descartes. We were in no condition to fend off those introduced by John D. Rockefeller, Henry Ford, and Alfred P. Sloan that promised speed, mobility, sexual adventure, and personal identity. Oil has undermined intelligence in at least six ways.

First, oil eroded our ability to think intelligently about community and the possibility of cooperation. Its nature is what game theorists call zero-sum: you have it or someone else does; you burn it or they do. Its possession set those who had it against those who did not: states against states; regions against regions; nations against nations; and the interests of one generation against those of generations to follow. Cheap oil and the automobile pitted community against community, suburban commuters against city neighborhoods. Money made from oil and oil-based technologies corrupted our politics, while our growing dependency corrupted our sense of proportion and scale. To guarantee our access to Middle Eastern oil we have declared our willingness to initiate Armageddon. We are now spending billions in fulfillment of this pledge even though a fraction of this annual bill would eliminate the need for oil imports altogether. The characteristics of oil and the way we have used it and have grown overly dependent on it have helped shape a mind-set that cannot rise above competition.

Second, oil has undermined our land intelligence by increasing the speed with which we move on it or fly over it. We no longer experience the landscape as a vital reality. Compare a trip by interstate highway from Pennsylvania to Florida with that taken by William Bartram in the eighteenth century. Where Bartram saw wonders and had the time to observe them carefully and be instructed and moved by them, modern travelers experience only a succession of homogenized images and sounds moving through an engineered landscape all tailored to the requirements of speed and convenience. As a result, our contact with land is increasingly abstract, measured as lapsed time and experienced as the dull exhaustion that accompanies jet lag or close confinement.

Third, oil has made us dumber by making the world more complicated but less complex. An Iowa cornfield is a complicated human contrivance resulting from imported oil, supertankers, pipelines, commodity markets, banks and interest rates, federal agencies, futures markets, machinery, asphalt highways, spare parts supply systems, and agribusiness companies that sell seeds, fertilizers, herbicides, and pesticides. In contrast, the forest or prairie that once existed in that place was complex, a highly resilient system consisting of a diversity of life-forms, ecological relationships, and energy flows. Complicatedness is the result of high energy use that destroys genetic and cultural information. With complicatedness has come specialization of knowledge and the "expert." Exit the generalist and the renaissance person. The result is a society and economy that no one comprehends—indeed, one that is beyond human comprehension. Complicatedness gives rise to unending novelty, surprise, and unforeseen consequences. As the possibility of foresight declines, the idea of responsibility also declines. People cannot be held accountable for the effects of actions that cannot be foreseen. Moreover, a high-energy society undermines our sense of meaning and our belief that our own lives can have meaning. It leads us to despair and to disparage the very possibility of intelligence.

Fourth, cheap oil and the automobile are responsible, in large measure, for the urban sprawl that has conditioned us to think that ugliness and disorder are normal or at least economically necessary. Where fossil energy was cheap and abundant, the idea of a land ethic based on the "integrity, stability, and beauty of the biotic community" has never taken firm hold. This is not just a problem of ethics; it is a deeper problem that has to do with how poorly we think about economics. Sprawling megalopolitan areas are not only an esthetic affront; they are sure signs of an unsustainable economy dominated by absentee corporations that vandalize distant places for "resources" and other places to discard wastes. A mind conditioned to think of ecological, esthetic, and social disorder as normal, which is to say a mind in which the categories of harmony and beauty have atrophied, is to that extent impoverished. It is rather like being able to see only half of the color spectrum. On the other hand, intelligence, I think, grows as the mind is drawn to the possibilities of creating order, harmony, beauty, stability, and permanence.

Fifth, oil has undermined intelligence by devaluing handwork and craftsmanship. To a great extent the history of high-energy civilization can be described by the shift in the ratio between labor and energy. Economic development is the process of substituting energy for labor, moving people from farms into cities and from craft professions into factories and eventually into "the service sector." This is not simply a matter of economic efficiency, as some argue; it is a problem of human intelligence. Thinking, doing, and making exist in a complex symbiotic relationship. The price we pay for the convenience and affluence of a service economy may well be paid in the coin of intelligence. The drift of high-energy civilization is to make the world steadily less amenable to the kind of thought that results from the friction of an alert mind's grappling with real materials toward the goal of work well done. To the modern mind, with its ghettos of costs and benefits, expertness, efficiency, built-in obsolescence, and celebration of technology that replaces manual skill, any alternative sounds hopelessly naive. However, we may find reason to reconsider, on the grounds of a larger efficiency and higher rationality, the reality that we are in fact *Homo faber* whose identity is defined by the close interplay of thought and making.

Finally, oil has undermined intelligence because it requires technologies that we are smart enough to build but not smart enough to use safely. This is the gap between knowing how to do something and knowing what one should do. Cheap oil has divided our capabilities from our sense of obligation, care, and long-term responsibility. Oil used at the rate of millions of barrels each day cannot be used responsibly. The *Exxon Valdez* oil spill in Prince William Sound, dozens of other large oil spills since, and now the BP eruption spreading disaster over the Gulf of Mexico are not accidents but the logical result of a system that operates on a scale that can only produce carelessness, corruption, and catastrophes. Our mistake is compounded by the belief that the catastrophes occur only when oil is spilled. It is, however, an equal, if more diffuse, catastrophe when oil emissions exit car engines, thereby causing air pollution and climate change. Oil has reduced our intelligence by dividing us between what we take to be realistic imperatives of economy and the commands of ethical stewardship. As a result we have become far less adept at thinking and acting ethically and far more adept at rationalizing and denying.

If oil has made us dumber, might water make us smarter about more things over a longer term? I think so. To this end, I suggest several things, beginning with an examination of contemporary curriculum to identify those parts that are based on the assumption of the permanence and blessedness of cheap energy. How much of the curriculum would stand if this assumption were removed? Education has generally prepared the young to live in a high-energy world. We have shaped whole disciplines around such assumptions without stopping to inquire about their validity or their larger effects. The belief in the permanence and felicity of high-energy civilization is found at the heart of most of contemporary economics, with its practice of discounting, development theory, marketing, business, political science, and sociology. The natural sciences have been largely directed toward manipulation of the natural world without any comparable effort to study impacts of doing so or alternative kinds of knowledge that work with natural systems. Behind a great deal of this is the belief that we can make an end run around nature and get away with it.

Second, water should be a part of every school curriculum. I would include, for example, Karl Wittfogel's (1956) study of the relationship between water management and despotic government, Donald Worster's (1985) study of the politics of water in the American West, and Charles Bowden's (1985) study of the relationship between water and the Papago people of Arizona. Water as part of our mythology, history, politics, culture, and society should be woven throughout curriculum, K through PhD.

Third, water should be the keystone in a new science of ecological design. Education in ecological design integrates a broad range of disciplines and design principles of resilience, flexibility, appropriate scale, sunlight, and durability. Biologist John Todd's work, as an example, is instructive in part because he has combined good engineering with ecology to create "living machines" that purify wastewater in a process that replicates the ecology of wetlands.

Fourth, water and water purification should be built into the architecture and the landscape of educational institutions. The very institutions that purport to induce the young into responsible adulthood often behave like vandals. This need not be. Institutional waste streams offer a good place to begin to teach applied (as opposed to theoretical) responsibility. Solar aquatic waste systems and similar approaches offer a way to teach the techniques of wastewater purification, biology, and closed-loop design. There are many reasons to regard resource and waste flows as a useful part of the curriculum, not merely a nuisance to be kept out of sight, mind, and curriculum.

Finally, I propose that restoration be made a part of the educational agenda. Every public school, college, and university is within easy reach of streams, rivers, and lakes that are in need of restoration. The act of restoration is an opportunity to move education beyond the classroom and laboratory to the outdoors, from theory to application and from indifference to healing. My proposal is for institutions to adopt streams or entire watersheds and make their full health an educational objective as important as, say, capital funds campaigns to build new administration buildings, parking decks, or athletic facilities.

What is the meaning of water? One might as well ask, "What does it mean to be human?" The answer may be found in our relation to water, the mother of life. When the waters again run clear and their life is restored, we might see ourselves reflected whole.

⁘ Chapter 6 ⁘

Gratitude

(2007)

AUTHOR'S NOTE 2010: *This was given as a baccalaureate talk at Oberlin with musical accompaniment by a band of students organized and directed by Andy Barnett.*

IN THE BEGINNING the Great Heart of God set the rhythms of the universe in motion—first the cymbal smash of the Big Bang . . . the beat heard through the still-expanding Creation and in the pulsations of energy and light that animate the cosmos. In the beginning was the Great Heart of God, and that rhythm drives the journeys of our little planet around its small star. Day follows night, one season follows another. The Great Heart of God beats in the Dance of Life, the ebb and flow of the tides, the migration of birds, the rhythms in our bodies, and the seasons of our lives. There is

> a time to be born, and a time to die; a time to plant, and a time to pluck up that which is planted; a time to kill and a time to heal; a time to break down and a time to build up; a time to weep, and a time to laugh; a time to mourn, and a time to dance; a time to cast away stones, and a time to gather stones together; a time to embrace and a time to refrain from embracing. (Ecclesiastes 3:2–5)

Break the rhythm and our little part of the cosmic dance stumbles to a halt. But in the beginning and forever: the rhythm of the Great Heart of God.

This article was originally published in 2007.

A fraction of a second ago, as geologists and ecologists measure time, another rhythm was begun. Some call this the "Fall." In one telling of the story, the cadence was changed by a snake and a woman (a libel against a perfectly fine life-form and one against all of womankind). More likely the discordant beat came from a few males who thought that an elite few could improve the Creation by changing the rhythm. C. S. Lewis once said that the intent was to control other men by seizing control of nature. Ecologically, control meant exploiting the vast pools of carbon—first the carbon-rich soils of the Fertile Crescent, later the carbon in the forests of Europe, and in our time the ancient carbon stored as coal and oil.

But it was not long before others, more sophisticated and clever, realized that a few could change the rhythm of Creation altogether. The heroes of disharmony, men like Francis Bacon, Descartes, and Galileo, taught us that we could and should conduct the symphony and in Bacon's words "put nature on the rack and torture her secrets out of her to the effecting of all things possible." And so in time we learned how to make things never made by nature, we learned to split the atom and to manipulate the code of life. Some are busy making devices that will be, they say, more intelligent than humans. In the conquest of nature and of other men, the rhythm changed to those of the business cycle, the product cycle, the electoral cycle, the seasons of fashion and style . . . the rhythms of commerce, greed, power, and violence. But we did not know what we were doing, as Wendell Berry once said, because we did not know what we were undoing.

Now we live in a time of consequences. Climate scientists have given us an authoritative glimpse of a literal hell not far in the future. Many scientists fear that we are fast approaching the threshold of runaway climate change . . . not just "global warming" but destabilization of the entire planet. A hotter time will change the seasons, the cycles of nature, the rhythms of life, and the great procession of evolution. The rhythm of the Great Heart of God has been drowned out by the cadence of hubris, greed, and violence. . . . And we should ask why.

After reflection I have come to believe that the great Jewish Rabbi Abraham Heschel had it right, that the source of dissonance is ingratitude: "As civilization advances," Heschel wrote, "the sense of wonder almost necessarily declines . . . mankind will not perish for want of information; but only for want of appreciation. The beginning of our happiness lies in the understanding that life without wonder is not worth living. What we lack is not a will to believe but a will to wonder." Heschel, here, connects appreciation with the sense of wonder and awe. The problem, as he defines it, is simply that as "a mercenary of our will to power, the mind is trained to assail in order to plunder rather than to commune in order to love" (Heschel 1951).

We were given the gift of paradise and thought that we could improve it—on our terms. We thought we could reduce the great mystery of life to a series of solvable problems each contained in one academic box or another. We thought that we could rid the world of reverence and so exorcise mystery, irony, and paradox. We thought that we might change the cadence of Creation and seize control of the great symphony of life with no adverse consequence.

But why is gratitude so hard for us? This is not a new problem. Luke tells us that Jesus healed the 10 lepers, but only one returned to say thank you. That's about average, I suppose. In our universities we teach a thousand ways to criticize, analyze, dissect, and deconstruct, but we offer very little guidance on the cultivation of gratitude—simply saying thank you.

And maybe we should not be grateful. In the spirit of pluralism, is there a case for ingratitude? Is gratitude merely a ploy that runs inversely proportional to favors not yet granted? One might suspect the psalmist of such. Or perhaps there is no cause for gratitude amidst the cares and trials of life. Shakespeare, for example, has Macbeth say that "life's but a . . . tale told by an idiot, full of sound and fury, signifying nothing." Political philosopher Thomas Hobbes similarly thought that life was full of peril and death—"nasty, brutish, and short." And many of us find our bodies, incomes, careers, and lives as less than we would like, whatever we might deserve.

But most of us, too, would find life without appreciation rather like a meal without flavor or living in a world without color, or one without music. So, we set aside one day of the year for Thanksgiving but mostly spend it eating too much and watching football. Gratitude comes hard for us for many reasons. For one thing, in psychologist Robert Emmons's words, "gratitude can be a bitter pill to swallow, humbling us and demanding as it does that we confront our own sense of self-sufficiency" (Emmons 2008, 15). For another, we spend nearly half a trillion dollars on advertising to cultivate ingratitude otherwise known as the seven deadly sins. The result is a cult of entitlement to have as much as possible for doing as little as possible. The pace of modern life leaves little time to be grateful or awed by much of anything. But there are deeper reasons for ingratitude. Gratitude does not begin in the intellect but rather in the heart. "Intellect," in

David Steindl-Rast's words, "only gets us so far" (Steindl-Rast 1984, 13). Our intellect should be alert enough to recognize a gift, but to acknowledge a gift as a gift requires an act of will and heart. To acknowledge a gift, however, is also "to admit dependence on the giver . . . but there is something within us that bristles at the idea of dependence. We want to get along by ourselves" (Steindl-Rast 1984, 15).

To acknowledge a gift, in other words, is to acknowledge an obligation to the giver. And herein is the irony of gratitude. The illusion of independence is a kind of servitude while gratitude—the acknowledgement of interdependence—sets us free. Only "gratefulness has the power to dissolve the ties of our alienation," as Steindl-Rast puts it. But "the circle of gratefulness is incomplete until the giver of the gift becomes the receiver; a receiver of thanks . . . and the greatest gift one can give is thanksgiving" (Steindl-Rast 1984, 17). Saying thank you is to say that we belong together—the giver and thanks giver—and it is this bond that frees us from alienation. And the gift must move. What is given must be passed on. In the end nothing can be held or possessed—a truth grasped by every culture that approaches what we've come to call sustainability.

Gratitude changes the rhythm. It restores the cycle of giver and receiver and back again. It extends our awareness back in time to acknowledge ancient obligations and forward in time to the far horizon of the future and to lives that we are obliged to honor and protect. Gratitude requires mindfulness, not just smartness. It requires a perspective beyond self. Gratitude is at once an art and a science, and both require practice. The arts and sciences of gratitude, which is to say, applied love, are flourishing in ironic and interesting ways. Businessman Ray Anderson has set his company on a path to operate by current sunlight and return no waste product to the Earth. Biologists are developing the science of biomimicry that uses nature's operating instructions evolved over 3.8 billion years to make materials at ambient temperatures without fossil fuels and toxic chemicals, rather like spiders that make webs from strands five times stronger than steel. The movement to power the civilization from the gift of sunshine and wind is growing at 40 percent per year worldwide. The American Institute of Architects and the U.S. Green Building Council have changed the standard for buildings to eliminate use of fossil fuels by 2030. Could we, in time, create a civilization that in all of its ways honors the great gift and mystery of life itself?

Can true gratitude transform our prospects? Can we harmonize the rhythms of this frail little craft of civilization with the pulse of the Great

Heart of God? I believe so, but gratitude cannot be legislated or forced. It will remain a stranger to any mind that lacks compassion. It must be demonstrated, but above all it must be practiced daily, and when we do, "gratitude elevates, it energizes, it inspires, it transforms" (Emmons 2008, 12).

Chapter 7

Orr's Laws

(2004)

FIFTEEN YEARS OR SO after its founding, the editorial staff of *Conservation Biology* decided to undertake an extensive review of the journal to assess its fitness as a medium of scientific communication. As part of the exercise, reviewers were asked to evaluate the 47 columns I'd written for the journal up to that time. Responses ranged from those saying such things as "I subscribe to *Conservation Biology* solely for David Orr's essays" and his "column is the first part I eagerly flip to" and "keep Orr as long as you can" to one saying that "folks such as David Orr are such hypocritical, unpatriotic blows as to make me sick. . . . He and his Saab-driving, white zinfandel–sipping bunch of pseudo-intellectual know-it-all, sanctimonious culls . . ." and so forth. Actually, at the time I drove a Ford Ranger pickup truck (I now drive a Prius and do, indeed, feel much more sanctimonious), and generally prefer a fine cabernet or pinot. Notwithstanding, I thought it a good time to reflect on what I'd learned over those years from comments from readers and editors alike, and from what I hoped suggested a process of thoughtful maturing, not merely aging.

That mulling-over retrospective resulted in a list of five "laws." There is nothing significant about the number five. Moses and his mentor arrived at an even ten, and some still think that too few. Alas, even that list is more honored than operational. Buddha offered four, but of course he had no proper theological training. An upstart, and unlicensed, rabbi later reduced it to two. I, neither so extensive nor so concise but similarly given

---

This article was originally published in 2004.

to making lists, settled on five. I put this forward fearful that some would regard the results as paltry enough to encourage me to take up bowling or a career in manual labor. But I assumed that a good offense is the best defense and stated these as "Orr's laws," thereby risking ever so slightly the possibility of enhancing my reputation in some circles as a "sanctimonious" and "Saab-driving, white zinfandel–sipping . . . know-it-all" but otherwise impressing the gullible and hopefully entertaining everyone else. So what gems did those 15 years yield? I humbly offered the following as "Orr's laws." (I should note in due modesty that since publication of Orr's laws, there has been no resulting upheaval in science, or literature, or law, or anything else of which I am aware. Apparently, it takes time for things to sink in.)

1. We pay for the conservation of biological diversity, climate stability, and environmental quality whether we get it or not.

In an age much devoted to the theology of free markets, the conventional wisdom holds that we cannot afford very much conservation, resource efficiency, preservation of landscapes, clean air and water. Based on incomplete and even selective accounting, that view is almost always wrong. Honest accounting requires that we keep the boundaries of consideration as wide as possible over the long term and deduct the collateral benefits that come from doing things right. Ignoring the costs of wars fought for "cheap" oil, or climate change, air pollution, and sprawl, my Ford Ranger was cheap enough. More-efficient transportation may have higher front-end costs, but price and cost should not be confused. We—or someone eventually—will pay for sustainability whether we get it or not. Economists operating strictly within the boundaries of the neoclassical paradigm cannot account for the true costs of impaired security, health, beauty, and spiritual comfort. But it is the height of folly to believe that we can dismantle forests, pollute, squander resources, erode soils, destroy biological diversity, remodel the biogeochemical cycles of the Earth, and create ugliness, human and ecological, without paying. The truth is that sooner or later, the full costs will have to be paid one way or another. The problem, however, is that the costs of environmental dereliction, as often noted, are diffuse and can be deferred to some other persons or to some later time, but they do not thereby disappear. The upshot is that much of our apparent prosperity is phony and so too the intellectual and ideological justifications for it. For anyone with concern for the future of humankind, there is no higher priority than to help put our understanding of economics on more solid ground.

Corollary 1: Selfishness is not equivalent to self-interest.
Establishing an ecologically solvent economics requires confronting errors that masquerade as realism. There is none more pernicious than the conventional belief that selfishness and self-interest are the same thing. Assuming that to be so, it is easy to conflate self-aggrandizement and selfishness as just different manifestations of the same drive, thereby explaining everything and nothing about human behavior. Doing so, however, confuses fundamentally different things, thereby making us more cynical and giving license to our lesser social instincts. We are unavoidably self-interested because we are self-aware, but we can choose not to be selfish. This leads to the often noted irony that increasingly the most self-interested thing we can do is to be more giving, caring, and altruistic, which requires expanding our sense of inclusiveness.

Corollary 2: Maximizing efficiency—measured as output for a given level of input—creates disorder, that is to say, inefficiency at higher levels.
The reasons are complex but have a great deal to do with our tendency to confuse means with ends. As a result efficiency often becomes an end in itself while the original purposes (prosperity, security, benevolence, reputation, etc.) are forgotten. The assembly line was efficient for the manufacturing firm, but its larger effects on workers, communities, and ecologies were often destructive, and the problems for which mass production was a solution have been compounded many times over. Neighborliness is certainly an inefficient use of time on any given day, but not when considered over months and years. For engineers, freeways are efficient at moving people up to a point, but they destroy communities, promote pollution, lead to congestion, change foreign policies, and eliminate better alternatives including design that eliminates some of the need for mobility. Walmart is an efficient marketing enterprise but eliminates its small competitors and many things that make for good communities. And, of course, nuclear weapons are wonderfully efficient as well. Hence, the higher-level efficiency of inefficiency underscores the need to reorient economic life toward ends, not means; the whole, not just parts; and the long term over the short term.

2. Problems of ecology are first and foremost political problems having to do with who gets what, when, and how.

As much as some might wish it otherwise, environmental protection, climate stability, and the conservation of biological diversity are unavoidably political. The environment is ultimately a mirror reflecting the collective decisions that we make about energy, forests, land, water, and resources. Environmentalists are often regarded as "liberal," and the rest are thought to be "conservative." This left-right perspective, too, is phony and obscures a great deal more than it reveals. The real political divisions are not between liberals and conservatives but between the present generation and its descendants. One can arrive at a decent regard for sustainability, ecological health, and the prospects of humankind as a true conservative or as a consistent liberal. Those are not opposing positions but different sides of a coin. Disguised as a kind of super patriotism, much that passes for "conservative" these days is merely reckless demagoguery that serves corporate interests. For their part, liberals have yet to reckon with the problem of how to limit human appetites without infringing freedom, which will require a higher definition of both. And neither has yet developed any plausible and decent view of a workable and sustainable global political order.

AUTHOR'S NOTE 2010: *In the years since this was written, our politics have gone from bad to worse to total gridlock.*

In 1989 as the Soviet Union was coming undone, the world seemed poised on the brink of a more promising time. In the years since, that promise has been squandered, and the reasons have much to do with the hold of corporate management and militarization on the U.S. political system and media. This cannot be unrelated to other things, such as the fact that the U.S. leads the world in the emission of greenhouse gases, debt, production of toxic substances, gun ownership, violent crime, television watching, the numbers of its population in jail, and obesity. Not only is the U.S. the least energy efficient of the "developed nations," we are also among the least socially progressive. I doubt our capacity to change significantly while we spend more than the next 21 nations combined on weapons and wars and refuse to come to grips with our energy gluttony.

The only conclusion to be drawn is that those who are concerned about the human future must become effective politically. The goal is a thorough rebuilding of the political system in order to recalibrate conservatism and liberalism alike to the inescapable realities of ecology and physics.

3. Humans are more ignorant than they are smart, and many seem to prefer it that way.

T. S. Eliot put it this way:

> Human kind
> Cannot bear very much reality. (Eliot 1971, 119)

From the publication of the *Global 2000 Report* in 1980 to the present, there is a veritable mountain of scientific evidence about human impacts on ecosystems and the biosphere and hence about our tenure on Earth. But our collective sleepwalk toward the edge of irreversible tragedy continues, suggesting that we are not so much rational creatures as we are creatures with a considerable talent for rationalizing. The reality is that we are coming to the end of a brief interlude in human history powered by fossil fuels. Had we been a truly wise lot, we would have burned little of this endowment and probably not have industrialized in the manner and to the degree that we did. Willed or otherwise, we did both out of ignorance.*

Wes Jackson (the Land Institute) and Bill Vitek (Clarkson University) once convened a small conference around the theme of "An Ignorance Based Worldview," aiming to lay the groundwork for a more accurate paradigm than that given us by apostles of smartness, beginning with Bacon, Galileo, and Descartes. In contrast to those believing ignorance to be a solvable problem, conferees agreed that there is lots that we cannot know and perhaps some that we should not know and that this admission requires a dramatically different worldview than that on which the industrial world was fashioned. But how, with the rapid growth of science, could we be irredeemably ignorant? One answer is that ignorance is built into the way science itself works. As knowledge grows, so too does the interface of the known with the unknown, which is to say that science is

*It is not insignificant that in these past 15 years the consolidation of media means more homogenization of news and information along an increasingly narrow bandwidth. We, in the United States, are the most media-saturated people in the world and arguably among the least well informed, hence the most adept at rationalizing in the face of evidence to the contrary.

not a zero-sum game by which ignorance retreats in direct proportion to every increment of knowledge.

In short, we are ignorant because of the vastness of what there is to be known relative to our intellectual and perceptual capacities. We are ignorant because we individually and collectively forget things that we once knew. We are ignorant because every human intervention and action changes the world that we aim to understand. We are ignorant because of our own limited intelligence and inability to make sense of it all. We are ignorant because we cannot know in advance the unintended effects of our actions in complex systems. We are ignorant even about the proper ends to which knowledge might be put. And, not the least, we are ignorant, as Eliot noted, because we choose to be.

The upshot for scientists is that their work is not finished when they have duly reported the research findings. Science will merely gather dust until it is incorporated into a larger and more accurate story of who and what we are relative to the mysteries of life, time, and space. That story of the human journey must be so compelling as to displace once and for all the myth that humans can be the lords and masters of Creation.

Corollary 1: Knowing is always accompanied by paradox, irony, and unintended consequences.
And that, too, is ironic.

Corollary 2: The amount of credulity in human societies remains constant over time but can only be seen in hindsight.
I was tempted to render this into a law rather like the first law of thermodynamics (both assume constancy in systems otherwise closed except to solar radiation and an occasional ray of insight), but I chickened out. It's one thing to be accused of being a Saab-driving, white zinfandel sipper but another entirely to be accused of either hubris or idiocy. Accordingly, I will merely suggest as a corollary to the law of ignorance the fact that humans, in all ages and times, are inclined to be as unskeptical, gullible, and deceivable as those living in any other. Only the causes of our gullibility change. People of previous ages read chicken entrails, relied on shamans, consulted oracles. We, on the other hand, use computer models, believe experts, and exhibit a touching faith in technology to fix virtually everything. But who among us really understands how computers or mathematical models work, the limits of expertise and its underlying theories, or the ironic ways in which technology "bites back"? Not one in

ten thousand! Has gullibility declined as science has grown more powerful? No, if anything, it is growing because science and technology are increasingly esoteric and specialized, hence removed from daily experience. People who understand less and less of either will believe almost anything. Gullibility feeds on mental laziness and stupidity and is magnified by ostracism, pressures for conformity, and penalties for deviance.

4. Humans are inescapably spiritual beings but only intermittently religious.

Philosopher Erazim Kohak once noted, "Humans can bear an incredible degree of meaningful deprivation but only very little meaningless affluence" (Kohak 1984, 170). In the former condition most of us tend to grow, harden, and mature, while we are undone in the latter. This is not a case for deliberately incurring misery, which tends to multiply with little assistance, but rather to underscore our inevitable spiritual nature that is like water bubbling upward from an artesian spring. Our only choice is whether that energy is directed to authentic purposes or not.

Too much of environmentalism is about data and numbers that have little resonance with the public. For conservationists this means that the case we make for preserving biotic systems must tap some deeper motivation than narrow self-interest. It must appeal to that early childhood fascination with living things and particular places that is often desiccated by formalized learning and professionalization of knowledge. It must be joined with the arts and humanities: music, poetry, drama, and literature. Shakespeare did not write text books on human psychology and ethics. He wrote plays that will move people as long as literature is read. Dry recitation of facts, data, and logic, however important for other purposes, will not in the end cause most of us to conserve much.

5. Stupidity is randomly distributed up and down the socio-economic-educational ladder.

In my lifetime I have known as many brilliant un-degreed people without much formal learning as those certified by a PhD. And there are likely as many thoroughgoing, fully degreed fools as there are un-degreed fools. This is an admission of some gravity, leading me as an "educator" to believe that the gift of intelligence and intellectual clarity can be focused and sharpened a bit but can be neither taught nor conjured. The numerous examples of the undereducated or those who were outright failures in the academic sense include Albert Einstein, Winston Churchill, and Frank Lloyd Wright. One should not conclude, however, that formal schooling is useless but that its effectiveness, for all of the puffery that adorns college catalogues and educational magazines, is less than advertised. And there are those, as lawyer John Berry once noted, who have been "educated beyond their comprehension," people made more errant by the belief that their ignorance has been erased by mere possession of facts, theories, and the sheer weight of learnedness.

For those engaged in the effort to preserve a habitable planet, the point is that we will have to enlist the ideas, efforts, and enthusiasm of millions of people who live, work, and think in close proximity to natural systems but who lack formal education. Such people are often tacitly dismissed because of an unspoken elitism attending the upper reaches of the learned scientific world—and elitism undercuts the larger effort to conserve natural systems. We are likely to find useful observation, workable ideas, and clearheadedness all along the socio-economic-educational spectrum, but not necessarily concentrated at the upper end.

Corollary: The intelligence of any organization or bureaucracy is always less than the sum total of its managers.

We understand human stupidity and dysfunction because we encounter it at a scale commensurate with our own. Faced with large-scale organizations, whether corporations, governments, or colleges and universities, we tend to equate scale, prestige, and power with perspicacity and infallibility. Nothing could be farther from the truth. The intelligence of large-scale organizations (if that is not altogether oxymoronic) is limited by the obligation to earn a profit, enlarge its domain, preserve entitlements, or maintain a suitable stockpile of prestige. Accordingly, the effort to preserve a habitable Earth should not become overly dependent on large-scale organizations and thereby be limited because of their pathologies. The alternative model, rather like ecosystems, favors working through networks of relatively small, agile, and highly effective organizations closer to the problems.

Of course these aren't my "laws" at all but a brief distillation drawn from the work of a remarkable worldwide group of scholars/activists/practitioners working against long odds over decades to redirect the course of a civilization bent on destruction. If there is one other "law" that I am tempted to add but lack the words to do so, it would have to do with the possibility that our prospects can never exceed the horizon of our hopes, and that gives us every reason to keep hope alive.

PART 2

On Sustainability

AUTHOR'S NOTE 2010

Salvation by better gadgetry is a peculiarly American faith, but it has taken root almost everywhere because of the power of the U.S. example and the mystique of technology. It hits us where we're weakest. Virtually every publication on the subject of sustainability prominently features glossy pictures of windmills, solar collectors, and exotic buildings, as if that were all there was to it. I'm not opposed to these things, or better gadgets for that matter, but I am not willing to become a technological fundamentalist or what longshoreman and writer Eric Hoffer once called a "true believer." The multiple challenges posed by the transition to what is called sustainability go deeper than technology. They are rooted in human nature, culture, history, and politics. In some ways technology is both a symptom of what ails us and sometimes a partial solution. But it is decidedly not a panacea. We certainly need lots of windmills, solar technology, hybrid vehicles, and LED lights—no argument. But having thoroughly deployed such things and others even more exotic, what will we have done? And how will we deal with the big issues that have to do with violence, fairness, justice, equality, and decency across the lines that separate us by tribes, nations, species, and generations? How we do so will decide whether we will be around for a while and under what conditions.

Chapter 8

Walking North on a Southbound Train

(2003)

A GREAT STORYTELLER I KNOW once told a story about a fox that appears at the edge of a clearing in which a dog is tethered to a pole. The fox begins to run in circles just outside the radius of the dog's tether, followed by the frantically barking dog. After a few laps the fox has managed to tie the dog to the tether, at which point he struts in to devour the dog's food while the helpless mutt looks on. Something like that has happened to all of us who believe climate stability and healthy ecosystems are worth preserving and that this is a matter of obligation, spirit, true economy, and common sense. But someone or something has run us in circles, tied us up, and is eating our lunch. It is time to ask who and why and how we might respond. Here is what we know:

1. Despite occasional success, overall we are losing the epic struggle to preserve the habitability of the Earth. The overwhelming fact is that virtually all important ecological indicators are in decline. The human population increased threefold in the twentieth century and will likely grow further before leveling off at 8–9 billion sometime around 2050. The loss of species continues and will likely increase in coming decades. Human-driven climatic change is now under way and is occurring more rapidly than many scientists thought possible even a few years ago. There is no political or economic movement

This article was originally published in 2003.

presently powerful enough to stop the process short of a doubling or tripling of the background rate of 280 ppm CO_2.

2. Even after the election of 2008, the forces of denial in the United States are more militant and brazen than ever before. Every day, millions in this country alone hear that those concerned about the environment are "wackos" or potential terrorists and no significant Washington politician utters any objection. Gun sales have soared because of paranoid fears about the imminent takeover of the U.S. government. Patently absurd views are propagated on FOX News and right-wing radio, many undergirded with fabricated data but repeated endlessly word for word until the next talking-points memo is distributed.
3. The movement to preserve a habitable planet is caught in the cross fire between fundamentalists of the corporate-dominated global economy and those of atavistic religious and political movements. It is far easier to see the latter than the former, but in a longer perspective those of perpetual economic expansion will be perceived to be at least as dangerous as are those of a purely religious sort. That danger is now magnified by a national policy dating to the Bush presidency that permits the U.S. to strike preemptively at any country deemed to be an enemy without resort to international law, morality, common sense, or public debate. In the words of one analyst, this is "a strategy to use American military force to permit the continued offloading onto the rest of the world of the ecological costs of the existing US economy—without any short-term sacrifices on the part of US capitalism, the US political elite or US voters" (Lieven 2002).
4. Fundamentalists of either kind require dependably loathsome enemies. For Osama bin Laden, the United States admirably serves that purpose. It is no less true that the foundering presidency of Mr. Bush in 2001 was revitalized by the activities of Mr. Bin Laden and subsequently by the less agreeable behavior of Saddam Hussein. Each is fulfilled and defined by an utterly vile enemy. Bin Laden, at this writing, is still on the loose, but his appeal as a suitable ogre is waning at the moment. Fundamentalists, accordingly, are inclined to select domestic targets such as President Obama, who for his part appears abnormal in needing no target on which to project his frustrations.
5. There has been a steep erosion of democracy and civil liberties in

the U.S., driven by what former president Jimmy Carter describes as "a core group of conservatives who are trying to realize long-pent-up ambitions under the cover of the proclaimed war against terrorism" (Carter 2002). There is a strong antidemocratic movement on the right wing of American politics that would limit voting rights, reduce access to information, prevent full disclosure of matters about the conduct of the public business and public control of military affairs, and gridlock policy-making well short of solving any of the major challenges facing the country.

6. In the decades before the economic collapse of 2008, massive amounts of wealth were transferred from the poor and middle classes to the richest by reducing taxes on wealth and corporate income. The result is that the wealth of the top 1 percent exceeds that of the bottom 95 percent. No good estimate exists of how much money is simply lost from the federal budget, given out to cronies, outsourced, or given away in tax breaks or taxes not collected.
7. For nearly a half century, government at all levels has been under constant attack by the extreme right wing with the clear intention of eroding our capacity to create collective solutions to national and global problems. The assumption is now common that markets are "moral" but that publicly created political solutions are not. The result is a continuation of what a Republican president, Teddy Roosevelt, once described as "a riot of individualistic materialism, under which complete freedom for the individual . . . turned out in practice to mean perfect freedom for the strong to wrong the weak" (Roosevelt 1913).
8. The strategy, once revealed by Ronald Reagan's director of the Office of the Budget, David Stockman, has been to cut taxes for corporations and the wealthy and increase military spending, thereby creating a severe fiscal crisis that requires cutting expenditures for health, education, mass transit, the environment, and cities.
9. Our problems are systemic and can only be solved by changing the system.
10. There are yet good possibilities to avert the worst of what may lie ahead.

In short, the movement to preserve the habitability of the Earth is in failure mode. The reasons can be found neither in the lack of effort or good intention by thousands of scientists, activists, and concerned citizens nor in a lack of information, data, logic, and scientific evidence. On these counts the movement has grown impressively, as has the quality and quantity of scientific evidence and rational discourse on which it rests. But we must look more deeply at how this is manifest in the larger arena in which public attitudes are formed and the way in which this influences the conduct of the public business.

We are in failure mode, first, because for 30 years or longer, we have tried to be reasonable on their terms, in the belief that we could persuade the powerful if we only offered enough reason, data, evidence, and logic. We have quantified the decline of species, ecosystems, and now planetary systems in exhaustive detail. We bent over backwards to accommodate the style and intellectual predilections of self-described "conservatives" and those for whom the corporate-dominated economy is far more important than the environment, in the belief that politeness and good evidence stated in their terms would win the day. Accordingly, we put the case for the Earth and coming generations in the language of economics, science, and law. With few exceptions we have been reasonable, erudite, clever, cautiously informative, *and*, relative to the magnitude of the challenges before us, ineffective. In short, we do science, write books, publish articles, develop professional societies, attend conferences, and converse learnedly. But they do politics, take over the courts, control the media, and incite and manipulate the fears and resentments endemic to a rapidly changing society.

The movement to preserve a habitable Earth is in failure mode, too, because it is fractured into different factions, groups, and arcane philosophies. In this respect it has come to resemble the nineteenth-century European socialist movement that became bitterly divided into warring factions, each more eager to be right than right and effective. When the world was finally ready for better ideas about how to decently organize industrial society, that movement delivered bolshevism, and the rest, as they say, is history. Historically progressives exhaust themselves in bloody internecine quarrels, the strategy, as David Brower once described it, of drawing the wagons into a circle and shooting inward. The Right generally suffers no such fracturing, in large part because their agenda is formed around less complicated aims having to do with power and pecuniary advantage.

Further, I think Jack Turner is right in saying that we are in failure mode because all too often we are complacent and lack passion. "We are," in his words, "a nation of environmental cowards . . . willing to accept substitutes, imitations, semblances, and fakes—a diminished wild. We accept abstract information in place of personal experience and communication" (Turner 1996, 21, 25). Effective protest, he continues, "is grounded in anger and we are not (consciously) angry. Anger nourishes hope and fuels rebellion, it presumes a judgment, presumes how things ought to be and aren't, presumes a caring. Emotion remains the best evidence of belief and value. Unfortunately, there is little connection between our emotions and the wild" (Turner 1996, 21–22). We are endlessly busy trading e-mails, blogging, doing research, writing papers, and attending conferences in exotic places but go into the wild less and less often. We are cut off from the source.

Finally, we are losing because we failed to appreciate the depth of human needs for transcendence and belonging. We have allowed those intending to pillage the last of nature to do so behind the cover of religion, national pride, community, and family. As a result, the majority of U.S. citizens—even those who regard themselves as "environmentalists"—see little conflict with the goals of human domination of nature and the perpetual expansion of the human estate on Earth. As Buddhists would have it, whatever we thought we were doing, we have built a system based on illusion, greed, and ill will.

What is to be done? To that question there can be no simple, easy, or definitive answer, but I do think there are some obvious places to begin. The first requires that we take back public words such as *conservative* and *patriot*, which have been co-opted and put to no good or accurate use. How is it, for example, that the word *conservative* came to describe those willing to run irreversible risks with the Earth? Intending to conserve nothing, they are not conservatives but vandals now working at a global scale. How have those driving their sport utility vehicles to the mall, sporting two American flags and a "God bless America" bumper sticker, come to regard themselves as patriots? They are not moved by authentic patriotism at all, but by self-indulgence. For that matter how has the great and noble word *liberal* been demeaned and slandered as the height of political and intellectual folly? Unable to defend the integrity of words, we cannot defend the Earth or anything else.

The integrity of our common language, however, depends a great deal on the cultivation of discerning intelligence in the public, and that requires better education than we now have. But education has been whittled down to smaller purposes of passing tests and ensuring large "lifetime earnings" in some part of the global economy. What passes for education has become highly technical and specialized, little of which is aimed to draw out the full human stature of young people. We've become a nation of specialists and technicians, not broadly educated and discerning people. Scholars have been too intent on developing "professional knowledge," arcane theories, complicated methodologies, instead of broad knowledge useful to the wider public. Consequently, we have fewer and fewer people who know history or how the world works as a physical system or the rudiments of the Constitution or have a respectable political philosophy. We are a people ripe for the plucking.

This leads me to say that we do not have an environmental crisis so much as we have a political crisis. A great majority of people still wish a decent and habitable world for their descendants, but those desires are thwarted by the machinery that ought to connect the popular will to public decisions but no longer does so. We will have to repair and perhaps reinvent the institutions of democratic governance for a global world, and that means dealing with issues that the founders of this republic did not and could not have anticipated. The process of political engagement at all levels has become increasingly Byzantine, confusing, and inaccessible. And in a mass consumption society, we have all become better consumers than citizens, which is to say, willing participants in our own undoing. The solution, however difficult, is to reconnect people with the political process and government at all levels.

Fourth, it is necessary to expose the mythology that surrounds what Marjorie Kelly calls "the divine rights of capital" and place democratic controls on corporations and the movement of capital (Kelly 2001). We once fought a revolutionary war to establish political democracy in Western societies but have yet to democratize the workplace and the ownership of capital. These are still governed by the same illogic of unquestioned divine right by which monarchies once ruled. The assumption that corporations are legal persons and thereby beyond effective public scrutiny, control, or law is foolishness and worse. The latest corporate scandals are only that: the latest in a recurring pattern of illegality, self-dealing, and political corruption surpassing even that of the robber baron era. The solution is to enforce corporate charters as public license to do business on behalf of the public that are revocable if and when the terms of the charters are violated. If private ownership is good thing, it should be widely extended, not restricted to the super wealthy. By the same logic, we must remove the corrupting influence of money from politics, beginning with corporate campaign contributions and the hundreds of billions of dollars of public subsidies for cars, highways, fossil fuels, and nuclear power that corrupt the democratic process and public policy.

Fifth, political reform requires an engaged, informed, and thoughtful citizenry. Compare, for example, the Illinois farmer-citizens who stood for hours to hear Lincoln and Douglas debate issues of slavery and sectionalism in 1858. Those debates were full of careful argument, eloquence, and wit. Those citizens applauded, laughed, and jeered, which is to say that they followed the flow of argument and heard what was being said. Later, some died for and because of those same arguments. They were citizens and were willing to sacrifice a great deal for that privilege. In our time, while the issues have grown to global scale with consequences that extend as far into the future as the mind dares to imagine, political argument is whittled down to sound bites fitted in between advertisements. The means whereby citizens are informed have been increasingly monopolized and manipulated. Only half or less of the citizenry bother to vote. Some believe public apathy and political incompetence to be good or at least tolerable. I do not. Unless we reverse course they will, in time, prove to be the undoing of democratic government and all that depends on a healthy democracy. The nature of what will replace it is already evident: an unconstrained managerial and well-armed plutocracy intent on global plunder.

Sixth, we need a positive strategy that fires the public imagination. The public, I think, knows what we are against, but not what we are for. And there are many things that should be stopped, but what should be started? The answer to that question lies in a more coherent agenda formed around what is being called ecological design as it applies to land use, buildings, energy systems, transportation, materials, water, agriculture, forestry, and urban planning. For three decades and longer, we have been developing the ideas, science, and technological wherewithal to build a sustainable society. The public knows of these things only in fragments but not as a coherent and practical agenda—indeed the only practical course available. That is our fault and we should start now to put a positive agenda before the public that includes the human and economic advantages of better technology, integrated planning, coherent purposes, and foresight.

Finally, we should expect far more of leaders than we presently do. Never has the need for genuine leadership been greater, and seldom has it been less evident. We cannot be ruled by ignorant, malicious, greedy, incompetent, and shortsighted people and expect things to turn out well.

If we are to navigate the challenges of the decades ahead, what E. O. Wilson calls "the bottleneck," we will need leaders of great stature, clarity of mind, spiritual depth, courage, and vision. We need leaders who see patterns that connect us across the divisions of culture, religion, geography, and time. We need leadership that draws us together to resolve conflicts, move quickly from fossil fuels to solar power, reverse global environmental deterioration, and empower us to provide shelter, food, medical care, decent livelihood, and education for everyone. We need leadership that is capable of energizing genuine commitment to old and venerable traditions as well as new visions for a global civilization that preserves and honors local cultures, economies, and knowledge.

So, imagine a world in which those who purport to lead us must first make a pilgrimage to ground zero at Hiroshima and publicly pledge "never again." Imagine a world in which those who purport to lead us must go to Auschwitz and the Killing Fields and pledge publicly "never again." Imagine a world in which leaders must go to Bhopal and say to the victims, "We are truly sorry. This will never happen again, anywhere." Imagine a world where the leaders of the industrial world publicly apologize to those in low-lying lands or island nations for making them climate refugees and work to stop climate destabilization. Imagine, too, those pilgrim leaders going to hundreds of places where love, kindness, forgiveness, sacrifice, compassion, wisdom, ecological ingenuity, and foresight have been evident.

Imagine a world in which those who purport to lead us must help identify places around the world degraded by human actions and help initiate their restoration. Some projects might take as long as 1000 years to restore, such as the Aral Sea, the ecology of the Harrapan region in India, the forests of Lebanon, soil fertility in the Middle East, the Chesapeake Bay, the North Atlantic cod fishery—the possibilities are many. Imagine a world in which those who intend to lead help lift our sights above the daily crisis to the far horizon of what could be.

Imagine, too, leaders with the kind of humility demonstrated by Czech president Vaclav Havel:

> In time I have become a good deal less sure of myself, a good deal more humble . . . every day I suffer more and more from stage fright; every day I am more afraid that I won't be up to the job . . . more and more often, I am afraid that I will fall woefully short of expectations, that I will somehow reveal my own lack of qualifications for the job, that despite my good faith

> I will make ever greater mistakes, that I will cease to be trustworthy and therefore lose the right to do what I do. (Havel 2002, 4)

Self-described "realists" will dismiss the idea of better leadership as muddleheaded or worse. Some will see in it some global conspiracy or another. Prospective leaders will profess sympathy but say that they do not have the time to improve themselves further. And those least qualified to lead will pay no attention at all. But it is not up to any of them to prescribe for us. We are now citizens of the Earth joined in a common enterprise with many variations. We have every right to insist that those who purport to lead us be worthy of the task. Imagine a time, not far off, when we might all be onboard a train heading north!

Chapter 9

Four Challenges of Sustainability

(2006)

The destiny of the human species is to choose a truly great but brief, not a long and dull career.
NICHOLAS GEORGESCU-ROEGEN

THE CONCEPT OF SUSTAINABILITY first came to public notice in Wes Jackson's work on agriculture in the late 1970s (Jackson 1980), Lester Brown's *Building a Sustainable Society* (Brown 1980), and *The World Conservation Strategy* (Allen 1980). The Brundtland Commission made it a central feature of its 1987 report, defining it as meeting the needs of the present generation without compromising the ability of future generations to do the same (World Commission on Environment and Development 1987). Their definition confused sustainable growth, an oxymoron, and sustainable development, a possibility. Ambiguities notwithstanding, the concept of sustainability has become the keystone of the global dialogue about the human future. But what exactly do we intend to sustain, and what will that require of us?

Such questions would have had little meaning to generations prior to, say, 1950, when nuclear annihilation became possible. Other than a collision between Earth and a large meteor, there was no conceivable way that civilization everywhere could have been radically degraded or terminated.

This article was originally published in 2006.

But now any well-informed high school student could make a long list of ways in which humankind could cause its own demise, ranging from whimpers to bangs. The dialogue about sustainability is about a change in the human trajectory that will require us to rethink old assumptions and engage the large questions of the human condition that not long ago were thought to have been solved once and for all.

The things that cannot be sustained are clear. The ongoing militarization of the planet, along with the greed and hatred that feeds it, are not sustainable. Sooner or later a roll of the dice will come up Armageddon, whether in the Indian subcontinent, in the Middle East, or by an accidental launch, acts of a rogue state, or an act of terrorism. A world with a large number of desperately poor cannot be sustained, because they have power to disrupt lives of the comfortable in ways that we are only beginning to appreciate, and it would not be worth sustaining anyway. The perpetual enlargement of the human footprint in nature cannot be sustained, because it will eventually overwhelm the capacity and fecundity of natural systems and cycles. The unrestrained development of any and all technology cannot be sustained without courting risks and adversity that we often see only in hindsight. A world of ever-increasing economic, financial, and technological complexity cannot be sustained, because sooner or later it will overwhelm our capacity to manage. A world divided by narrow, exclusive, and intense allegiances to ideology or ethnicity cannot be sustained, because its people will have too little humor, compassion, forgiveness, and wisdom to save themselves. Unrestrained auto-mobility, hedonism, individualism, and conspicuous consumption cannot be sustained, because they take more than they give back. A spiritually impoverished world is not sustainable, because meaninglessness, anomie, and despair will corrode the desire to be sustained and the belief that humanity is worth sustaining. But these are the very things that distinguish the modern age from its predecessors. Genuine sustainability, in other words, will come not from superficial changes but from a deeper process akin to humankind growing into a fuller stature.

The question then is, not whether we change, but whether the transition is done with more or less grace and whether the destination is desirable or not. The barriers to a graceful transition to sustainability, whatever form it may take, are not so much technological as they are social, political, and psychological. It is possible that we will be paralyzed by information overload leading to a kind of psychic numbness. It is possible that we will suffer what Thomas Homer-Dixon calls an "ingenuity gap" in which problems outrun our problem-solving capacities (Homer-Dixon 2000). It is possible that the sheer scale and complexity of human systems will become utterly unfathomable, hence unmanageable. It is possible that we will fail to comprehend the nature of nature sufficiently to know how to live well on the Earth in large numbers. It is possible that we will fail to make a smooth transition, because of political ineptitude and a lack of leadership and/or because power is co-opted by corporations and private armies. It is possible that we will fail because the powers of denial and wishful thinking cause us to underestimate the magnitude of our problems and overlook better possibilities. And it is possible that we might fail because of what can only be called a condition of spiritual emptiness. The challenges of sustainability come hard on the heels of a century in which perhaps as many as 200 million people were killed in wars, ethnic conflicts, and extermination camps, taking a psychic toll that we dimly understand.

On the other hand it is possible, and I think likely, that the challenge of survival is precisely what will finally bring humankind together in the realization of the fragility of civilization and the triviality of most of our causes relative to the one central issue of survival. The overall challenge of sustainability is to avoid crossing irreversible thresholds that damage the life systems of Earth while creating long-term economic, political, and moral arrangements that secure the well-being of present and future generations. We will have to acknowledge that the Enlightenment faith in human reason is, in some measure, wrong. But this does not mean less enlightenment, but rather a more enlightened enlightenment tempered by the recognition of human fallibility—a more rational kind of reason. In this light the great discovery of the modern era is not how to make nuclear fire, or alter our genes, or communicate 24/7 at the speed of light but, rather, the discovery of our interconnectedness and implicatedness in the web of life (Capra 1996, 2002). What Thomas Berry calls the "Great Work" of the twenty-first century will be to comprehend what that awareness means in every area of life in order to calibrate human demands with what the Earth can sustain. Broadly speaking, the transition to sustainability poses four challenges.

First, we need more accurate models, metaphors, and measures to describe the human enterprise relative to the biosphere. We need a compass that defines true north for a civilization long on means and short on direction. On the one hand the conventional wisdom describes us as masters of the planet, destined to become ever more numerous and rich without explaining how this is possible or why it might be desirable. In contrast, Howard and Elisabeth Odum argue, for example, "that many, if not all, of the systems of the planet have common properties, organize in similar ways, have similar oscillations over time, have similar patterns spatially, and operate within universal energy laws" (Odum and Odum 2001, 5). From the perspective of systems ecology, the efflorescence of humanity in the twentieth century is evidence of a natural pulsing. But having exhausted much of the material basis for expansion (Odum and Odum 2001, 85), like other systems, we are entering a down cycle, a "long process of reorganizing to form a lesser economy on renewable resources," before another upward pulse (Odum and Odum 2001, 8). The pattern of growth/retreat they find in all systems stands in marked contrast to the rosy assumptions of perpetual economic growth. For the Odums smart policy would include plans for a prosperous descent, to avoid an otherwise catastrophic collapse. The specific tasks they propose are to "stabilize capitalism, protect the Earth's production of real wealth, and develop equity among nations" (Odum and Odum 2001, 133).

Archeologist Joseph Tainter (1988) proposes a similar model based on the rise and collapse of complex societies. Collapse eventually occurs when "investment in sociopolitical complexity . . . reaches a point of declining marginal returns" (Tainter 1988, 194). In Tainter's view, this is "not a fall to some primordial chaos, but a return to the normal human condition of lower complexity" (Tainter 1988, 198). Patterns of declining marginal returns he believes are now evident in some contemporary industrial societies in areas of agriculture, minerals and energy production, research, health care, education, and military and industrial management. Like the Odums, Tainter regards expansion and contraction as parts of a normal process. But how do we know whether we are in one phase or the other? The answer requires better accounting tools that relate human wealth generation to some larger measure of biophysical health. The Odums propose the concept of *emergy*, or what they define as "the available energy of one kind that has to be used up directly and indirectly to make a product or service" (Odum and Odum 2001, 67). By their accounting, the amount of embodied energy in solar equivalent units gives a more accurate picture of our relative wealth than purely financial measures. Others are developing different tools to the same purpose of including natural capital otherwise left out of purely economic accounting.

Second, the transition to sustainability will require a marked improvement and creativity in the arts of citizenship and governance (Carley and

Christie 2000). There are some things that can be done only by an alert citizenry acting with responsive and democratically controlled governments. Only governments moved by an ethically robust and organized citizenry can act to ensure the fair distribution of wealth within and between generations. Only governments prodded by their citizens can act to limit risks posed by technology or clean up the mess afterward. Only governments and an environmentally literate public can choose to adopt and enforce standards that move us toward a cradle-to-cradle materials policy. Only governments acting on a public mandate can license corporations and control their activities for the public benefit over the long term. Only governments can create the financial wherewithal to rebuild ecologically sound cities and dependable public transportation systems. Only governments acting with an informed public can set standards for the use of common property resources, including the air, waters, wildlife, and soils. And only governments can implement strategies of resilience that enable the society to withstand unexpected disturbances. Resilience means dispersed, not concentrated, assets, control, and capacity. A resilient society, for instance, would have widely dispersed manufacturing, many small farms, many small cities and towns, greater self-reliance, and few if any technologies vulnerable to catastrophic failure, acts of God, or human malice. Sustainability, in short, constitutes a series of public choices that require effective institutions of governance and a well-informed and politically engaged citizenry.

The third challenge, then, is to inform the discretion of the public through greatly improved education. The kind of education needed for the transition to sustainability, however, has little to do with improving SAT or GRE scores or advancing skills necessary to an expansionist phase of human culture. "During growth," in the Odums' words, "emphasis was on getting new information . . . but as resource availability declines, emphasis [will be] on efficiency in teaching information that we already have" (Odum and Odum 2001, 258). They suggest a curriculum organized around the study of the relationships between energy, environment, and economics and how these apply across various scales of knowledge. Students of all ages will need the kind of education and skills appropriate to building a society with fewer cars but more bicycles and trains, fewer large power plants but more windmills and solar collectors, fewer supermarkets and more farmers' markets, fewer large corporations and more small businesses, less time for leisure but more good work to do, and less public funding but more public spirit. The rising generation, then, must restore natural capital of soils, forests, watersheds, and wild areas; clean up the toxic messes from the expansionist phase; build habitable cities; relearn the practices of good farming; and learn the arts of powering civilization on efficiency and sunlight. Education appropriate to their future, not the least, will require the courage to provide "intellectual leadership for the Long Run" based on a clear understanding of where we stand relative to larger cycles and trends (Odum and Odum 2001, 262).

Fourth, it is easy to offer long lists of solutions and still not solve the larger problem. The difficulty, once identified by E. F. Schumacher, is that human problems, like those posed by the transition to sustainability, are not solvable by rational means alone. These are what he called "divergent" problems formed out of the tensions between competing perspectives that cannot be solved, but can be transcended (Schumacher 1977, 120–33). In contrast to "convergent" problems that can be solved by logic and method, divergent problems can only be resolved by higher forces of wisdom, love, compassion, understanding, and empathy. The logical mind does not much like divergent problems, because it operates more easily with "either/or, or yes/no . . . like a computer" (Schumacher 1977). Recognizing the challenge of sustainability as a series of divergent problems leads to the fourth and most difficult challenge of all.

The transition to sustainability will require learning how to recognize and resolve divergent problems, which is to say a higher level of spiritual awareness. By whatever name, something akin to spiritual renewal is the sine qua non of the transition to sustainability. Scientists in a secular culture are often uneasy about matters of spirit, but science on its own can give no reason for sustaining humankind. It can, with equal rigor, create the knowledge that will cause our demise or that necessary to live at peace with each other and nature. But the spiritual acumen necessary to solve divergent problems posed by the transition to sustainability cannot be a return to some simplistic religious faith of an earlier time. It must be founded on a higher order of awareness that honors mystery, science, life, and death.

Specifically, the kind of spiritual renewal essential to sustainability must enable us to forgive the terrible wrongs at the heart of the bitter ethnic and national rivalries of past centuries and move on. There is no convergent logic or scientific solution that will enable us to transcend self-perpetuating hatreds and habitual violence. The only solution to this divergent problem is a profound sense of forgiveness and mercy that rises above the convergent logic of justice. The spiritual renewal necessary for the transition must provide convincing grounds by which humankind can justify the project of sustainability. We are, in Lynn Margulis's words, "upright mammalian weeds" (Margulis 1998, 149). But is this all that we are or all that we can be? If so, we have little reason to be sustained beyond our sheer will to live. Perhaps this is enough, but I doubt it. A robust spiritual sense may not mean that we are created in the image of God, but it must offer hope that we may grow into something better than a planetary plague. A robust spirituality must help us go deeper in order to resolve what Ernest Becker once described as the "terror of death" (Becker 1973, 11) that "haunts the human animal like nothing else" (Becker 1973, ix). The effort, to deny the reality of our death, he believed, serves as "a mainspring of human activity" including much that we now see cannot be sustained. "Modern man is drinking and drugging himself out of awareness or he spends his time shopping, which is the same thing" (Becker 1973, 284). "Taking life seriously," he wrote, "means that whatever man does on this planet has to be done in the lived truth of the terror of creation, of the grotesque, of the rumble of panic underneath everything." In words written shortly before his own death Becker concluded, "The urge to cosmic heroism, then, is sacred and mysterious and not to be neatly ordered and rationalized by science and secularism" (Becker 1973, 284). No culture has gone farther than our own to deny individual mortality, and in the denying, it is killing the planet. A spirituality that allows us to face our own mortality honestly without denial or terror contains the seeds of the daily heroism necessary to preserve life on Earth. Instead of terror, a deeper spirituality would lead us to a place of gratitude and celebration.

Chapter 10

The Problem of Sustainability

(1992)

THREE CRISES LOOM DEAD AHEAD. The first is a food crisis evident in two curves that intersect in the not too distant future: one showing worldwide soil losses of 24 billion tons, the other a rapidly rising world population. The second crisis is that caused by the era of cheap fossil energy and its conclusion. We are in a race between the exhaustion of fossil fuels, global warming, and the policy requirements necessary to transition to a new era based on efficiency and solar energy. The third crisis, perhaps best symbolized by the looming prospect of a global climate change, has to do with ecological thresholds and the limits of natural systems. We can no longer assume that nature will be either bountiful or stable or that the Earth will remain hospitable to civilization. These three crises feed upon one another. They are interactive in ways that we cannot fully anticipate. Together they constitute the first planetary crisis, one that will either spur humans to a much higher state or cause our demise. It is not too much to say that the decisions about how or whether life will be lived in the next century are being made now. We have a few decades, perhaps, in which we must make unprecedented changes in the way we relate to each other and to nature.

In historical perspective, the crisis of sustainability appeared with unprecedented speed. Very little before the 1960s prepared us to understand the dynamics of complex interactive systems and the force of exponential

---

This article was originally published in 1992.

growth. A few prescient voices, including those of George Perkins Marsh, John Muir, Paul Sears, Fairfield Osborn, Aldo Leopold, William Vogt, and Rachel Carson, warned of resource shortages and the misuse of nature. But their warnings went largely unheard. Technological optimism, economic growth, and national power are deeply embedded in the modern psyche. The result is an enormous momentum in human affairs without as yet any good end in sight.

The crisis is unique in its range and scope including energy, resource use, climate, waste management, technology, cities, agriculture, water, biological resilience, international security, politics, and human values. Above all else it is a crisis of spirit and spiritual resources. We have it on high authority that without vision people perish. We need a new vision, a new story that links us to the planet in more life-centered ways. The causes of the crisis are related to those described by the early critics of modernity such as Marx, Weber, Durkheim, Dostoevsky, Freud, and Gandhi. But they dealt principally with the social effects of industrialization, not with its biophysical effects. It is our challenge to see both as parts of a single system. The anomie, rootlessness, and alienation of the modern world are part of a larger system of values, technologies, culture, and institutions which also produce acid rain, climate change, toxic wastes, terrorism, and nuclear bombs.

From one perspective these represent a set of problems, which by definition are solvable with enough money, the right policies, and technology. From another they are more accurately regarded as dilemmas for which there can be no purely technical solution. Put differently, can the values, institutions, and thrust of modern civilization be adapted to biophysical limits, or must we begin the task of creating something different? The answer hinges on what we believe to be the causes of unsustainability, which is to say, where and how we went wrong. What problems are we attempting to solve? How do these mesh with different policies, technologies, and behavior now proposed as solutions?

Five possibilities stand out. The crisis can be interpreted as a result of one or more social traps; it may stem from flaws in our understanding of the relation between the economy and the Earth; it could be a result of the drive to dominate nature evident in our science and technology; it may have deeper roots that can be traced to wrong turns in our evolution; or finally, it may be due to sheer human perversity. I am inclined to believe that any full explanation of the causes of our plight would implicate all five. They are like the layers of an onion: peel one off and you discover yet another below. In the intellectual peeling, asking why leads to the next layer and to deeper levels of causation. I will consider these from the "outside in," from the most apparent and, I think, least problematic causes to deeper ones that become harder to define and more difficult to resolve.

The Crisis as a Social Trap

The crisis of sustainability, in Robert Costanza's words, is in part the result of rational behavior in "situation(s) characterized by multiple but conflicting rewards. . . . Social traps draw their victims into certain patterns of behavior with promises of immediate rewards and then confront them with consequences that the victims would rather avoid" (Costanza 1987). Arms races, traffic jams, cigarette smoking, population explosions, and overconsumption are traps in which individually rational behavior in the near term traps victims into long-term destructive outcomes. With each decision, players are lured into behavior that eventually undermines the health and the stability of the system. In Garrett Hardin's famous essay "The Tragedy of the Commons," the villager rationally decides to graze an additional cow on an already overgrazed commons because the system rewards him for doing so. He can ignore the costs to others and eventually to himself because the system rewards individual irresponsibility. Similarly, the dynamic of technological competition, such as arms races, creates pressures to deploy a new device or weapon, only to be matched or overmatched by others, thereby raising the costs of deadlock and increasing the risks of system failure. In both cases the rewards are short-term and the costs are long-term and paid by all.

To the extent that the crisis of sustainability is a product of social traps in the way we use fossil energy, land, water, forests, minerals, and biological diversity, the solutions must in one way or another change the timing of payoffs so that long-term costs are paid up front as part of the purchase price. This is the rationale behind proposals for carbon taxes and life-cycle costing. Hardin's villager would be deterred from grazing another cow by having to pay the full cost of additional damage to the commons. The Pentagon's weapons addiction might be reversed by something like a tax on all weapons that could be used offensively in direct proportion to their potential destructiveness. In these and other instances, honest bookkeeping would deter entry into social traps.

The theory is entirely plausible. No rational decision maker willingly pays higher costs for zero net gain, and no rational society rewards members to undermine its existence. To the contrary, rational societies would reward decisions that lead to long-term collective benefits and punish the contrary. A sustainable society, then, will result from the calculus of self-interest. This approach requires minimal change in existing values, and fits most of our assumptions about human behavior derived from economics.

The theory is vulnerable, however, to some of the same criticisms made of market economics. Do we have, or can we acquire, full information about the long-term costs of our actions? In most cases the answer is "no." Consumers who used freon-charged spray cans in the 1960s, thereby contributing to ozone depletion, could not be charged because no one knew the long-term costs involved. Given the dynamism of technology and the complexity of most human/environment interactions, it is not likely that many costs can be predicted in advance and assigned prices to affect decisions in a timely way. Some may not even be calculable in hindsight. But assuming complete information, would we willingly agree to pay full costs rather than defer costs to the future and/or to others? There is a peculiar recalcitrance in human affairs known to advertisers, theologians, and some historians. It has the common aspect of preference for self-aggrandizement in the short term, devil take the hindmost in the long term. People who still choose to smoke or who refuse to wear seatbelts persist, not because they are rational, but because they can rationalize. Some who risk life and livelihood for others do so not because these represent "rational" choices, as that word is commonly understood, but because of some higher motivation.

Efforts to build a sustainable society on assumptions of human rationality must be regarded as partial solutions and first steps. But recognition of social traps and designing policies to avoid them would constitute important steps in building a sustainable society. Why we fall into social traps and generally find it difficult to acknowledge their existence—that is, to behave rationally—leads to the consideration of deeper causes.

The Crisis as a Consequence of Economic Growth

A second and related cause of the crisis of sustainability has to do with the propensity of all industrial societies to grow beyond the limits of natural systems. Economic growth is commonly regarded as the best measure of government performance. It has come to be the central mission of all developed and developing societies. In the words of political scientist

Henry Teune, "an individually based secular morality cannot accept a world without growth" (Teune 1988, 111). Growth, he asserts, is necessary for social order, economic efficiency, equitable distribution, environmental quality, and freedom of choice. In the course of his argument we are instructed that agribusiness is more efficient than family farms, which is not true, that forests are doing fine, which is not true, and that we are all beneficiaries of nuclear power, which deserves no comment. Nowhere does Teune acknowledge the dependence of the economy on the larger economy of nature, or the unavoidable limits set by that larger economy. For example, humans now use, directly and indirectly, 40 percent of the net primary productivity of terrestrial ecosystems on the planet and are changing climate, exterminating species, and toxifying ecosystems. How much more of nature can we take without undermining the biophysical basis of civilization, not to mention growth? Professor Teune does not say.

The most striking aspect of arguments for unending growth is the presumption that it is the normal state of things. Nothing could be further from the truth. The growth economy along with much of the modern world is, in a larger view, an aberration. For perspective, if we compare the evolutionary history of the planet to a week's time, as David Brower proposes, the industrial revolution occurred 1/40th of a second before midnight on the seventh day, and the explosive economic growth since 1945 occurred in the last 1/500th of a second before midnight. In the words of historian Walter Prescott Webb, the years between 1500 and 1900 were "a boom such as the world had never known before and probably never can know again" (Webb 1975, 13). The discovery of a "vast body of wealth without proprietors" in the New World radically altered ratios of resources to people. But by the time Frederick Jackson Turner announced the closing of the American frontier in 1893, these ratios were once again what they had been in the year 1500. Technology, for Webb, offered no way out: "On the broad flat plain of monotonous living [he was from Texas] we see the distorted images of our desires glimmering on the horizons of the future; we press on toward them only to have them disappear completely or reappear in a different form in another direction" (Webb 1975, 282). Webb would not have been surprised either by the frantic expectations raised by various technological magic bullets or the ways in which they fail to meet overblown expectations. For him, the inexorable facts were the ratios of people to land and resources.

Twenty-two years later, a team of systems scientists at MIT armed with computer models came to similar conclusions about the limits to growth

(Meadows 1972). Their results showed that population and resource use could not continue to grow exponentially without catastrophic collapse in the later decades of the twenty-first century. Marked increases in resource efficiency and pollution control did not appreciably alter the results. Catastrophe in exponentially growing systems is not necessarily evident until it is too late to avert.

The assumption of perpetual growth raises fundamental questions about the theoretical foundations of modern economics. Growth does not happen without cause. It is in large part the result of a body of ideas and theories that inform, motivate, and justify economic behavior. In the twentieth century the world economy expanded by 1300 percent, but can growth continue at this pace in the next century? Mainstream economists are evidently still in agreement with conclusions reached by Harold Barnett and Chandler Morse in 1963:

> Advances in fundamental science have made it possible to take advantage of the uniformity of matter/energy—a uniformity that makes it feasible, without preassignable limit, to escape the quantitative constraints imposed by the character of the earth's crust. . . . Science, by making the resource base more homogeneous, erases the restrictions once thought to reside in the lack of homogeneity. In a neo-Ricardian world, it seems, the particular resources with which one starts increasingly become a matter of indifference. The reservation of particular resources for later use, therefore, may contribute little to the welfare of future generations. (Daly 1980, 8)

Or as Harvard economist Robert Solow once said, "The world can, in effect, get along without natural resources." For Julian Simon, resources "are not finite in any economic sense" (Simon 1980, 17). Human ingenuity is "the ultimate resource" (the title of Simon's book) and will enable us to overcome constraints that are merely biophysical.

Nonetheless, a different economics is emerging, rooted in the fact that "the economic process consists of a continuous transformation of low entropy into high entropy, that is, into irrevocable waste" (Georgescu-Roegen 1971, 281). The laws of thermodynamics, which say that we can neither create nor destroy energy and matter and that the process goes from ordered matter, or "low entropy," to waste, or "high entropy," set irrevocable limits to economic processes. We burn a lump of coal, low entropy, and create ashes and heat, high entropy. Faster economic growth only increases the rate at which we create high entropy in the form of waste, heat, garbage, and disorder. The destiny of the human species, according to Georgescu-Roegen, "is to choose a truly great, but brief, not a long and dull, career" (Georgescu-Roegen 1971, 304).

Economic growth is the sum total of what individuals make, grow, buy, sell, and discard. And at the heart of conventional growth economics, which purports to explain all of this, one meets a theoretical construct that economists have named "economic man," a proudly defiant moral disaster programmed to maximize his utility, which is whatever he is willing to pay for. By all accounts this includes a great many things and services that used to be freely included as a part of the fabric of life in societies with village greens, front porches, good neighbors, sympathetic saloon keepers, and competent people. Economic man knows no limits of discipline, or obligation, or satiation, which may explain why the growth economy has no logical stopping point, and perhaps why good neighbors are becoming harder to find. Psychologists identify this kind of behavior in humans as "infantile self-gratification." When this kind of behavior is manifested by entire societies, economists describe it as "mature capitalism."

In a notable book in 1977, economist Fred Hirsch described other limits to growth that were inherently social (Hirsch 1976). As the economy grows, the goods and services available to everyone theoretically increase, except for those that are limited, like organizational directorships and lakeside homes, which Hirsch calls "positional goods." After basic biological and physical needs are met, an increasing portion of consumption is valued because it raises one's status in society. But, "if everyone in a crowd stands on tiptoe," as Hirsch writes, "no one sees better." Rising levels of consumption do not necessarily increase one's status. Consumption of positional goods, however, gives some the power to stand on a ladder. The rest are not necessarily worse off physically but are decidedly worse off psychologically. The attendant effects on economic psychology "become an increasing brake" on economic growth. Growing numbers of people whose appetites have been whetted by the promise of growth find only social congestion that limits leadership opportunities and status. Hirsch puts it this way:

> The locus of instability is the divergence between what is possible for the individual and what is possible for all individuals. Increased material resources enlarge the demand for positional goods, a demand that can be satisfied for some only by frustrating demand by others. (Hirsch 1976, 67)

The results, which he describes as the "economics of bad neighbors," include a decline in friendliness, the loss of altruism and mutual obligation, increased time pressures, and indifference to public welfare. Moreover, the pursuit of private and individual satisfaction by corporations and consumers undermines the very moral underpinnings—honesty, frugality, hard work, craftsmanship, and cooperation—necessary for the system to function. In short, after basic biological needs are met, further growth both "fails to deliver its full promise" and "undermines its social foundations" (Hirsch 1976).

The economist Joseph Schumpeter once made a similar argument. Capitalism, he thought, would ultimately undermine the attitudes and values necessary to its stability. "There is in the capitalist system," he wrote in 1942, "a tendency for self-destruction" (Schumpeter 1962, 162). Robert Heilbroner argues similarly that business civilization will decline, not only because of pollution and "obstacles of nature," but also because of the "erosion of the 'spirit' of capitalism" (Heilbroner 1976, 111). A business civilization inevitably becomes more "hollow" as material goods fail to satisfy deeper needs, including those for truth and meaningful work. Its demise will result from the "vitiation of the spirit that is sapping business civilization from within" (Heilbroner 1976, 115). At the very time that the system needs the loyalty of its participants most, they will be indifferent or hostile to it.

If the evidence suggests that economic growth is ecologically destructive, and soon to be constrained by biophysical and/or social limits, why do most economists want even more of it? A common answer is that growth is necessary to improve the situation of the poor. But this has not happened as promised. Rapid growth between 1980 and the market collapse of 2008 dramatically increased the concentration of wealth in the United States. The same pattern is evident worldwide, as the gap between the richest and poorest has widened from 3:1 in 1800 to perhaps 25:1 or more at present. Within poor countries, the benefits of growth predominantly go to the wealthiest, not to those who need them most. The importance of growth to the modern economy cannot be justified empirically on the grounds that it creates equity. Growth serves other functions, one of which is the avoidance of having to face the issue of fair distribution. As long as the total pie is growing, absolute but not relative wealth can be increased. If growth stops for any reason, the questions of distribution become acute. Political scientist Volkmar Lauber has made a good case that "the main motivation of growth . . . is not the pursuit of material gratification by the masses but the pursuit of power by elites" (Lauber 1978, 200). His case rests in part on analysis of public opinion polls in Europe and the United States showing only indifferent support for economic growth and much stronger support for quality of life improvements. In other words, economic growth occurs, not because people demand it, but because elites do. Growth makes the wealthy more so, but it also gives substantial power to government and corporate elites who manage the economy, its technology, and all of its side effects.

From the perspective of physics and ecology, the flaws in mainstream economics are fundamental and numerous. First, the discipline lacks a concept of optimal size for the economy. Second, as Daly argues, it mistakenly regards an increasing gross national product as an achievement, rather than as a cost required to maintain a given level of population and artifacts. Third, it lacks an ecologically and morally defensible model of the "reasonable person," helping to create the behavior it purports only to describe. Fourth, growth economics has radically misconceived nature as a stock to be used up. The faster a growing volume of materials flows from mines, wells, forests farms, and oceans through the economic pipeline into dumps and sinks, the better. Depletion at both ends of this stream explains what Wendell Berry calls the "ever-increasing hurry of research and exploration" driven by the "desperation that naturally and logically accompanies gluttony" (Berry 1987, 68). Fifth, growth economics assumes that the human economy is independent of the larger economy of nature, with its cycles and ecological interdependencies, and of the laws of physics that govern the flow of energy.

The prominence of the economy in the modern world, and that of growth economics in the conduct of public affairs, explains, I think, why we fall into social traps. The cultivation of mass consumption through advertising promotes the psychology of instant gratification and easy consumer credit, which create pressures that lead to risky technological fixes, perhaps the biggest trap of all. The discipline of economics has taught us little or nothing of the discipline imposed on us by physics and by natural systems. To the contrary, these are regarded as minor impediments to be overcome by substitution of one material for another, more ingenious technology, and the laws of supply and demand. But economics is, in turn, a part of a larger enterprise to dominate nature through science and technology.

The Crisis as the Result of the Urge to Dominate Nature

At a deeper level, then, the crisis of sustainability can be traced to a drive to dominate nature that is evident in Western science and technology. But what is the source of that urge? One possibility, according to historian

Lynn White, is that the drive to dominate nature is inherent in Judeo-Christian values (White 1967). The writers of Genesis commanded us to be fruitful, multiply, and have dominion over the Earth and its creatures. We have done as instructed. And this, according to White, is the source of our problems. But the Bible says many things, some of which are ecologically sound. Even if it did not, there is a long time between the writing of Genesis and the onset of the problems of sustainability. An even larger gap may exist between biblical commandments generally and human behavior. We are enjoined, for example, to love our enemies, but as yet without comparable results. Something beyond faith seems to be at work. That something is perhaps found in more proximate causes: capitalism, the cult of instrumental reason, and industrial culture.

Lewis Mumford attributes the urge to dominate nature to the founders of modern science: Bacon, Galileo, Newton, and Descartes. Each, in Mumford's words, "lost sight of both the significance of nature and the nature of significance" (Mumford 1970, 82). Each contributed to the destruction of an organic worldview and to the development of a mechanical world that traded the "totality of human experience . . . for that minute portion which can be observed within a limited time span and interpreted in terms of mass and motion" (Mumford 1970, 57).

Similar themes are found earlier in writings of Martin Heidegger and Alfred North Whitehead and in the recent work of Carolyn Merchant, William Leiss, Morris Berman, Jacques Ellul, and nearly all critics of technology. With varying emphases, all argue that modern science has fundamentally misconceived the world by fragmenting reality, separating observer from observed, portraying the world as a mechanism, and dismissing nonobjective factors, all in the service of the domination of nature. The result is a radical miscarriage of human purposes and a distortion of reality under the guise of objectivity. Beneath the guise, however, lurks a crisis of rationality in which means are confused with ends and the domination of nature leads to the domination of other persons. C. S. Lewis said:

> At the moment, then, of man's victory over nature, we find the whole human race subjected to some individual men, and individuals subjected to that in themselves which is purely "natural"—to their irrational impulses. Nature, untrammelled by values, rules the Conditioners and, through them, all humanity. (Lewis 1947, 79–80)

The crisis of rationality of which Lewis wrote is becoming acute with the advent of nuclear weapons and genetic engineering. In a remarkable article entitled "The Presumptions of Science" in the journal *Daedalus* in 1978, biologist Robert Sinsheimer asked, "Can there be forbidden or inopportune knowledge?" (Sinsheimer 1978, 23–35). *Frankenstein* was Mary Shelley's way of asking a similar question 160 years earlier: is there knowledge for which we are unwilling or unable to take responsibility? It is common to believe that all knowledge—whatever its effects—is good and all technology unproblematic. These articles of faith rest, as Sinsheimer notes, on the belief that "nature does not set booby traps for unwary species" and that our social institutions are sufficiently resilient to contain the political and economic results of continual technological change. He recommends that "we forgo certain technologies, even certain lines of inquiry where the likely application is incompatible with the maintenance of other freedoms" (Sinsheimer 1978).

The idea that science and technology should be limited on grounds of ecological prudence or morality apparently struck too close to the presumptions of establishment science for comfort. Sinsheimer's article was met with a thundering silence. Science and technology are religion in Western culture. Research, adding to society's total inventory of undigested bits of knowledge, is now perhaps as holy a calling as saving the heathen was in other times. Yet the evidence mounts that unfettered scientific exploration, now mostly conducted in large, well-funded government or corporate laboratories, can sometimes add to the difficulties of building a durable society. Weapons labs create continual upward pressures on the arms race, independent of political and policy considerations. The same is true in the economy where production technologies displace workers, threaten the economies of whole regions, and introduce a constant stream of environment-threatening changes (for example, thousands of new chemicals introduced each year; synthetic fabrics substituted for cotton and wool; plastics for leather and cellulose; detergents for soap; chemical fertilizers for manure; fossil or nuclear energy for human, natural, or animal energy). In each case, the reason for the change has to do with economic pressures and technological opportunities. In historian Donald Worster's words, the problem posed by science and technology lies "in that complex and ambitious brain of *Homo sapiens*, in our unmatched capacity to experiment and explain, in our tendency to let reason outrun the constraints of love and stewardship" (Worster 1987, 101). For Worster, as for Sinsheimer, we need "the most stringent controls over research."

On the other side of the issue is the overwhelming majority of scientists, engineers, and their employers who regard science and technological innovation as inherently good and essential, either to surmount natural constraints or to develop energy and resource efficiency necessary for sustainability. These two positions differ, not on the importance of knowledge, but over the kind of knowledge necessary. On the minority side are those seeking what Erwin Chargaff calls "old and solid knowledge," which used to be called wisdom. It has less to do with specialized learning and the cleverness of means than with broad, integrative understanding and the careful selection of ends. Such knowledge, in Wendell Berry's phrase, "solves for pattern." It does not result, for example, in the expenditure of millions of federal research dollars to develop genetically derived ways to increase milk production at the same time that the U.S. Department of Agriculture is spending millions to slaughter dairy herds because of a milk glut.

No one, of course, is against wisdom. But while we mass-produce technological cleverness in research universities, we assume that wisdom can take care of itself. The results of technical research are evident and most often profitable. Wisdom is not so easy—what passes for wisdom may be only eloquent foolishness. Real wisdom may not be particularly useful. The search for integrative knowledge would probably not contribute much to the gross national product, or to the list of our technological achievements, and certainly not to our capacity to destroy. As often as not, it might lead us to stop doing a lot of things that we are now doing, and to reflect more on what we ought to do.

But any attempt to control scientific inquiry and technology runs into three major problems. The first is that of separating the baby from the bathwater. Research needs to be done, and appropriate technologies will be important building blocks of a sustainable world. In this category, I would include research into energy efficiency and solar technologies, materials efficiency, the restoration of damaged ecosystems, how to build healthy cities and to revitalize rural areas, how to grow food in an environmentally sound manner, and the conditions of peace. These are things on which our survival, health, and prosperity depend. Without much effort, we could assemble another list of research that works in the opposite direction. The challenge before us is to learn how to make distinctions between knowledge that we need from that which we do not need, including that which we cannot control. This distinction will not always be clear in advance, nor would it always be enforceable. What is possible, however, is to clarify the relationship between technology, knowledge, and the goals of sustainability and to use that knowledge to shift public research and development expenditures accordingly.

A second problem is the real possibility that controls will undermine freedom of inquiry and First Amendment guarantees. Sinsheimer argues that freedom of inquiry should be balanced against other freedoms and values. Freedom of inquiry, in short, is not an absolute but must be weighed against other values, including the safety and survival of the system that makes inquiry possible in the first place. A third concern is the effectiveness of any system of controls. Sinsheimer proposes that limits be placed on funding and access to instruments, while admitting that past efforts to control science have often given license to bigots and book burners. Part of the difficulty lies in our inability to predict the consequences of research and technological change. Most early research is probably innocent enough, and only later does research become dangerous when converted into weapons, reactors, toxic chemicals, and production systems. Even these cannot automatically be regarded as bad without reference to their larger social, political, economic, and ecological context. If one society successfully limits potentially dangerous scientific inquiry, however, work by scientists elsewhere continues unless similarly proscribed. The logic of the system of research and technological development operates by the same dynamics evident in arms races or Hardin's tragedy of the commons. Failure to pursue technological developments, regardless of their side effects, places a corporation or a government at a potential disadvantage in a system where competitiveness and survival are believed to be synonymous.

There are no easy answers to issues posed by technology and science, but there is no escape from their consequences. At every turn, the prospects for sustainability hinge on the resolution of problems and dilemmas posed by that double-edged sword of unfettered human ingenuity. At the point where we choose to confront the effects of science and technology, we will discover no adequate philosophy of technology to light our path. Technology has expanded so rapidly, and initially with so much promise, that few thought to ask elementary questions about its relation to human purposes and prospects. Intoxication replaced prudence.

There is another way to see the problem. Perhaps much of our technology is not taking us where we want to go anyway. The thrust of technology has almost always been to make the world more effortless and efficient. The logical end of technological progress, as George Orwell once put it, was to "reduce the human being to something resembling a brain in a bottle" or "to make the world safe for little fat men." Our goal, Orwell thought, should be to "make life simpler and harder instead of softer and more complex" (Orwell 1958, 201, 210). Making life simpler, however, requires only a fraction of the technology now available.

Technological extravagance is most often justified because it makes our economy more competitive, that is, it enables us to grow faster than other economies. In doing so, however, we find ourselves locked into behavior patterns that impose long-term costs for short-term gains. Beyond social traps, growth economics, and the drive to dominate nature are more distant causes having to do with human evolution and the human condition.

The Crisis as the Result of an Evolutionary Wrong Turn

Perhaps in the transition from hunter-gatherer societies to agricultural and urban cultures we took the wrong fork in the road. That primitive hunter-gatherer societies more often than not lived in relative harmony with the natural world is of some embarrassment to the defenders of the faith in progress, and as anthropologist Marshall Sahlins reports, they did so at a high quality of life, with ample leisure time for cultural pursuits and with high levels of equality (Sahlins 1972). The designation of hunter-gatherers as "primitive" is a useful rationalization for cultural, political, and economic domination. In spite of vast evidence to the contrary, we insist that Western civilization should be the model for everyone else, but for most anthropologists there is no such thing as a superior culture, hence none that can rightly be labeled as primitive. Colin Turnbull concluded in *The Human Cycle* that in many respects hunter-gatherer tribes handled various life stages better than contemporary societies (Turnbull 1983). In Stanley Diamond's words, the reason "springs from the very center of civilization, not from too much knowledge but from too little wisdom. What primitives possess is the immediate and ramifying sense of the person, and . . . an existential humanity—we have largely lost" (Diamond 1981, 173).

If civilization represents a mistaken evolutionary path, what can we do? Human ecologist Paul Shepard once proposed a radical program of cultural restructuring that would combine elements of hunter-gatherer cultures with high technology and the wholesale redesign of contemporary civilization (Shepard 1973). Later, he argued for a more modest course that required rethinking the conduct of childhood and the need to connect the psyche with the Earth in the earliest years. Contact with earth, soil, wildlife, trees, and animals, he thought, is the substrate that orients adult thought and behavior to life. Without this contact with nature, maturity is spurious, resulting in "childish adults" with "the world's flimsiest identity structures" (Shepard 1982, 124).

For all of the difficulty in translating the work of Sahlins, Diamond, Shepard, and others into a coherent strategy for change, they offer three perspectives important for thinking about sustainability. First, from their work we know more about the range of possible human institutions and economies. In many respects, the modern world suffers in comparison with earlier cultures from a lack of complexity, if not complicatedness. This is not to argue for a simple-minded return to some mythical Eden of the sort described by Rousseau but an acknowledgement that earlier cultures were not entirely unsuccessful in wrestling with the problems of life, nor we entirely successful. Second, from their work, we know that aggressiveness, greed, violence, sexism, and alienation are in large part cultural artifacts not inherent in the human psyche. Earlier cultures did not engender these traits nearly as much as mass-industrial societies have. Riane Eisler reinterprets much of the prehistorical record and concludes that the norm prior to the year 5,000 was peaceful societies that were neither matriarchal nor patriarchal (Eisler 1987). Third, the study of other cultures offers a tantalizing glimpse of how culture can be linked to nature through ritual, myth, and social organization. Our alienation from the natural world is unprecedented. Healing this division is a large part of the difference between survival and extinction. If difficult to embody in a programmatic way, anthropology suggests something of lost possibilities and future potentials. A fourth possibility remains to be considered, having to do with the wellsprings of human behavior.

The Crisis of Sustainability and the Human Condition

In considering the causes of the crisis of sustainability, there is a tendency to sidestep the possibility that we are a flawed, cantankerous, willful, perhaps fallen, but certainly not entirely planet-broken race. These traits, however, may explain evolutionary wrong turns, flaws in our culture and science, and an affinity for social traps which describe the human condition. In psychologist Ernest Becker's words, "we are doomed to live in an overwhelmingly tragic and demonic world" (Becker 1973, 281). The demonic is found in our insatiable restlessness, greed, passions, and urge to dominate, whether fueled by Eros, Thanatos, fear of death, or the echoes of our ancient reptilian brain. At the collective level, there may be what John Livingston calls "species ambition" that stems from our chronic insecurity. "The harder we struggle toward immortality," he writes, "the fiercer becomes the suffocating vise of alienation" (Livingston 1982, 79). We are caught between the drive for Promethean immortality, which likely takes us to extinction, and what appears to be a meaningless survival in the recognition that we are only a part of a larger web of life. Caught between the prospect of a brief, exciting career and a long, dull one, the anxious animal chooses the former. In this statement of the problem we can recognize a variant of Gregory Bateson's description of a double bind from which there is no purely logical escape.

Can we build a sustainable society without seeking first the Kingdom of God or some reasonable facsimile thereof? Put differently, is cleverness enough, or will we have to be good in both the moral and ecological sense of the word? And if so, what does goodness mean in an ecological perspective? The best answer to this question, I believe, was given by Aldo Leopold: "A thing is right when it tends to preserve the integrity, stability, and beauty of the biotic community. It is wrong when it tends otherwise" (Leopold 1966, 224–25). The essence of Leopold's land ethic is "respect for his fellow members, and also respect for the (biotic) community as such" (Leopold 1966). Respect implies a sense of limits, things one does not do, not because they cannot be done but because they should not be done. But the idea of limits, or even community, runs counter to the Promethean mentality of technological civilization and the individualism of laissez-faire economics. At the heart of both, David Ehrenfeld argues, is an overblown faith in our ability "to rearrange the world of nature and the affairs of men and women." But "in no important instance," he writes, "have we been able to demonstrate comprehensive successful management of our world, nor do we understand it well enough to be able to manage it in theory" (Ehrenfeld 1979, 105). Even if we could do so, we could never outrun all of the ghosts and fears that haunt Promethean men.

All theological explanations, then, lead to proposals for a change in consciousness and deeper self-knowledge that recognize the limits of human rationality. In Carl Jung's words, "we cannot and ought not to repudiate reason, but equally we must cling to the hope that instinct will hasten to our aid" (Jung 1965, 341). The importance of theological perspectives in the dialogue about sustainability lies in their explicit recognition of persistent and otherwise inexplicable tragedy and suffering in history, and in history to come—even in a world that is otherwise sustainable. This realism can provide deeper insight into human motives and potentials, and an antidote to giddy and breathless talk of new ages and paradigm shifts. Whatever a sustainable society may be, it must be built on the most realistic view of the human condition possible. Whatever the perspectives of its founders, it must be resilient enough to tolerate the stresses of human recalcitrance. Theological perspectives may also alert us to the need to cultivate qualities of compassion and tolerance in the certainty that a sustainable society will require a great deal of it. They also alert us to the desirability of scratching where we itch. If we can fulfill all of our consumer needs, desires, and fantasies, as cornucopians like Julian Simon or devotees of technology and efficiency predict, there may be other nightmares ahead, of the sort envisioned by Huxley in *Brave New World* or that which afflicted that early student of paradox, King Midas. There is good reason not to get everything we want, and some reason to believe that in the act of consumption and fantasy fulfillment, we are scratching in the wrong place. But it is difficult to link these insights into a program for change; indeed the two may be antithetical. Jung, for one, dismissed the hyper intellectuality found in most rational schemes in favor of the process of metanoia arising from the collective unconscious. After a lifetime of reflection on these problems, Lewis Mumford could only propose grassroots efforts toward a decentralized, "organic" society based on "biotechnics" and "something like a spontaneous religious conversion . . . that will replace the mechanical world picture with an organic world picture" (Mumford 1970, 413).

Conclusion: Causes in Historical Perspective

The crisis of sustainability is without precedent, as is the dream of a sustainable global civilization. In attempting to build a durable social order, we must acknowledge that efforts to change society for the better have a dismal history. Societies change continually—but seldom in directions hoped for, for reasons that we fully understand, and with consequences that we can anticipate. Nor, to my knowledge, has any society planned and successfully moved toward greater sustainability on a willing basis. To the contrary, the historical pattern is, as Chateaubriand said, for forests to precede civilization, deserts to follow. The normal response to crises of carrying capacity has not been to develop a careful and thoughtful response meshing environmental demands with what the ecosystem can sustain over the long run. Rather, the record reveals either collapse of the offending culture, or technological adaptation that opens new land (new sources of carbon), water, or energy (including slave labor to contemporary use of fossil fuels). Economic development has largely been a crisis-driven process that occurs when a society outgrows its resource base (Wilkinson 1973).

The argument, then, that humankind has always triumphed over adversity in the past and will therefore automatically meet the challenges of the future has the distinction of being at once bad history and irrelevant. Optimists of the "ultimate resource" genre neglect the fact that history is a tale written by the winners. The losers, including those who violated the commandments of carrying capacity, disappeared mostly without writing much. We know of their demise, in part, through painstaking archeological reconstruction that reveals telltale signs of overpopulation, desertification, deforestation, famine, and social breakdown—what ecologists call "overshoot."

Even if humankind had always triumphed over challenges, the present crisis of sustainability would be qualitatively different, without any historical precedent. It is the first truly global crisis. It is also unprecedented in its sheer complexity. Whether by economics, policy, passion, education, moral suasion, or some combination of the above, advocates of sustainability propose to remake the human role in nature, substantially altering much that we have come to take for granted, from Galileo to Adam Smith to the present. Most advocates of sustainability recognize that it will also require sweeping changes in the relations between people, societies, and generations. And all of these must, by definition, have a high degree of permanence.

Still, history may provide important parallels and perspectives, beginning with the humbling awareness that we live on a planet littered with ruins that testify to the fallibility of our past judgments and foresight. Human folly will undoubtedly accompany us on the journey toward sustainability, which further suggests something about how that journey should be made. This will be a long journey. The poet Gary Snyder writes of a 1000-year process. Economists frequently write as if several decades will do. Between the poet's millennia and the economist's decades, I think it is reasonable to expect a transition of at least several centuries. But the major actions to stabilize the vital signs of Earth and stop the hemorrhaging of life must be made much sooner.

History, however, gives many examples of change that did not occur, and of other changes that were perverted. The Enlightenment faith in reason to solve human problems ended in the bloody excesses of the French Revolution. Historian Peter Gay said:

> The world has not turned out the way the philosophes wished and half expected that it would. Old fanaticisms have been more intractable, irrational forces more inventive than the philosophes were ready to conjecture in their darkest moments. Problems of race, of class, of nationalism, of boredom and despair in the midst of plenty have emerged almost in defiance of the philosophes' philosophy. We have known horrors, and may know horrors, that the men of the Enlightenment did not see in their nightmares. (Gay 1977, 567)

So to the extent that the faith in reason survives, it is applied to narrow issues of technology. The difference, in Leo Marx's words, "turns on the apparent loss of interest in, or unwillingness to name, the social ends for which the scientific and technological instruments of power are to be used" (Marx 1987, 71). Similarly, Karl Marx's vision of a humane society became the nightmare of Stalin's Gulags.

In our own history, progressive reforms far more modest than those necessary for sustainability have run aground on the shoals of corporate politics. The high democratic ideals of late-nineteenth-century populism gave way to a less noble reality. One historian put it this way:

> A consensus thus came to be silently ratified: reform politics need not concern itself with structural alteration of the economic customs of the society. This conclusion, of course, had the effect of removing from mainstream reform politics the idea of people in an industrial society gaining significant degrees of autonomy in the structure of their own lives. . . . Rather, . . . the citizenry is persuaded to accept the system as "democratic"—even as the private lives of millions become more deferential, anxiety-ridden, and less free. (Goodwyn 1978, 284)

A similar process is apparent in the decline of the reforms of the 1960s, which began with the high hopes of building "participatory democracy" described in the Port Huron Statement, only to tragically fall apart in chaos, camp, racism, assassinations, domestic violence, FBI surveillance, and a war that never should have been fought.

History is a record of many things, most of which were not planned or foreseen. And after Auschwitz, Hiroshima, the H-bomb, gulags, and killing fields we know that at best it is only partially a record of progress. It is easy at this point to throw up one's hands and conclude with the Kentucky farmer who informed the lost traveler that "you can't get there from here." That conclusion, however, breeds self-fulfilling prophecies, fatalism, and resignation—perhaps in the face of opportunities, but certainly in the face of an overwhelming need to act. We also have the historical examples of Gandhi, Martin Luther King, and Albert Schweitzer that suggest a different social dynamic, one that places less emphasis on confrontation, revolution, and slogans and more on patience, courage, moral energy, humility, and nonpolarizing means of struggle. And we have the wisdom of E. F. Schumacher's admonition to avoid asking whether we will succeed or not and instead to "leave these perplexities behind us and get down to work" (Schumacher 1977, 140).

Finally, the word *crisis*, based on a medical analogy, misleads us into thinking that after the fever breaks, things will revert to normal. This is not so. As long as anything like our present civilization lasts, it must monitor and restrain human demands against the biosphere. This will require an unprecedented vigilance and the institutionalization (or ritualization) of restraints through some combination of law, coercion, education, religion, social structure, myth, taboo, and market forces. History offers little help, since there is no example of a society that was or is both technologically dynamic and environmentally sustainable. It remains to be seen how and whether these two can be harmonized.

Chapter 11

Two Meanings of Sustainability

(1988)

A SUSTAINABLE SOCIETY, as commonly understood, does not undermine the resource base and biotic stocks on which its future prosperity depends. In the words of Lester Brown, Christopher Flavin, and Sandra Postel, "a sustainable society is one that satisfies its needs without jeopardizing the prospects of future generations" (Brown et al. 1990, 173). To be sustainable means living on income, not capital. The word *sustainable*, however, conceals as much as it reveals. Hidden beneath the rhetoric are assumptions about growth, technology, democracy, public participation, and human values. The term entered wide public use with Lester Brown's book *Building a Sustainable Society* and with the International Union for Conservation of Nature's *World Conservation Strategy*, both of which appeared in 1980. In 1987, the Brundtland Commission adopted "sustainable development" as the pivotal concept in its report *Our Common Future*. As defined by the Brundtland Commission, development is sustainable if it "meets the needs of the present without compromising the ability of future generations to meet their own needs" (World Commission on Environment and Development 1987, 43). Sustainable development requires "more rapid economic growth in both industrial and developing countries." The commission, therefore, politely appeased both sides of the debate. The word *sustainable* pacifies

This article was originally published in 1988.

environmentalists, while *development* has a similar effect on businessmen and bankers.

The phrase *sustainable development* raises as many questions as it answers. It presumes that we know, or can discover, levels and thresholds of environmental carrying capacity, which is to say, what is sustainable and what is not. But a society could be sustainable in a number of technology, population, and resource configurations. To be sustainable, for example, a larger population would have to live with less of almost everything per capita than a smaller society drawing on the same resource base. The phrase also deflects consideration about the sustainability and resilience of political and economic institutions, which certainly have their own limits. Third, the phrase seems to imply social engineering on an unlikely scale. Finally, the phrase suggests agreement about the causes of unsustainability, which does not exist. The dialogue about environment and development is mostly centered on discussion about policy adjustments or technological fixes of one sort or another. The deeper causes discussed in the previous chapter are seldom mentioned, perhaps because they raise the possibility that we are in much more dire straits than most care to believe.

In effect, the commission hedged its bets between two versions of sustainability, the first of which I will call "technological sustainability," the second, "ecological sustainability." In the most general terms, the difference is whether a society can become sustainable within the modern paradigms through better technologies and more accurate prices, or whether sustainability requires the transition to a culture that transcends the individualism, anthropocentrism, consumerism, nationalism, and militarism of modern societies. If regarded as successive stages, these are not necessarily mutually exclusive. To the contrary, I consider both to be necessary parts of a sustainable world. To use a medical analogy, the vital signs of the heart attack victim must be stabilized first or all else is moot. Afterward comes the longer-term process of dealing with the causes of the trauma, which have to do with diet and lifestyle. If these are not corrected, however, the patient's long-term prospects are bleak. Similarly, technological sustainability is about stabilizing planetary vital signs. Ecological sustainability is the task of finding alternatives to the practices that got us in trouble in the first place; it is necessary to rethink agriculture, shelter, energy use, urban design, transportation, economics, community patterns, resource use, forestry, the importance of wilderness, and our central values. These two perspectives are partly complementary, nonetheless their practitioners tend to have very different views about the extent of our plight, technology, centralized power, economics and economic growth, social change and how it occurs, the role of public participation, and the importance of value changes and ultimately very different visions of a sustainable society.

Technological Sustainability

Advocates of technological sustainability tend to believe that every problem has either a technological answer or a market solution. There are no dilemmas to be avoided, no domains where angels fear to tread. Resource scarcity will be solved by materials substitution, or genetic engineering. Energy shortages will be solved by more efficiency improvements, better technology, and, for some, nuclear power. The belief in technological sustainability rests on the following beliefs.

The first and most important of these is the assertion that humans, as Herman Kahn once said, should be "numerous, rich, and in control of the forces of nature." The goal of sustainable development in this sense is familiar to devout readers of the dominion passage in Genesis and to acolytes of Francis Bacon. From Bacon we found justification for the union of science and power that, in his words, would "command nature in action." Bacon sought, not truth as such, but a particular kind of truth that would lend itself to specific outcomes. His means of "vexing" nature were aimed to "squeeze and mould" her in ways more desirable to her interrogators and molders. Bacon's legacy is found in our time in the belief that nature can be "managed" by understanding and manipulating natural processes. The goal is to manage all "assets," whether human or natural, to promote economic growth. This assumes a great deal about human management abilities. For advocates of technological sustainability, ecology provides the scientific underpinnings for a system of planetary management. Technological sustainability is the total domination of nature plus population control. It is Gifford Pinchot with high technology.

Advocates of technological sustainability, second, believe that humans are best described by the model of economic man, who knows no limits of sufficiency, satiation, or appropriateness. Economic man maximizes gains and minimizes losses according to an internal schedule of preferences that does not distinguish between right and wrong. These assumptions are familiar to students of sociobiology and behaviorist psychology. In varying ways, both assume that humans are products of their neurological structure, conditioning, genes, and appetites, not free choice informed by considerations of ethics and morality. This view, in Clifford Geertz's words, "is the moral equivalent of fast food, not so much artlessly neutral as skillfully impoverished" (Schwartz 1986, 325). The issue is not whether people are capable of being greedy or selfish—they most certainly are—but whether human nature makes them inescapably so, and whether society rewards such behavior or not. After reviewing what passes for scientific literature about human nature drawn from economics, sociobiology, and behavioral psychology, psychologist Barry Schwartz concludes that "each discipline is importantly incomplete or inaccurate even within its own relatively narrowly defined domain. . . . Even if we accept what the disciplines have to say within their own domains, there is no reason to accept their principles as a general account of what people are" (Schwartz 1986, 317).

The society created in the belief that people are incapable of rising above narrow self-interest will differ from one in which other assumptions prevail. In other words, our beliefs about our nature tend to become self-fulfilling prophecies which produce the behavior they purport only to describe.

Arguments for technological sustainability rest heavily on beliefs that humans as economic maximizers are incapable of the discipline implied by limits, even though they are somehow capable of the wisdom and good judgment necessary to manage all of the Earth's resources in perpetuity. This deeply pessimistic view of human potentials assumes that we cannot control our appetites, act for the common good, or wisely direct our collective energies.

Advocates of technological sustainability, moreover, believe that economic growth is essential. The World Commission on Environment and Development, for example, calls for a "new era of growth," by which they mean "more rapid economic growth in both industrial and developing countries, freer market access for the products of developing countries, lower interest rates, greater technology transfer, and significantly larger capital flows" (World Commission on Environment and Development 1987, 89). The commission plainly regards growth as the engine for sustainable development everywhere. James Gustave Speth, president of the World Resources Institute, in a more resigned fashion believes that "economic growth has its imperatives; it will occur." He cites a projection of a "five-fold expansion in world economic activity." Instead of the radical disbelief such numbers should elicit, he is "excited" by the prospects for "greening" technology, as he puts it, and for the transformation of industry, eventually permeating "the core of the economies of the world" with ecological good sense (Speth 1989, 3–5).

AUTHOR'S NOTE 2010: *Speth has since come to a more pessimistic and, I think, realistic view. See Speth, 2008.*

This view raises several questions. First, since growth and environmental deterioration have occurred in tandem, how could they now be disassociated? It is not easy to envision sustainable growth in the main sectors of the industrial economy—energy, chemicals, automobiles, and the extractive industries. Newer parts of the economy, such as genetic engineering, remain unproven; they may spawn entirely new threats to the habitability of the planet. They will also lead to vast new concentrations of wealth with all that portends for democracy. And growth in the industrial world has not consistently helped the poor at home or abroad; to the contrary, the gap between the richest and the poorest is mostly widening. Why would growth in the developed world in even more precarious times lead to different results?

AUTHOR'S NOTE 2010: *"From 1979 to 2005 incomes for the highest earners increased almost fourfold, while the median income went up only 12 percent" (*New York Times, *March 24, 2010, p. A19).*

Second, advocates of technological sustainability are not clear on what it is that is being sustained: development, a new concept, or growth as more of the same with greater efficiency. The Brundtland Commission compounded the confusion by defining sustainable development as economic growth. Sustainable growth, in economist Herman Daly's words, "implies an eventual impossibility" of unlimited growth in a finite system (Daly 1988). Sustainable development, implying qualitative change, not quantitative enlargement, might be sustainable. The distinction is fundamental and usually overlooked. Because growth cannot be sustained in a universe governed by the laws of thermodynamics, we must confront issues of scale and sufficiency. "We need something like a Plimsoll line," Daly writes, "to keep the economic scale within ecological carrying capacity" (Daly 1988, 3). Carrying capacity, the total population times resource-use level that a given ecosystem can maintain, cannot be specified with precision. But neither can we be absolutely clear about other concepts in economic theory, such as time and money. Daly proposes three criteria to determine optimal scale: (1) it must be sustainable over the long term;

(2) there must be limits to human appropriation of global net primary productivity, which is now 25 percent or 40 percent of terrestrial primary productivity; and (3) from the work of Charles Perrings, "the economy [must] be small enough to avoid generating feedbacks from the ecosystem that are so novel and surprising as to render economic calculation impossible" (Daly 1988).

A related ambiguity concerns the relationships between developed and less developed economies. For example, growth in the developed economies depends on a steady flow of food, energy, and raw materials from the less developed world. The acres from which such food, timber, minerals, and materials are extracted and on which industrial economies depend constitute "ghost acreage," the land and resources outside national boundaries which supply the difference between consumption and resources. The use of ghost acres creates two problems. First, an imbalance is created by the price differential between exports of raw materials and imports of finished goods. Second, sellers of raw materials are highly vulnerable to price fluctuations and materials substitution. Together, they give ample reason for developing countries to selectively disengage from the global economy and chart alternative strategies for meeting basic needs. For theorists of sustainability, they raise practical and ethical questions. To what extent must population and resource use stay within the limits of regional or national carrying capacity? What level of imports of which commodities constitutes unsustainability? The Japanese, for example, have preserved their remaining forests at the expense of those in Alaska, Brazil, and Southeast Asia. In Daly's words, "a single country may substitute man-made for natural capital to a very high degree if it can import the products of natural capital from other countries which have retained their natural capital to a greater degree" (Daly 1988, 26). Either some must agree to remain undeveloped while others develop, or the structural disparity between developed economies and less developed economies must be rectified.

Advocates of technological sustainability often assume that the problems are those of inaccurate pricing and poor technology. Sustainability merely means getting the policy right, adjusting prices to reflect true scarcity and real costs, and developing greater efficiency in the use of energy and resources. And who will do this? For advocates of technological sustainability, the answer is policy makers, scientists, corporate executives, banks, and international agencies. Advocates rarely mention citizens, citizen groups, or grassroots efforts around the world. This perspective perhaps explains why the poor are often regarded as the cause of problems.

The authors of the World Resources Institute's (1985) study of tropical deforestation, for example, state that "it is the rural poor themselves who are the primary agents of destruction," none of whom were included as "task force members." Not surprisingly, those who control decisions about land tenure, or those who have systematically uprooted and undermined village economies that were once sustainable, were not mentioned. This perspective may reflect an inordinate desire to appear "reasonable," or it may come from the parochialism that enfogs (a new word) too many conferences in expensive settings that exclude people with calloused hands. Technological sustainability is largely portrayed as a painless, rational process managed by economists and policy experts sitting in the control room of the fully modern, totally computerized society, coolly pulling levers and pushing buttons. There is little evidence that its proponents understand democratic process or comprehend the power of an active, engaged, and sometimes enraged citizenry. This may also explain the near total neglect of environmental education in the Brundtland Commission report and other policy reports coming regularly from Washington think tanks. If sustainability is a top-down process, then an active, ecologically competent citizenry is irrelevant, and the effort to create such a citizenry through education is a diversion of scarce funds.

Ecological Sustainability

A second approach to the issues of sustainability holds that we will not get off so easily. Wendell Berry, for example, writes, "We must achieve the character and acquire the skills to live much poorer than we do. We must waste less, we must do more for ourselves and each other" (Berry 1989, 19). This, however, has less to do with policy levers than it does with general moral improvement in society, which may not otherwise care to find policy levers. Ivan Illich similarly regards the goals of development as a fundamental mistake:

> The concept implies the replacement of widespread, unquestioned competence at subsistence activities by the use and consumption of commodities; the monopoly of wage labor over all other kinds of work; redefinition of needs in terms of goods and services mass-produced according to expert design; finally, the rearrangement of the environment in such fashion that space, time, materials and design favor production and consumption while they degrade or paralyze use-value oriented activities that satisfy needs directly. (Illich 1981, 15)

According to Wolfgang Sachs, "eco-developers" (his term for advocates of technological sustainability) "transform ecological politics from a call for new public virtues into a set of managerial strategies." Without questioning the economic worldview, Sachs argues, one cannot question the "notion that the world's cultures converge in a steady march toward more material production" (Sachs 1989, 16–19). The alternative he proposes is one that regards development as a cultural process in which needs and their satisfaction arise from a vernacular culture. Ecological sustainability can be portrayed in terms of four characteristics.

First

Humans, they argue, are limited, fallible creatures. Wendell Berry, for example, writes:

> We only do what humans can do, and our machines, however they may appear to enlarge our possibilities, are invariably infected with our limitations.... The mechanical means by which we propose to escape the human condition only extend it. And further: No amount of education can overcome the innate limits of human intelligence and responsibility. We are not smart enough or conscious enough or alert enough to work responsibly on a gigantic scale. (Berry 1989, 22)

Berry describes two different kinds of limits: those on our ability to coordinate and comprehend things beyond some scale and those inherent in our nature as creatures with a limited sense of the good and willingness to do it. Even if the first could be overcome, the second limit would remain to infect the results. In other words, we cannot escape our creaturehood, and we can compound our problems many times over in the attempt to do so.

Second

A second component of ecological sustainability has to do with the role of the citizen in the creation of a sustainable future. The modern world is one in which the corporation and the state are dominant over the small enterprise and the community. People in the modern world have become increasingly passive in their roles as consumers and employees. Sustainability in the postmodern world will rest on different foundations that require an active, competent citizenry. Lewis Mumford, writing in 1938, described this task, or what he called "regional development," in these words:

> We must create in every region people who will be accustomed, from school onward, to humanist attitudes, co-operative methods, rational controls. These people will know in detail where they live and how they live: they will be united by a common feeling for their landscape, their literature and language, their local ways, and out of their own self-respect they will have a sympathetic understanding with other regions and different local peculiarities. (Mumford 1938, 386)

His approach to regional planning was based on the need to "educate citizens; to give them the tools of action, to make ready a background for action, and to suggest socially significant tasks to serve as goals" (Mumford 1938). Political scientist John Friedmann proposes a similar "escape" from our plight which involves the

> re-centering of political power in civil society, mobilizing from below the countervailing actions of citizens and recovering the energies for a political community that will transform both the state and the corporate economy from within. (Friedmann 1987, 314)

His approach to "radical planning" is premised on the belief that

> the great strength of American radicals is the self-organizing capacity of the American people on a local level, and the bastion of the national state is too powerful and too remote from the centers of radical practice to become an arena in its own right. This is not to say that the struggle cannot occasionally be carried to Washington, but in this huge country, America, the political life that holds promise is, for the time being, better concentrated in the diversity of its many local communities and the fifty states of the Union. (Friedmann 1987, 374)

Friedmann proposes to center the political life of the community on "restructured households that have shed their passivity and embraced the 'production of life' as their central concern." While acknowledging that the interdependent global economy will not unravel anytime soon, Friedmann, along with Daly and Cobb, proposes a selective de-linking of the economics and politics of local communities from those of the larger world (Friedmann 1987, 375–82). It is important to note that none of these advocate a return to parochial and closed communities or nations. Rather, they propose a process of rebuilding from the bottom up, seeing an active and competent citizenry as the foundation for a world appropriately linked.

Wendell Berry's comments about the "futility of global thinking" must be understood in this context. For Berry, global problems begin in the realm of culture and character, for which there can be no national or international solutions separate from those that begin with competent, caring, and disciplined people living artfully in particular localities. Biologist Garrett Hardin similarly argues that most "global problems" are, in fact, aggregations of national or local problems, for which effective solutions can only occur at the same level. Even if this were not the case, top-down solutions are often inflexible, destructive, and unworkable. Even if this were not true, the best policies in the world will not save ecologically slovenly, self-indulgent people who are not likely to tolerate such policies in any case. In other words, the constituency for global change must be created in local communities, neighborhoods, and households from people who have been taught to be faithful first in little things.

Proponents of ecological sustainability, then, aim to restore civic virtue, a high degree of ecological literacy, and ecological competence throughout the population. This, in contrast to recent conservatism, begins by conserving people, communities, energy, resources, and wildlife. It is rooted in the Jeffersonian tradition of an active, informed, competent citizenry. A citizenry capable of conservation is a product of good homes, good farms, good communities, good churches and synagogues, good schools, and right livelihood. There is a synergy between an active, competent citizenry and visionary leadership. A country made up of good communities will tend to foster and support leadership, and real leaders will empower citizens and communities.

Third

Ecological sustainability is rooted as much in past practices, folkways, and traditions as in the creation of new knowledge. Anthropologist Michael Redclift, for example, writes that "if we want to know how ecological practices can be designed which are more compatible with social systems, we need to embrace the epistemologies of indigenous people, including their ways of organizing their knowledge of their environment" (Redclift 1987, 151). One of the conceits of modern science is the belief that knowledge can be applied everywhere in the same manner. Traditional knowledge is mostly specific to a particular place and evolved over centuries. It is rooted in a local culture and serves as a source of community cohesion, a framework that explains the origins of things (cosmology), and provides the basis for preserving fertility, controlling pests, and conserving biological diversity and genetic variability. Knowledge is not separated from the multiple tasks of living well in a specific place over a long period of time. The crisis of sustainability has occurred only when and where this union between knowledge, livelihood, and living has been broken and knowledge is used for the single purpose of increasing productivity. It may be, as Redclift says, that the "question is whether 'we' [the "developed" nations] are prepared for the cultural adaptation that is required of us" (Redclift 1987). For the most part, we have systematically uprooted both the kind of traditional knowledge of this sort and the people who created and preserved it. The loss of traditional knowledge, economist Richard Norgaard argues, is directly related to increased species extinction and the risks inherent in the rise of a single knowledge-economic system controlling agriculture worldwide. He writes:

> The patchwork quilt of traditional agroeconomies consisted of social and ecological patches loosely linked together. The connections between beliefs, social organization, technology, and the ecological system were many and strong within each patch for these things co-evolved together. Between patches, however, linkages were few, weak, and frequently only random. The global agroeconomy, on the other hand, is tightly connected through common technologies, and international crop, fertilizer and pesticide, and capital markets. (Norgaard 1987)

For the present system, any failure of knowledge, technology, research, capital markets, or weather can prove highly destabilizing or fatal. Disruptions of any sort ripple throughout the system. Not so for traditional agroeconomic systems. A failure of one does not threaten others.

Finally, Norgaard notes that the "global exchange economy" treats all parts of the world the same regardless of varying ecological conditions. Since "the diversity of the ecological system is intimately linked to the diversity of economic decisions people make," there is a steady reduction of biological diversity. Biological diversity is a factor in social risks, since "agroeconomic systems with many components have more options for tinkering and happening upon a stable combination or for learning and systematically selecting combinations with stabilizing negative feedbacks" (Norgaard 1987).

Ecological sustainability will require a patient and systematic effort to restore and preserve traditional knowledge of the land and its functions. This is knowledge of specific places and their peculiar traits of soils, microclimate, wildlife, and vegetation, as well as the history and the cultural practices that work in each particular setting. Sustainability will not come primarily from homogenized top-down approaches but from the careful adaptation of people to particular places. This is as much a process of rediscovery as it is of research.

Fourth

Proponents of ecological sustainability regard nature not just as a set of limits but as a model for the design of housing, cities, neighborhoods, farms, technologies, and regional economies. Sustainability depends upon replicating the structure and function of natural systems. John and Nancy Todd, for example, propose nine design precepts (Todd and Todd 1984, 18–92):

- The living world is the matrix for all design.
- Design should follow the laws of life.
- Biological equity must determine design.
- Design must reflect bioregionality.
- Projects should be based on renewable energy sources.
- Design should integrate living systems.
- Design should be coevolutionary.
- Building and design should heal the planet.
- Design should follow a sacred ecology.

Ecology is the basis for their work on the design of bioshelters (houses that recycle waste, heat and cool themselves, and grow a significant portion of the occupants' food needs) and in the design and construction of solar aquatic systems for purifying wastewater. In the design of solar aquatic waste systems, John Todd asks how nature would deal with organic wastes. The answer, he believes, lies in the creation of "living machines," ensembles of plants that perform specific functions necessary to remove human wastes, heavy metals, and toxics from water. Three working models confirm the theory at costs and performance levels superior to standard waste systems that require great amounts of energy and chemicals. Todd's living machines "are engineered with the same design principles used by nature to build and regulate its great ecologies in forests, lakes, prairies, or estuaries. Their primary energy source is sunlight. Like the planet they have hydrological and mineral cycles."

Todd sees the world as a "vast repository of biological strategies and components that might be integrated into a more coherent science and into economies wrapped in the wisdom of the natural world" (Todd 1990).

Amory and Hunter Lovins, cofounders of the Rocky Mountain Institute, similarly draw on ecology for the design of resilient technological systems. Resilience implies the capacity of technological systems to withstand external disturbances and internal malfunctions. Resilient systems absorb shock more gracefully and forgive human error, malfeasance, or acts of God. Resilience does not imply a static condition, but rather flexibility that permits a system "to survive unexpected stress; not that it achieve the greatest possible efficiency all the time, but that it achieve the deeper efficiency of avoiding failures so catastrophic that afterwards there is no function left to be efficient." Resilient systems exhibit certain qualities, including (Lovins and Lovins 1982)

- modular, dispersed structure;
- multiple interconnections between components;
- short linkages;
- redundancy;
- simplicity;
- loose coupling of components in a hierarchy.

Like the process of evolution, designers of resilient systems tend to follow the old precepts, such as, keep it simple stupid; if it ain't broke, don't fix it; you don't put all your eggs in one basket; and if anything can go wrong, it will. Resilience implies small-scale, locally adaptable, resource-conserving, culturally suitable, and technologically robust solutions whose failure does not jeopardize much else.

Wes Jackson uses the prairie as a model for ecologically complex farms that do not rely on tillage and chemical fertilizers. Ecologically and esthetically, they would resemble the original prairie that once dominated the Great Plains. For Wes Jackson, "the patterns and processes discernible in natural ecosystems still remain the most appropriate standard available to sustainable agriculture . . . what is needed are countless elegant solutions keyed to particular places" (Jackson and Piper 1989). Jackson's work follows that of Sir Albert Howard, who once proposed the forest as the model for agriculture:

> Mother earth never attempts to farm without livestock; she always raises mixed crops; great pains are taken to preserve the soil and to prevent erosion; the mixed vegetable and animal wastes are converted into humus; there is no waste; the processes of growth and the processes of decay balance one another; ample provision is made to maintain large reserves of fertility; the greatest care is taken to store the rainfall; both plants and animals are left to protect themselves against disease. (Howard 1979, 4)

The case for regarding nature as a model for farms, housing, cities, technologies, and economies rests on three beliefs. First, the biosphere is a catalogue recorded over millions of years of what works and what does not, including life-forms and biological processes. The sudden intrusion of new technologies, chemicals, and other massive human impacts disrupts established patterns and introduces novel elements for which nature has no adaptive experience. In other words, human activity will be disruptive unless it is designed to fit within ecological processes and the carrying capacity of natural systems.

Second, ecosystems are the only model we have of stable organization in a world of change. The energy efficiency, closed loops, redundancy, and decentralization characteristic of ecosystems allow them to swim upstream against the force of entropy. Industrial systems, on the contrary, assume linearity, perpetual growth, and progress which increase entropy and decrease stability.

A third argument has overtones of mysticism and theories of vitalism. The Todds' "sacred ecology," for example, reflects the belief in an underlying structure which connects the human and natural worlds in an unknowable "metapattern." Similar interpretations are often made of the biosphere as portrayed in the Gaia hypothesis of James Lovelock and of Teilhard de Chardin's "noosphere," in which human intelligence and communications technology are presumed to be something like a planetary nervous system in the making.

Advocates of ecological sustainability use nature as a model, but they do not necessarily agree on how the model should be used. Does sustainable development require the restoration of natural systems as authentically as possible, or only the imitation of their structure and ecological processes? Restoration ecology is the best example of the former, while Wes Jackson's efforts to breed perennial polycultures that resemble prairies exemplifies the latter. Attempts to mimic nature and ecological processes may in time come to resemble Baconian science with its goal of total mastery. If, on the other hand, sustainability is interpreted to mean the restoration (and/or preservation) of natural systems as authentically as possible, letting natural selection do most of the work, then its advocates must develop a clear understanding of what is natural, what is not, and why the difference is important.

Among the most important implications of using nature as a model for human systems are issues of scale and centralization. If ecology is the model, should society be more decentralized? Surface-to-volume ratios limit the size of biological organisms and physical structures. Are there similar principles of optimum size for cities, nations, corporations, and technologies? Leopold Kohr, E. F. Schumacher, and other proponents of decentralization supported decentralization and appropriate scale on three grounds, the first of which has to do with human limits to understand and manage complex systems. Wendell Berry similarly argues that the ecological knowledge and level of attention necessary to good farming limits the size of farms. Beyond that limit, the "eyes to acres" ratio is insufficient for land husbandry. At some larger scale it becomes harder to detect subtle differences in soil types, changes in plant communities and wildlife habitat, and variations in topography and microclimate. The memory of past events like floods and droughts fades. As scale increases, the farmer becomes a manager who must simplify complexity and homogenize differences in order to control. Beyond some threshold, control requires power, not stewardship. Grand scale creates islands of ignorance, small things that go unnoticed, and costs that go unpaid.

Is the same true of things other than farms? I think so, even if we cannot prescribe the ideal size of a city or corporation any more than we can define the exact number of acres one person can farm responsibly. To know the optimum farm size requires that we know the farmer's intelligence, skill, depth of motivation, energy level, age, state of marriage, type of land, and so forth. Appropriate scale is not an absolute but a continuum, bounded by the limits of nature and those of the mind. Disorder, breakdown, ugliness, and disease suggest that these limits have been transgressed. In the transition from Plato's ideal polis of 5000 to global cities of 20 million, neighborhoods unravel, pollution overwhelms local ecosystems, public health deteriorates, transportation becomes congested, civility declines, crime increases. But not all of these things happen at once. As scale increases, good things happen as well. Growing cities support symphony orchestras, but when they continue to grow, people are mugged leaving the symphony and acid rain dissolves the exterior marble of the civic auditorium. So we can speak only of a ratio of good to bad that gradually or precipitously declines as scale crosses some unknown threshold.

When obscure place names—Seveso, Bhopal, Three Mile Island, Chernobyl, Love Canal, Times Beach, Prince William Sound—become synonymous with disasters, a similar dynamic is at work in technological systems. In each case, large scale, complexity, improbability, and human error led in due time to what Charles Perrow describes as "normal accidents," that is, events which are entirely predictable given enough time (Perrow 1984).

The thread connecting all questions of appropriate scale from farms to technological systems, then, has to do, first, with the human limits to comprehend and manage beyond some threshold of scale and complexity. Increasing scale increases the number of things that must be attended to and the number of interactions between components. Rising scale also increases the costs of carelessness. Preoccupation with quantity replaces the concern for quality: the farm becomes an agribusiness, the city become a megalopolis, the shop becomes a corporation, tools become complicated technologies, the legitimate concern for livelihood becomes an obsession with growth, and weapons become instruments of total destruction.

The second ground for decentralization and appropriate scale is that centralization and large scale undermine the potential for ethical action and increase the potential for mischief. As scale increases, it becomes easier to separate costs and benefits, creating winners and losers who are mostly strangers to each other. Ethical responsibility means paying the full costs for one's actions, or mutually agreed-upon compensation to those whose lot is it to pay them initially. Ethical behavior seems most likely when the decision maker's own hide is at stake. It still works fairly well if costs are levied against friends, neighbors, and relatives encountered face to face. The likelihood of ethical behavior, however, decreases with distance in time and space between beneficiaries and losers.

Scale can also make it difficult to assign responsibility. Who can be blamed for acid rain, human-induced climate change, species extinction, or "normal" accidents? In each case the costs are widely distributed while responsibility is diffused among political leaders, utilities, corporations, government agencies, and the consuming public.

Leopold Kohr argues, as the third ground for decentralization, that large scale, whether in nations or social organizations, provides the impetus for imperialism, war, and aggression: "For whenever a nation becomes large enough to accumulate the critical mass of power . . . it will become an aggressor" (Kohr 1978, 35). He draws the conclusion that wickedness derives from bigness and that "no misery on earth can be handled, except on a small scale" (Kohr 1978, 79). Smallness is nature's principle of health, bigness the principal cause of disease.

The paradigm of ecological sustainability has evolved an epistemology of sorts around the concept of interrelatedness. This epistemology involves what Gregory Bateson called the "pattern that connects." This pattern always includes both observer and observed, subject and object. "We are not outside the ecology for which we plan," he says, "we are always and inevitably a part of it" (Bateson 1975, 1979). The search for interrelatedness is a revolt from Cartesian logic, reductionism, and the fragmentation characteristic of modern science, conventional economics, and even some of modern ecology. It also recognizes that the world is paradoxical and that our understanding of it will always be incomplete. We are makers of and participants in reality, not just observers. Where science has dismantled nature, we must study whole systems, linkages, processes, patterns, context, and emergent properties at higher systems levels. "Holistic science" cannot be conducted through the reductionist methods characteristic of much science. We cannot reach valid knowledge of nature simply by taking it apart and studying the pieces any more than we could understand human behavior from the study of anatomy.

The recognition of interrelatedness leads to equally radical changes in the conduct of human affairs. Conflict has often been essential to the existence of nations, churches, movements, and ideologies that identify themselves in opposition to something else. The tendency is to presume one's side to be the sole possessor of truth. But truth is no less uncertain, incomplete, relative, and paradoxical in human affairs than it is in the physical world described by Heisenberg or Einstein. It is a truism to say that we become that which we hate, but life is often like a dance of opposites, each necessary to the other. "Truth," in William Irwin Thompson's words, "cannot be expressed except in relationships of opposites" (Thompson 1987). We cannot fathom the unconscious drives and purposes which create irony and counterintuitive effects; anything like total truth is beyond our comprehension. We intend one thing and do the opposite. From this we can learn humility in the fact of unfathomable mystery and paradox. We can make no absolute distinctions between the self and the world. Treating others as we would have them treat us isn't just good for them; it's also in our own self-interest whether we like it or not. Goodness, mercy, justice, and ecological prudence have both survival value and spiritual rewards. Before rushing out to do good, however, we might reflect on how much of the world's misery began with good intentions. Competence in doing good is still an underdeveloped art.

The Limits of Metaphors

As with all concepts and metaphors, we must ask where that of ecological sustainability applies and where it does not. Two categories are particularly problematic. Cities will always be something of an exception to the model of natural systems. Under the best conditions, large urban areas will import substantial amounts of food, energy, water, and materials, and they will export roughly equivalent amounts of sewage, garbage, pollution, and heat. Many of these impacts could be reduced by better mass transit, careful urban planning that includes parks, systematic use of solar energy, urban-regional agriculture, urban reforestation, laws (like bottle bills) reducing material flows, and biological treatment of organic wastes. Nevertheless, although these measures significantly reduce environmental damage, they do not make cities "sustainable" such that the net environmental impact of urban concentrations is within the absorptive and healing capacities of the surrounding natural systems. The sheer concentration of large numbers of people will reduce environmental resilience, encroach on wildlife habitat, and impose significant ecological costs elsewhere. Urban concentrations must be justified on their contributions to intellectual, economic, and cultural life, not their sustainability. I do not think that cities have to be as ugly, formless, inhuman, and inefficient as we have made them. But given that we have urbanized badly, and cannot quickly undo what we have done, urban conglomerations cannot easily be made a harmonious part of a sustainable society. This is not an argument against cities, but rather one against megapolitan areas without plan or form. It is also one for "green cities" with greenbelts, urban parks, urban agriculture, and urban wilderness preserves.

Another and increasingly problematic area is that of technology. The cumulative effects of technology extend human power over nature so that we can transcend the limits of gravity, space, time, and biology and now, with computers, those of mind. In the process, we remove ourselves further and further from the natural conditions, both good and bad, that previously constrained human development. In a society that worships technology, questions of this sort are heresy. Technology is our declaration of independence from nature. As a user of airplanes, automobiles, computers, cell phones, and more, I am a cosigner. These things allow me to avoid a great many things about nature that I do not like. But this may be a Pyrrhic victory of convenience over substance. It may also reflect the domination of technology over free choice, since many of the technologies

I use I do so out of necessity. I would much prefer to travel by train, for example, but the passenger rail service is virtually nonexistent. Regardless, there can be no question that the use of technology is now the preeminent fact in modern societies. Whether it can be controlled and harnessed to the long-term benefit of humanity is the question of our civilization. If so, the goal of a sustainable society based on the model of natural systems is not necessarily antithetical to technology. The question then becomes what kind of technology, at what scale, and for what purposes. But we lack a philosophy of technology that could help us decide such things, and without much clarity, we are prone to what Langdon Winner has called "technological somnambulism," a "willing sleepwalk," a passive acceptance of whatever technologies are thrust upon us by whomever for whatever purposes. Because artifacts do have politics, in Winner's words, any decent philosophy of technology will be a political philosophy that clarifies the effects of technology on the distribution of power and control in society. It will also be a philosophy of nature because technological choices often have sweeping effects on ecosystems. An alternative, postmodern technology, in philosopher Frederick Ferre's view, would aim to optimize rather than maximize, to cultivate rather than manipulate, and to differentiate rather than centralize. The beginnings of postmodern technology are evident in solar technologies, in the development of regenerative farming practices, and, perhaps, in computers. Future advances in ecological technology will combine artifice and nature in subtle and ingenious ways representing a radically new departure that is neither a rejection of technology nor a sleepwalk along the edge of catastrophe.

The modern world has failed; a decent alternative world is still to be born. Transitions such as this are times of both promise and peril. The promise comes from the opportunity driven by necessity to reconsider, rethink, reform, restore, and rebuild our world and worldview. This process raises old issues and some new ones having to do with the balance between centralization and decentralization, urban and rural, freedom and order, individual and community, sacred and secular, organic and mechanical. The peril comes from the urgency, scope, and sheer numbers of problems coming down on us. The question is whether we can muster the intellectual clarity, goodwill, and moral power needed to make wise choices about the issues having to do with whether and how humanity survives.

∴ Chapter 12 ∴

Leverage

(2001)

Leverage has less to do with pushing levers than it does with disciplined thinking combined with strategically, profoundly, madly letting go.
Donella Meadows

I once asked a class to explain the dead zone, which is roughly the size of New Jersey, in the Gulf of Mexico, the fact that one-third of U.S. teenagers are overweight or obese, and the possible relationships between the two. After an hour, they had filled the blackboard with boxes and arrows that included federal farm subsidies, U.S. tax law, chemical dependency, feedlots and megafarms, the rise of the fast-food industry, declining farm communities, corporate centralization, advertising, a cheap food policy, research agendas at land-grant institutions, urban sprawl, the failure of political institutions, cheap fossil energy, and so forth. Most of the things described by those boxes, however, resulted from decisions that were once thought to be economically rational or at least within the legitimate self-interest of the parties involved. But collectively they are an unfolding continental-scale disaster affecting the health of people and land alike.

The same connect-the-dots kind of exercise could be done to explain urban decay and land sprawl, a defense policy that undermines true security, a de facto energy policy that promotes inefficiency, transportation gridlock, and the failure to provide universal health care. Our individual and collective failure to comprehend and act on the connectedness of

This article was originally published in 2001.

things is pervasive and systemic and threatens our health and long-term prosperity. It deserves urgent national attention but is scarcely noticed. Why is this so?

First, we have organized our national affairs to create persistent gridlock reflecting the founders' fear of excessive government and true democracy. As a result authority is divided among local, state, and national governments and then between executive, judicial, and legislative branches. At the federal level, dozens of congressional and senatorial committees and subcommittees oversee the nation's water, air, wildlife, lands, and resources. Within the executive branch, environmental policy is a continual negotiation between various cabinet agencies and subagencies with competing agendas. As a result we have no national environmental policy for agriculture affecting 700 million acres and the largest U.S. source of water pollution. After decades of talk, we have no comprehensive and farsighted national energy policy. Instead in these and other areas we have a patchwork of laws and regulations inconsistently enforced at various levels of government, often working at cross-purposes, and in jeopardy to a hostile administration and court system. Almost without exception these laws and regulations operate after environmental damage has occurred, and most conflict with other statutes that aim to promote economic growth. Further, laws governing pollution tend to move pollutants from one medium to another. So, for example, we scrub SO_2 from power plants, only to dispose toxic sludge on land. We "clean" water only to disperse toxic-laced solids on farmland or landfills. Pollution control becomes a kind of giant shell game by which we move pollutants between air, water, groundwater, and land.

Similarly, the hodgepodge of laws and regulations that govern chemical pollution are easily corrupted and constitute no effective protection to human or ecosystem health. Of some 75,000 chemicals in common use, only a few have been tested for a full range of health effects. Such tests do not include how one chemical interacts with others, even though it is known that those interactions can sometimes increase toxicity by orders of magnitude. Nothing in the law and little in our political habits so far causes us to seek out better alternatives to the use of hazardous chemicals. So the debate tends to revolve around the rate at which we can legally poison each other.

It did not—and does not—have to be this way. In 1969 the National Environmental Policy Act described a different course intended to develop a systemic, unified, and long-term approach that would "use all practicable means and measures . . . to create and maintain conditions under which man and nature can exist in productive harmony." Although widely emulated in other countries, the promise of NEPA in the U.S. has not been realized. The prospect of a systemic and farsighted approach to environmental policy was undermined from the beginning by the inherent limitations of fragmented government, by entrenched economic interests with easy access to the White House and members of Congress, by hostile courts, and by the tendency of industries to "capture" the agencies created to regulate them. Relative to the goals set forth in the NEPA, U.S. environmental policy, despite its strengths, is a nickel solution for a dollar problem.

Its failure, moreover, is excused by our attitudes toward government generally. For three decades or longer we've stewed in the odd notion that "government is the problem." Sometimes it is, but corporations, state governments, citizen apathy, talk radio, and fanaticism of all kinds can be problems as well. Extreme interpretations of individualism and property rights combined with a pervasive suspicion of government, for example, continually undermine the idea that public problems can and ought to be solved publicly, hence the possibility of actually solving problems as citizens acting together. Candidates, funded by interests wanting less public scrutiny, run for office on antigovernment platforms, pledging to do their best to limit governmental power while leaving that of corporations unimpeded. The U.S. has a long tradition of libertarian attitudes that serve to justify the ecological cacophony of sprawl, pollution, and waste. "What is missing from American environmental policy," in political scientist Richard Andrews' words, "is a coherent vision of common environmental good that is sufficiently compelling to generate sustained public support for government action to achieve it" (Andrews 1999, 370).

There is a second reason for failure inherent in the limitations of the reigning theories of economics. It is old news to the alert that, relative to our real wealth, the practice of mainstream economics and contemporary methods of accounting conceal as much as they reveal. They do not, for example, account for the "services" of natural systems such as pollination, water purification, or recycling of organic matter. Nor do they properly account for the loss of "natural capital" such as soils, forests, or species diversity (Hawken et al. 1999). The theory of economics, in either its classical or neoclassical version, followed industrialization and closely mirrored the reality thus created. It was a theory derived in full innocence of how the world works as a physical system and why this might be important, even for the economy.

In no instance is this more evident than in the way we allocate investment by discounting future outcomes back to some purported net present value. A dollar in hand today, in other words, is worth more than $1.10 a year from now. The practice is geared to maximize short-term benefits, often at a substantial long-term cost. Discounting the future means that we place little value on the possibilities of severe loss or even catastrophe a few decades hence. More shopping malls now trump concern for the decline in biodiversity; more highways now trump the need for farmland that could be needed by midcentury; more fossil energy now trumps concern about climatic change ahead. And so it goes, with the interests of our grandchildren discounted to zero. The practice of discounting, in economist Colin Price's words, "cannot be justified" (Price 1993, 345). What can be justified is a "cold and rational altruism, driven by a belief in the propriety of sharing with later times the things we have valued, in the time which has been given to us" (Price 1993, 345–47).

That much is well documented but widely ignored by defenders of doctrinal purity or acolytes of the myth of unfettered free markets. Beneath the formal theory there are crucial assumptions about what makes humans tick. Economic theory describes you and me as simple maximizers of our self-interest and further assumes that we know well enough what we want and how best to get it. But little in the human record supports such simplistic notions. We are, surely, far more complicated, as students of advertising and propaganda well know. We are moved by lots of things beyond simple calculations of economic self-interest: fame, glory, power, group protection, nationalism, values, the prospect of salvation, sex, virtue, duty, obsession, and sometimes by plain orneriness and at other times by the better angels of our nature. The failure of economics to account for such things, too, is both widely remarked and mostly ignored.

From such assumptions it is believed to be the height of rationality to ask, first, the cost of a thing, not whether it is a good thing to make or to do or how it fits with other priorities and values. The proper answer to "how much does it cost" is "relative to what and to whom and over what period of time?" Since most are unable or unwilling to ask such questions, supposed cost is the most frequent reason given for not doing something otherwise good or necessary. In this intellectual and moral vacuum, we lose sight of the simple fact, as the late Donella Meadows put it, that "we don't get to choose which laws, those of the economy or those of the Earth, will ultimately prevail" (Meadows 2001). The result is that short-term wealth for a few is purchased at the cost of long-term prosperity for all.

There are certainly other reasons that we fail to perceive systemic causes and act accordingly, including overspecialization, reductionism, discipline-centric education, and the "dumbing down" of lots of things. But the point is that we've created a global system that may be an ecological absurdity, but it is neither accidental nor incomprehensible. It is, rather, the result of decisions and choices we've made about how we conduct public affairs and how we evaluate our success. More to the point, are there places in the system where different decisions and relatively small reforms can produce large environmental results? For example, we spend, by one estimate, $1.4 trillion in "perverse" subsidies worldwide that "exert adverse effects of both environmental and economic sorts over the long run" in fishing, agriculture, mining, road building, logging, and energy extraction and use. Perverse subsidies are the sort described by Paul Hawken, by which "the government subsidizes energy costs so that farmers can deplete aquifers to grow alfalfa to feed cows that make milk that is stored in warehouses as surplus cheese that does not feed the hungry" (Myers 1998). Eliminating such subsidies would go a long way to reducing overfishing, excessive road building, energy inefficiency, deforestation, and the loss of biodiversity.

In the U.S., specifically, estimates of subsidies for the automobile range from $400 billion to $730 billion. Levied as a tax on gasoline, this amounts to something between $3.75 and $7.00 per gallon. Moreover, the Pentagon spends billions to ensure U.S. access to Persian Gulf oil. Traffic congestion in 68 major cities costs another $78 billion in lost productivity, 6.8 billion gallons of wasted fuel, and 4.5 billion hours and a fair amount of sanity for drivers sitting in long lines of blue haze. Ending the large array of subsidies for automobiles over, say, 10 years would help other modes of transportation become more viable and help us to rethink the design of urban areas to minimize the need for transport in the first place.

Economist Robert Frank has identified another leverage point—a tax on consumption aimed to eliminate the proliferation of luxury goods. "Our houses are bigger and our automobiles are faster . . . than ever before," he writes, but "we have less time for family and friends, and less time for sleep and exercise . . . our streets are dirty and congested. Our highways and bridges are in disrepair, placing countless lives in danger. And the misery in our inner cities continues unabated" (Frank 1999, 267). Although not primarily an environmental tax, a tax on luxury goods would save the resources that now fuel consumption races. His argument is roughly parallel to that given by economists to curb pollution by taxing it. At the same time, he proposes to remove taxes from savings. Others propose more sweeping environmental taxes aimed to remove taxation from income and put it on things we do not want, such as pollution and inefficiency. There are proposals for specific taxes, called "feebates," that set energy performance standards above or below which buyers are given rebates or charged fees. Other countries, notably the Netherlands and New Zealand, have developed practical national "Green Plans" with targeted reductions in energy and resource use and life-cycle cost accounting.

The point is simply that there are good ways to improve our situation, reduce costs, increase fairness within and between the generations, eliminate a great deal of environmental damage, and create a sustainable prosperity while appealing to both liberals and conservatives. There are acceptable possibilities, in other words, to solve environmental problems in ways that promote the larger good without undue sacrifice now. What we lack is the mind-set to see connections between things. In a real sense we do not have environmental problems, we have perceptual problems, and what we've failed to see is the human enterprise and our little enterprises connected in space and time in more ways and at more levels than we could ever count. Once we've fully absorbed the reality of our interdependence in space and time, the rest is a great deal easier.

Is it possible to organize our public affairs and private lives in ways that honor the integrity of the whole over the long term? No one I know thought more deeply or creatively about this question than Donella Meadows. Toward the end of a brilliant and vibrant life, she concluded that to change paradigms, "you keep pointing at the anomalies and failures in the old paradigm, you come yourself, loudly, with assurance from the new one, you insert people with the new paradigm in places of public visibility and power. You don't waste time with reactionaries; rather you work with active change agents and with the vast middle ground of people who are open-minded." There are no "cheap tickets to system change," she wrote. "You have to work at it." And you have to "madly let go" of fear, greed, narrowness, and sometimes, yes, stupidity (Meadows 1997).

Chapter 13

Shelf Life

(2009)

Those of us who have looked to the self-interest of lending institutions to protect shareholder's equity, myself included, are in a state of shocked disbelief. . . . Yes, I have found a flaw. I don't know how significant or permanent it is. But I've been very distressed by that fact.
Alan Greenspan

The self-confidence of learned people is the comic tragedy of civilization.
Alfred North Whitehead

For a long time to come, economists, pundits, and politicians will be wondering what to do about the largest and deepest economic crisis since the Great Depression of the 1930s and how it happened, yet again. Overlooked is the fact that we are simultaneously running two intertwined deficits with very different time scales, dynamics, and politics. The first deficit is short-term and has to do with money, credit, and how we create and account for wealth, which is to say a matter of economics. However difficult, it is probably repairable in a matter of a few years, give or take, most likely in the time-honored fashion of stimulating even more consumer spending and causing greater environmental disorder. The second is an ecological deficit. It is permanent, in significant ways unrepairable, and potentially fatal to civilization. The economy, as Herman Daly has pointed out for decades, is a subsystem of the biosphere, not the other way around. Accordingly, there are short-

This article was originally published in 2009.

term solutions to the first deficit that might work for a while, but they will not restore longer-term ecological solvency and will likely make it worse. Ecological debt cannot be remedied so easily, and in the case of climate destabilization, no matter what we do, it is a steadily—perhaps rapidly—worsening condition with which humankind will have to contend for a long time to come. University of Chicago geophysicist David Archer puts it this way:

> The climatic impacts of releasing fossil fuel CO_2 to the atmosphere will last longer than Stonehenge. Longer than time capsules, longer than nuclear waste, far longer than the age of human civilization so far. Each ton of coal that we burn leaves CO_2 gas in the atmosphere. The CO_2 coming from a quarter of that ton will still be affecting the climate one thousand years from now at the start of the next millennium. And that is only the beginning. (Archer 2009, 1)

In short, we have already bought disaster but not necessarily anything like a final catastrophe.

How could we have come so rapidly to the brink of extinction with hardly a twitch of apprehension? Assuming that we don't go over the cliff, it is a question that doubtless will provide much fodder for several millennia of academic conferences and dissertations and, for the sensitive, much deep thinking and anguished soul-searching. As we mull it over, we will rediscover, as conservative philosopher Richard Weaver once said, that "ideas have consequences." And in our case some really bad ideas of the last half century are leaving a legacy of very bad consequences. Weaver's 1948 book was an extended argument for conservatism, beginning with the recognition of knowledge higher than our own and the importance of such things as virtue, character, craftsmanship, enduring quality, civility, and, above all, piety. Applying this to nature, Weaver argued for a "degree of humility" such that we might avoid meddling "with small parts of a machine of whose total design and purpose we are ignorant" (Weaver 1984, 173). "Our planet," he wrote, "is falling victim to a rigorism, so that what is done in any remote corner affects—nay, menaces—the whole. Resiliency and tolerance are lost" (Weaver 1984, 173). Weaver regarded the modern project to reconstruct nature as an "adolescent infatuation." One can reasonably imagine what he would have said about the exhibition of thievery and stupidity leading to our present circumstances.

Weaver's idea that ideas have real consequences, alas, had less consequence than one might wish. It is honored mostly among a small band of true conservatives, the uncommon sort who actually value the conservation of tradition, law, custom, nature, culture, and religion and who take ideas and their real-world implications seriously. Other than for the title of his book, however, Weaver is presently unknown to the wider public and probably not at all to the faux conservatives who daily hissy up on FOX News. Unfortunately, ideas, whatever their consequences, seldom "yield to the attack of other ideas," in John Kenneth Galbraith's words, "but to the massive onslaught of circumstances with which they cannot contend" (Galbraith 2001, 30). That appears to be true in our own time, in which the pecuniary imagination was given such full reign. The convenient idea that foxes could be persuaded to reliably guard the henhouse, for example, derived from free marketeers like Milton Friedman, libertarians like George Gilder, supply-side economists like Arthur Laffer, and "long-boomers" like Peter Schwartz, did not voluntarily surrender to superior reason, logic, or evidence. Rather it was an idea whose consequences turned out to be bad both for the hens and a bit later for the starving foxes, some of whom, now professing different ideas, stand in line for public bailouts. Roadkill on the highway called reality. What might be called the shelf life of such ideas will turn out to be brief as such fads go; the consequences, however, will last a long time.

When delusion is popular, however, durable ideas are unpopular or more likely forgotten altogether. But in the present wreckage we have no choice but to search for more durable ideas with more benign or even positive consequences. When we find truly durable ideas, they are mostly about limits to what we can do or should do—but restraint, prudence, and caution are "oh mah God, sooo not cool," as one of my students thoughtfully expressed it. Accordingly, such things are put on the shelf, where they gather dust until necessity strikes again, and then they are called back into use as we try once again to find our bearings amidst the debris of popular delusions gone bust.

In this regard, the Greek poet Archilochus left a fragment of a manuscript with the words "the fox knows many things; the hedgehog knows one big thing." Like the hedgehog, advocates for the environment, animals, biological diversity, water, soils, landscapes, and climate stability know one big thing, as biologist Garrett Hardin once put it, which is that "we can never do just one thing" (Hardin 1972, 38). In other words, there are many unforeseen consequences from what we do, and so there are limits to what we can safely do. Since consequences are not only unpredictable but often remote in time and distant from the cause, we are often ignorant of the victims of our actions, and so there are moral limits on what we should do as well. To think about consequences over time requires, further, that we know how things are linked as systems and understand that small actions can have large consequences, many of which are unpredictable.

Around the first Earth Day in 1970 there was an efflorescence of brilliant thinking along these lines. In different ways, it was mostly about the things we could not do. Rachel Carson's *Silent Spring* (1962), for example, launched the modern environmental movement with the simple message that we could not carelessly spread toxic chemicals without causing damage to animals and eventually to ourselves. The book was attacked by proponents of what she called "Neanderthal biology," most of whom—then and now—with a great deal of money and/or reputation invested in the petrochemical business.

Other books making the same point addressed unlimited growth of population, economies, technology, and scale. However prescient and true, most of the wisdom was quickly forgotten. Now, however, we live in "the age of consequences" and have good reason to rethink many ideas and systems of ideas called paradigms. Perhaps this is what educators describe as a "teachable moment" or inventors refer to as the "aha moment." Assuming that it may be so, I have some suggestions for those assigned to rebuild the U.S. and global economies.

The first is the old idea that we cannot build a durable economy that is so utterly dependent on trivial consumption. In 2007, for instance, Americans spent $93 billion on tobacco and another $83 billion on casino gambling, but only $46 billion on books. Another example is from the recent SkyMall catalog, found in the seat pockets of commercial airplanes, which announces that it is "going beyond the ordinary." To do so, it offers those burdened with money and credit such items as a "startlingly unique" 2-foot-high representation of Bigfoot, to be placed in the garden where it will no doubt amaze and delight, available for only $98.95. "The keep your distance bug vacuum" equipped with a 22,400 rpm motor is available for $49.95. And for just $299.99 cat lovers can buy a marvel of advanced technology: "the 24/7 self-cleaning, scoopfree litter box!" Technology-oriented catalogues regularly offer dozens, nay, hundreds of devices that digitally amaze, ease, simplify, gratify, sort, store, scratch, waken, warn, multiply, compute, freshen, check, sanitize, and personalize. It may be possible, one day, to live in a digital, stainless steel nirvana of the sort George Orwell once said would make "the world safe for little fat men."

An economy so dependent on ephemeralities is a fraud because it cannot satisfy the desires that it arouses. It is a lie because it purports to solve by trivial consumption what can only be solved by better human relations. It is immoral because it takes scarce resources from those who still lack the basics and gives them to those with everything who are merely bored. It is unsustainable because it creates waste that destroys climatic stability and ecosystems. It is unintelligent because it redirects the mental energies of producers and consumers alike to illusion, not reality, which makes us stupid. And because of such things an economy organized to promote fantasy will eventually collapse of its own weight.

In *The Memory of Old Jack*, Wendell Berry describes the main character as "troubled and angered in his mind to think that people would aspire to do as little as possible, no better than they are made to do it, for more pay than they are worth" (Berry 1974). The masters of the recently imploded financial universe, who made millions while destroying much of the economy with the chief executive officers of any number of corporations from Enron to General Motors, would have appropriately aroused Old Jack's fury, as they should ours.

I have a second suggestion which is simply that we ought to build a slower economy. It is also an old idea embodied in aphorisms such as "the race is not to the swift" and "haste makes waste" and ancient in practices of fallowing and seventh-year sabbaticals. In the age of hustle, cell phones, twittering, and instant everything, we are inclined to forget that lots of worthwhile things can only be done slowly. It takes time for all of us, economists included, to think clearly. It takes time to be a good parent or friend. It takes time to create quality. It takes time to make a great city. It takes time to restore soil. It takes time to restore one's soul. In each case, speed distorts reality and destroys the harmonies of nature and society alike. Hurry certainly changes society for the worse.

Ivan Illich's provocative 1974 book *Energy and Equity* makes the case that "high quanta of energy degrade social relations just as inevitably as they destroy the physical milieu" (Illich 1974, 3). "Beyond a critical speed," Illich argued, "no one can save time without forcing another to lose it" (Illich 1974, 30). But any proposal to limit speed "engenders stubborn opposition . . . expos[ing] the addiction of industrialized men to consuming ever higher doses of energy" (Illich 1974, 55). Even assuming nonpolluting energy sources, the use of massive energy of any sort "acts on society like a drug . . . that is psychically enslaving" (Illich 1974, 6). The irony, Illich exposes, is that the time it takes to earn the money to travel at high speed divided into the average miles traveled per year gives a figure of about 15 mph—about the average speed of travel in the year 1900 but at considerably higher cost.

What does a slow economy look like? Woody Tasch, chairman of Investors' Circle, offers one view:

> It would be driven by . . . the imperatives of nature rather than by the imperatives of finance. Its first principle would be, I suppose, the principle of carrying capacity, embedded in the process of *nurturing*. (Tasch 2008, 175)

A slow money economy would change the way we invest and discipline the expectations of quick returns to capital to, say, 5 to 8 percent per year, which now sounds pretty good. It would require buyers, for example, to hold stock for, say, 6 months before they could sell. Tasch calls this "patient capital," but by any name, it involves the recalibration of money and finance to the pace of nature. And that would be a revolution.

My third suggestion is to build an economy on ecological realities, not on the belief that we are exempt from the laws of ecology and physics. In his classic 1980 book *Overshoot*, William Catton writes, "The alternative to chaos is to abandon the illusion that all things are possible" (Catton 1980, 9). He goes on to say that "we need an ecological worldview; noble intentions and a modicum of ecological information will not suffice" (Catton 1980, 12). Each ratchet upward of human population and dominance required the diversion of "some fraction of the earth's life-supporting capacity from supporting other kinds of life to supporting our kind" (Catton 1980, 27). Eventually, the method of enlarging our estate by expanding into unoccupied lands gave way to industrialization and drawing down ancient ecological capital. "The myth of limitlessness dominated people's minds" to the point where we have nearly trapped ourselves (Catton 1980, 29). For Catton we are caught in an irony of epic proportions: "The very aspect of human nature that enabled *Homo sapiens* to become the dominant species in all of nature is also what made human dominance precarious at best, and perhaps inexorably self-defeating" (Catton 1980, 153).

Nobel Prize–winning chemist and economic theorist Frederick Soddy (1877–1956) made a similar point in *Wealth, Virtual Wealth, and Debt* (1926):

> Debts are subject to the laws of mathematics rather than physics. Unlike wealth, which is subject to the laws of thermodynamics, debts do not rot

> with old age and are not consumed in the process of living. On the contrary, they grow at so much per cent per annum, by the well-known mathematical laws of simple and compound interest. (Daly 1996, 178)

Debt," in Herman Daly's words, "can endure forever; wealth cannot, because its physical dimension is subject to the destructive force of entropy" (Daly 1996, 179). As a result, Daly continues, "the positive feedback of compound interest must be offset by counteracting forces of debt repudiation, such as inflation, bankruptcy, or confiscatory taxation, all of which breed violence" (Daly 1996, 179). The growth of the money economy in reality represented the expansion of claims (debt) against a stable or now diminishing stock called nature. As the economy grew, what we call wealth represented only a growing number of claims against a finite stock of soil, forests, wildlife, resources, and land and hence was the source of long-term inflation and ruin.

Although he apparently did not know of Soddy's work, economist Nicholas Georgescu-Roegen later made many of the same points about the relation of entropy to economic growth in his monumental but widely ignored *The Entropy Law and the Economic Process* (1971). "Every Cadillac produced at any time," he wrote, "means fewer lives in the future." Given our expansive nature and the laws of physics, our fate, he concluded, "is to choose a truly great but brief, not a long and dull, career" (Georgescu-Roegen 1971, 304). Or as John Ruskin once put it more poetically: "the rule and root of all economy—that what one person has, another cannot have; and that every atom of substance, of whatever kind, used or consumed, is so much human life spent" (Ruskin 1968, 192).

~

We are running two deficits simultaneously, and we must solve them together. If we fail to do so, nature will take its course. This will require a great deal of rethinking, and it will not be easy. But the present economic collapse is too far-reaching and the threat of climate disaster too real to do otherwise. Taken together, they indicate that we are nowhere near as rich as we once presumed. We have been living far beyond our means by drawing down natural capital, rather like a corporation selling off assets in a fire sale, and calling the proceeds profit. The housing bubble, dishonest accounting, and the use of unaccountable financial instruments like derivatives are merely the tip of a far larger problem that includes the failure to account for carbon emissions and the loss of species diversity.

The ideas that lead us to the brink are the equivalent of junk bonds and derivatives, unsecured by real assets and ungrounded in reality. There are better ideas by which to order our economic and ecological affairs based on the principle that "for every piece of wise work done, so much life is granted; for every piece of foolish work, nothing; for every piece of wicked work, so much death" (Ruskin 1968, 202).

Chapter 14

The Constitution of Nature

(2003)

TURMOIL IN THE MIDDLE EAST, Africa, and South Asia has plunged the world into yet another era of nation building. The U.S. is now engaged in the difficult task of reconstructing Afghanistan and Iraq, reportedly along democratic lines. Beneath all of the rhetoric it is assumed that democracy is a useful model for severely conflicted Muslim countries with no experience of it and, further, that ours is an adequate framework in which to conduct the public business of any country in the twenty-first century. The first assumption has been challenged as premature or even naive (Zakaria 2003). But it is the second, and more important, of the two that I intend to question—and in particular the constitutional framework within which our own politics occur.

The U.S. Constitution, ratified in 1788, reflects the opinions that originated in the Enlightenment era about many things, not the least of which is that ordinary people—within limits—are capable of self-governance. It had the virtues of flexibility and ambiguity that have allowed it to frame U.S. political life from that time to our own through civil war and the transition from an agrarian society to an industrial and technologically advanced behemoth. It has provided, as Robert Dahl notes, a model of sorts for more than 100 other nations, but few have adopted its core assumptions about the actual organization of power (Dahl 2002). In our own history the transition of the Constitution from "charter into scrip-

---

This article was originally published in 2003.

ture" did not occur until sometime in the late nineteenth century (Jones 1973, 37; Kammen 1987). Since 1791 it has been amended 17 times but without substantial alteration of the overall document. That fact alone suggests caution about changing something that has worked so well for so long. Or has it?

"Compared with other democratic countries," in Robert Dahl's words, the "performance [of the U.S. Constitution] appears, on balance, to be mediocre at best" (Dahl 2002, 118). His judgment is based on political criteria, but there are other and broader ways by which we might judge the Constitution. How well, for example, has it worked as a framework for protecting the waters, land, forests, soils, wildlife, and ecological integrity of the United States? A thorough reading of the evidence indicates serious decline in virtually every category (Heinz Center 2002; Abell et al. 2000; Ricketts et al. 1999; U.S. Geological Survey 1998). Dead zones, extinctions, toxic pollution, soil erosion, radioactivity, urban sprawl, smog, industrial sacrifice areas, and changing climate are the ecological hallmarks of economic development in the United States. But do such things reflect failures of the Constitution or broader failures in our political system, or some combination of the two?

Such questions would not have been intelligible to the framers. For them the conquest of nature by science and technology was an unmixed blessing. In our time, we can see the limits of nature, some say its end. We know what they could not have known: that nature is an intricate web of causes and effects often widely separated in space and time and that small changes can have very large effects. We know, too, that what we mean by nature is complicated by our being bound up in it in ways that are hard to fathom. And we know, or ought to know, that we could bring it and ourselves crashing down gradually or quickly. The framers of the U.S. Constitution could not have foreseen this, although James Madison and Thomas Jefferson came to believe that the experiment with democracy might not last beyond the time of cheap land (Matthews 1995, 210). We, however, know the ecological history of the intervening years and, arguably, have a better capacity to comprehend the future (McNeill 2000). All of this is to say that we can judge the Constitution and the political life it framed in an ecological perspective that the framers did not have. From this vantage point, three issues are seen as particularly important: the inclusiveness of constitutional protection, the applicability of due process, and the fragmentation of political power.

Inclusiveness

Although they began with the words "we the people," the framers did not include women, Native Americans, or African Americans. The omissions were rectified by the 13th, 14th, 15th, 19th, and 24th amendments. But no such protection has yet been granted to future generations, even though we know that the decisions and actions of the present generation cast a long shadow on their prospects in ways that could not have been known in the eighteenth century. Of the founders, Jefferson is notable for his worries about the intergenerational effects of debt, but no one could have known about intergenerational ecological debt and such things as the extinction of species, climatic change, and toxic pollution. "We the people" meant we the present generation with the caveat that the framers intended to "secure the blessings of liberty to ourselves and our posterity." To do so meant getting the framework issues right enough to balance interests, avoid the tyranny of either minority or majority, provide democratic representation, create national institutions, and establish a creditworthy government. But the framers placed no restrictions on the rights of the living relative with respect to those of subsequent generations. It would be a mistake, however, to infer that the framers had no further regard for posterity. To the contrary, I think that they did but assumed that obligations to the future had been discharged by the creation of a durable national government. But many now believe that future generations need more explicit protection.

In 1986 the Supreme Court of the Philippines, for example, upheld the standing of children to litigate in order to stop deforestation on behalf of future generations' rights to "a balanced and healthy ecology." To acknowledge standing, the court drew from no specific textual reference, saying only, "These basic rights need not even be written in the Constitution for they are assumed to exist from the inception of humankind" (Ledewitz 1998, 605). The proper question is then, not whether succeeding generations have legitimate rights to a balanced and healthy ecology, but how those rights would be determined and enforced in the present. But given the intergenerational reach of technology, the issues go much past the protection of resources. Do our descendants, for example, have a legitimate claim to a genetic heritage stable within definable limits (McKibben 2003)? Ought their interests to be weighed in decisions, say, about genetic enhancement of human intelligence and extension of the human life span that would become permanent features or perhaps even the beginnings of a new species? If so, how would we know their preferences or best interests, a different thing? Should their probable wishes or interests be considered in other decisions having to do with the development of nanotechnologies and artificial intelligence that might diminish their prospects or foreclose them altogether? If such rights are extended to the future generations, who will speak for them and how will those rights be honored in practice? Regarding the former, there are instructive precedents in trusteeship and court-appointed guardians for those unable to defend their own interests. And there are a variety of public policy tools to protect future generations, including prices that include true ecological costs, depletion quotas or severance taxes that slow the drawdown of resources, taxes on pollution, land trusts, and the police power of the state through regulation. Difficulties in applying these or other methods should not be used to override the fact that no good argument can be made for the right of one generation to deprive subsequent generations of the ecological requisites necessary to pursue life, liberty, and property.

Going further, the Constitution deals solely with humans and their affairs, which is to say that it is purely anthropocentric. But there is a broader way to think about constitutions. French sociologist Bruno Latour, for one, proposes that we distinguish between "the full constitution" and "the constitution of jurists" (Latour 1993, 13–15; Ledewitz 1998, 233). The former includes the unstated assumptions underlying the latter and accounts for "the distribution of powers among human beings, gods, nonhumans; the procedures for reaching agreements; the connections between religion and power; ancestors; cosmology; property rights; plant and animal taxonomies" (Latour 1993, 14). This larger constitution "defines humans and nonhumans, their properties and their relations, their abilities and their groupings" (Latour 1993, 15). This is, I think, what Aldo Leopold had in mind when he described humans as "plain members and citizens of the land-community" (Leopold 1987, 204, 223). But nowhere in the U.S. Constitution are the other members of the land community acknowledged. As Latour notes, the framers assumed that nature and society were entirely separate and that humans were "free to reconstruct [nature] artificially" (Latour 1993, 139). Lacking any constitutional recognition or protection, nature was there for the taking and it was taken. In the words of historian Howard Mumford Jones, "there was a continent to ravage" and Americans took a

> fierce, adolescent joy in smashing things—in stripping mountains to get at the ore, laying forests waste for their better timber, plowing up the plains whether normal crops could grow on them or not, slaughtering millions of bison . . . scarifying whole counties with the poisonous fumes of smelters, polluting rivers with sludge from oil wells, slaughterhouse, and city sewage. (Jones 1973, 107)

If the Constitution could, in time, be broadened to rectify past human wrongs through the amendments that extended the rights of citizenship to African Americans and women, no such thing has been done relative to the members of the land community. And some are still caught up in the adolescent joy of smashing things.

Latour proposes "a different democracy . . . [one] extended to things" (Latour 1993). Aldo Leopold similarly believed that inclusion of "soils, waters, plants, and animals, or collectively: the land" in our definition of community was both "an evolutionary possibility and an ecological necessity" (Leopold 1966, 253; Stone 1974). Leopold never wrote about the legal implications of this idea, assuming that the beginning point for law and policy was first to enlarge the boundaries of ethical consideration, and law would someday follow. If and when it does, there are difficult issues to resolve about how rights and duties pertain across species boundaries. Is it logical or practical to include the rights of species? Ought we to consider the rights of ecosystems, as Leopold proposed, and what does this mean? How are we to discern the interests of nonhuman entities or consider ourselves obliged when reciprocity is not possible? If these could be decided affirmatively, how might they be integrated into our complicated systems of politics and jurisprudence? Again, complexities should not be used as an excuse to dismiss the issues and thereby the possibilities of extending constitutional protections in important and novel ways. Leopold believed that an ecological comprehension of our own self-interest would lead us, in time, to see that our well-being was inextricably tied to the health of the land community. Said differently, human interests and the efficacy of law would be markedly diminished in a ruined ecological system. What could it possibly mean for Americans to have the rights guaranteed in the Constitution in a land with a diminished biota, despoiled landscapes, polluted air and water, little topsoil, ravaged forests, and a climate growing more severe decade by decade? Rights in such conditions would be no better than having legal entitlement to an apartment in a demolished building.

Due Process

The Fourth Amendment protects "the right of people to be secure in their persons, houses, papers, and effects." Yet these people so secured have dozens or hundreds of chemicals in their bloodstreams and fatty tissues from exposure to the thousands of chemicals in our food, air, water, and materials (Thornton 2000). The privacy of the body has been invaded mostly without our knowledge or permission, and with little accountability by those responsible. The ubiquity of pollution means that responsibility is difficult to ascertain, and it is still more difficult to determine which of hundreds or thousands of chemicals mixing in ways beyond our comprehension caused exactly what pathology. Our knowledge of such things is inescapably general. We know that some of these substances, singly or in combination, undermine health, reproductive potential, intelligence, ability to concentrate, and emotional stability, hence the capacity to pursue and experience life, liberty, and happiness. But it is nearly impossible to know exactly which ones, in what combinations, and at what specific levels. We know that children are more vulnerable to chemicals and heavy metals than adults and that some physical and mental effects are permanent. But we cannot know in advance which ones are most susceptible. We know, however, that the liberty of some to make and disperse toxic chemicals and heavy metals conflicts with the rights and liberties of those exposed. In some cases the effects will manifest far into the future, placing perpetrators beyond the reach of law and leaving their victims without remedy. What, then, does it mean that we cannot "be deprived of life, liberty, or property," including property of the body, without "due process of law," as stated in both the 5th and 14th amendments?

The framers could not have known about carcinogenic, mutagenic, or endocrine-disrupting substances or radioactivity, but we do. For many toxic substances we know that there is no safe threshold of exposure, and none at all for radioactivity. Chemicals that disrupt the endocrine system do their work in parts per billion, wreaking havoc on the development and immune systems of children. Had they known what we now know about the ubiquity of chemicals and their effects, would the framers have extended the protections of due process to include the fundamental right of bodily integrity? And should such protections be extended more broadly to include deprivation of other ecologically grounded requisites of life and liberty?

E. O. Wilson, for example, describes our affinity for nature as "biophilia," which he defines as an innate "urge to affiliate with other forms of life" (Wilson 1984, 85). "We are human in good part," he writes, "because of the particular way we affiliate with other organisms" (Wilson 1984, 139). Nature, then, is not something "out there" but, rather, something that has been inscribed in us, and after several million years of evolution it would be surprising were it otherwise. Environmental psychologists, similarly, describe nature as experienced in childhood as a kind of substrate of our consciousness and emotions. Could it be that the disruption of natural processes diminishes the possibility of affiliation with nature? Does ugliness, in all of its modern forms, diminish the human psyche and, thereby, the capacity for biophilia? Could it be that the diminished possibility for affiliation with a healthy nature reduces the quality of life? A growing body of scientific research suggests that that chain of reasoning is more than just plausible. If so, the constitutional protections of due process ought to be broadened someday to protect those aspects of life and liberty uniquely and irrevocably grounded in the experience of nature.

Fragmented Power

The framers created a system aimed to check ambition, competing interests, and the possibility of tyranny from a highly centralized government. To these ends, the Constitution divides power between the legislative, executive, and judicial branches and further between federal and state governments. Over time the fragmentation of government powers has increased with the growth of federal agencies, departments, and programs. The result is that, relative to environmental policy, the right hand of government often knows not what the left hand is up to. The Department of Commerce, for example, promotes economic expansion while the Environmental Protection Agency is expected to clean up the resulting messes. The Department of Energy promotes an energy plan with more nuclear power plants that, were it implemented, the Department of Defense could not conceivably defend from terrorists. The system of checks and balances, further, limits the ability of the federal government to anticipate, plan, and respond to systemic problems or, better yet, avoid them altogether. But the larger problem is the mismatch between the way nature works in highly connected and interactive ecosystems and the fragmentation of powers built into the Constitution. Nature is a unified mosaic of ecosystems, functions, and processes. Government, on the other hand, was conceived by the founders as a limited and fractured enterprise.

In the intervening years, government programs have grown as disjointed and incremental responses intended to solve particular problems. Not infrequently, a solution to one problem becomes the cause of later problems. The Clean Air Act of 1970, for example, required scrubbing of power plant emissions, but the substances so removed were deposited on land, becoming a land-use problem. The effect, in this and other cases, has been a kind of shell game by which problems are not solved but moved from air, to water, to land, and back again. Governments commonly deal with the coefficients of problems, not with the system that created problems in the first place. Reduced automobile pollution is a worthy goal, but the problem is systemic, having to do with the lack of anything like an intelligent transportation system that would include fast intercity trains, urban light-rail, bike trails, walking paths, and highways. Environmental laws seldom prevent or solve environmental problems. At best they render them somewhat more manageable while providing fertile ground for legal wrangling over the permissible rates by which the citizenry is poisoned and the land degraded.

The intent of the framers to limit and divide power, in other words, has become an impediment to the creation of effective environmental policy. "The Madisonian model," in Steven Kelman's words, "make[s] it more difficult to produce government action of any sort" (Kelman 1988, 49). Relative to environmental matters and the rights of future generations, gridlock is now the default setting of U.S. government. Consequently, since the 1970s there has been virtually no advance in our ability to protect or enhance environmental quality. At best, air and water quality are in a holding pattern while other, and more serious, problems worsen.

⁓

Even though "it is time—long past time—to invigorate and greatly widen the critical examination of the constitution and its shortcomings," in Robert Dahl's words, "public discussion that penetrates beyond the Constitution as a national icon is virtually nonexistent" (Dahl 2002, 154–56). Dahl believes the Constitution to be insufficiently democratic. It is also insufficiently ecological, and these are, I think, related issues. The framers' worldviews were a complicated mosaic of European and Scottish philosophy, agrarianism, frontier practicality, and Native American wisdom. And they were businessmen with an eye to pecuniary advantage, as Charles

Beard observed long ago (Beard 1913). They had, as Dahl and others note, mixed opinions about democracy. But they did not know and could not have known how the world works as an ecological system and that the unfettered advance of technology would someday cast a dark shadow on a distant posterity. The question is whether that lack can be remedied within the constitutional framework and its amending process or will require broader political change, or a combination of the two.

Either way, the essence of the remedy requires that we act conservatively in cases where the risks of widespread, severe, and irreversible harm are high or simply unknown (Raffensperger and Tickner 1999). The burden of proof ought to be placed on the generator of risk, not those involuntarily put at risk, now or later. The truth is that we are ignorant about many of the unanticipated adverse effects of technology and growth that accumulate by stealth or manifest by surprise. But precaution is as commonplace in daily affairs as it seems radical in the realm of present-day law, economics, and public policy. As individuals we buy insurance, have annual physical exams, and wear seat belts, which is to say that we exercise caution for reasons so obvious as to require no explanation. In medicine the principle of precaution is widely accepted in the words "first do no harm." In the realm of public policy, however, we have not acknowledged a comparable logic even though the risks we now incur may be catastrophic and our ignorance far exceeds our knowledge. It is one thing for individuals to incur risks to themselves and another thing entirely for some few to risk the welfare of the many, including future generations, who have no say in the matter. The present situation privileges the rights of the elite who cannot be held accountable if and when things turn out badly. The benefits of risk are, in effect, privatized while the risks are socialized across generational lines.

Is it possible to legislate precaution within and between generations? The most notable example to that end was the National Environmental Policy Act of 1970, which was intended "to improve and coordinate Federal plans, functions, and programs ... to the end that the nation may fulfill the responsibilities of each generation as trustee of the environment for succeeding generations." NEPA is an eloquent statement of a national environmental policy that had and still has great potential. The act mandated environmental impact statements for "Federal actions significantly affecting the quality of the human environment." But it has foundered, in the opinion of its principal author, on the shoals of presidential indifference, judicial misinterpretation, public apathy, broad incomprehension of the environment, and the lack of "a great unifying goal," none of which are specifically matters of the Constitution (Caldwell 1998, 147). Much of the same can be said of the effectiveness of the Endangered Species Act of 1973, which also required an ecologically literate Congress and executive and informed public. NEPA notwithstanding, the United States has no effective environmental policy and none whatsoever relative to energy use, land use, transportation, or agriculture. Instead we have a hodgepodge of poorly enforced laws, regulations, and practices, some of which work at cross-purposes, and no one of which prevents environmental degradation in the first place. Whatever the intentions of Congress, environmental laws and regulations have been watered down for the convenience of major economic interests. Despite ardent rhetoric about devotion to our children and theirs, we remain in thrall to economic expansion and resource exploitation, devil take the hindmost.

"It is in regard to the future," legal scholar Bruce Ledewitz writes, "that our policies are most clearly heedless . . . today's generation may be thought of as the majority using its political muscle to permanently disadvantage future generations" (Ledewitz 1998, 587, 591). In his view "there is no impediment in the political Constitution to the derivation of expansive constitutional rights particularly at a time in which the future of humankind may be at stake" (Ledewitz 1998, 620). Despite the failure of earlier attempts to amend the Constitution in 1967, 1968, and 1970, Ledewitz believes that the time will come when the environmental rights of future generations will be protected by law. Doing so will require, among other things, careful definition of what the words *healthy environment* mean, establishment of standing for future generations, and acceptance of the growing body of international legal opinion. The obstacles, in his view, are not problems of logic, law, or the intent of the founders but, rather, the embarrassing ecological obsolescence of U.S. constitutional law, a legal community ignorant of the scale of environmental problems, and the possibility that "the current generation may prefer its own wealth and convenience over that of future generations," which is a political problem (Ledewitz 1998, 631). But despite the power of the idea and the urgency of the situation, Ledewitz concedes that "the time is not ripe" for expanding the scope of the law or passing a constitutional amendment. Something more is necessary.

⁓

The idea that the relationship between government and the governed ought to be defined and limited in a written constitution was a novel and powerful idea in 1788—the culmination of decades of political reasoning, pamphleteering, and a revolutionary war. In the intervening years most nations have adopted constitutional frameworks defining procedures, rights, and structure between government and the people. In truly democratic societies, constitutions ought to promote fairness and public deliberation, replacing arbitrary authority with what Cass Sunstein calls "a republic of reasons" (Sunstein 1993, 347; Sunstein 2001). But in U.S. experience, constitutional law has also been a battleground between those wishing to preserve the status quo and others intent on promoting fairness and the flexibility to meet changing circumstances. Not infrequently the courts have used the Constitution to stop change, "often without adequate justification" (Sunstein 1993, 348; Sunstein 2001, 68–92). And the Supreme Court is presently silent regarding the ecological prerequisites for life, liberty, and property for the present and succeeding generations. That silence allows the living to deprive their posterity without due process and with little or no prospect of fair compensation.

The great virtue of the U.S. Constitution, however, is its "extraordinary capacity for self-revision" (Sunstein 1993, 354). It is, as Ledewitz reminds, "an open and revolutionary document . . . [and] need not be interpreted to stand mute while the environment and the interests of the future are sacrificed" (Ledewitz 1998). Our situation is no less revolutionary than that of the framers' time. Indeed, it is far more so. The stakes involved in climate change, loss of species, destruction of ecosystems, and tropical deforestation are much higher than those of the framers' era because they are global and permanent and threaten to destroy the ecological foundations of civilized societies. Looming on the horizon are technologies that, once deployed, could fundamentally and irrevocably alter what it means to be human and the role of humankind in a world of machines designed to be smarter than people and capable of self-replication. The time has come for a more thorough consideration of law, the rights of property, the public trust, and the human prospect.

In their time the framers recognized that tyranny could be remote, that those affected by decisions ought to be represented, that all are equal before the law, that those who govern ought to be held accountable and could be replaced through regular, fair, and open elections. However imperfectly executed, those principles are as revolutionary as ever. And there are still older principles contained in the Public Trust Doctrine holding that the commons of air, water, and lands ought to be managed for the public good, not private gain. But what do such ideas and precedents mean in our time, and what will they mean to posterity?

First, they mean acknowledging and eliminating the sources of remote tyranny in our time. In contrast to the framers' expectations, power has steadily gravitated from the people and elected governments to corporations, an entity they did not anticipate. Subsequently, the courts have been wonderfully kind to corporations, making them fictionalized persons protected by the rights accorded to real people in the Bill of Rights. Congress and various administrations have become highly indebted to them, thereby giving monied interests an undeserved advantage over the public interest. The public is losing, or has already lost, control over much of the public commons, capital, information, airwaves, land, health care, employment, genetic information, and, if the acolytes of free trade have their way, the power to control our own economic affairs. Further, we the people are excluded from fundamental decisions about war and peace, nuclear weapons policy, and the growing number of decisions about technology in which there is some probability of infinite disaster. Once, we became exercised about "taxation without representation," but the present reality is more akin to "extermination without representation."

Second, the shadow cast by ecologically profligate generations is the source of tyranny far less amenable to remedy. If that intergenerational remote tyranny is to be avoided, the present generation will have to do it. If it is true that "all men [including posterity] are created equal, that they are endowed by their Creator with certain unalienable rights, that among these are Life, Liberty, and the Pursuit of Happiness," then no generation has a license to diminish the unalienable rights of subsequent generations by changing the biogeochemical systems of Earth or impairing the stability, integrity, and beauty of biotic systems—the consequences of which are a form of tyranny stretching across generations. Ignorance can no longer serve as a plausible defense for actions that compromise the legitimate rights of present and future generations.

Accordingly, a truly conservative and revolutionary reading of the U.S. Constitution would build on the idea that we are trustees poised between our forebears and our posterity. In trust we are obliged by decency, fairness, justice, and affection to protect, preserve, and honor the ecological prospects of existing life and that yet to be. Without that guarantee, other purely legal rights can have little meaning. This obligation will require us to extend rights across generational lines, hold power accountable, and restrain the advance of technology where it impinges on fundamental rights to life, liberty, and property now and in the future. It is absurd to believe that the framers, seven generations ago, would have wished us to preserve the letter of the Constitution of 1788 while permitting the destruction of the very ground on which that document and life itself depend.

Chapter 15

Diversity

(2003)

I don't mean only insects and bacteria. There's far too much life of every kind about, animal and vegetable. We haven't really cleared the place yet.
C. S. Lewis, *That Hideous Strength*

By all credible accounts the diversity of species is in sharp decline, headed toward what Richard Leakey and Robert Lewin (1995) call "the sixth extinction." The causes of species decline include population growth, economic expansion, pollution, climate change, mining, logging, urban sprawl, overfishing, and the displacement of indigenous peoples. The legal and administrative protections placed between endangered species and eternity work, at best, in a limited fashion for a time. Our successes in preservation, as David Brower once noted, are temporary while our failures are permanent. But the problem is not limited to the decline of biological diversity. Many of the same forces that erode biological diversity jeopardize diversity of all kinds, including that of languages and culture. The modern world, it seems, is at war with difference even while professing devotion to it.

The loss of diversity cannot be attributed, on the whole, to ignorance. Considerable effort has been and is being made to document the decline, but how much such information reaches the public or particular decision makers is hard to say. What is apparent, for those who wish to see, is an increasingly detailed and discouraging picture. E. O. Wilson believes that we could lose a quarter of the world's biodiversity over the next century,

This article was originally published in 2003.

or with foresight and a bit of luck we might hold the loss to 10 percent (Wilson 1992, 342). Nor are we ignorant, for the most part, of the many reasons why diversity should be preserved. Self-interest, cultural prudence, and arguments in defense of the intrinsic rights of species converge on the same goal of saving all that can be saved. We know that it would be foolish to wantonly eliminate the many services provided by species and healthy ecosystems. Beyond arguments grounded in utility, however, we have powerful moral reasons to preserve species that provide no useful service to *Homo sapiens*. But neither argument has changed much relative to what needs to be changed if biological and cultural diversity is to be preserved, and so species, languages, and indigenous cultures continue to spiral downward. One suspects that the effort to preserve diversity runs against deeper currents. We profess great devotion to that which we seem incapable of protecting, in large part, I think, because the very logic of modern culture aims everywhere for uniformity and control.

Modernization from its earliest beginnings formed around ideas such as progress, economic growth, and human superiority over nature and the goals of standardization, legibility, efficiency, and control. Lewis Mumford begins his magisterial history of technology, for example, with a discussion of the clock which he describes as "the foremost machine of modern technics" (Mumford 1934, 15). The widespread adoption of the clock in the fourteenth century led to the quantification of time and "a new medium of existence" in which "one ate, not upon feeling hungry, but when prompted by the clock: one slept, not when one was tired, but when the clock sanctioned it" (Mumford 1934, 17). The irregularities of actual days and seasons as well as those inherent in the human organism were swept aside in favor of time that could be measured, counted, and made to count. The clock was followed in rapid succession by geometrically precise maps, double entry bookkeeping, the art of perspective in painting, and the marriage of vision and quantification that, in Alfred Crosby's words, "snap the padlock—reality is fettered" (Crosby 1997, 229). Ability to quantify visual space enabled Europeans to extend control in hitherto unimaginable ways and envision things yet to be invented giving rise to still greater control and uniformity.

The same trend toward greater control became evident in European philosophy, science, and politics in the turbulent seventeenth century. "From 1630 on," in philosopher Stephen Toulmin's words, "the focus of philosophical inquiries has ignored the particular, concrete, timely and local details of everyday human affairs: instead, it shifted to a higher, stratospheric plane, on which nature and ethics conform to abstract, timeless, general, and universal theories" (Toulmin 1990, 35). The dream of reason aimed to establish rational methods, an exact language, and a unified science, in other words, "a single project designed to purify the operations of the Human Reason by de-situating them: that is, divorcing them from the compromising association of their cultural contexts" (Toulmin 2001, 78). This was nothing less than the triumph, temporary perhaps, of rationality over reasonableness, calculation over emotion, generality over particularity. The result, in Toulmin's words, was "injury to our commonsense ways of thought [and] confusion about some highly important questions" (Toulmin 2001, 204).

The search for certainty extended past the abstractions of philosophy and science to change how people related to physical reality as well. Early and inaccurate means of surveying land, for example, surrendered to precise land measurement based on Edmund Gunter's 22-yard-long surveying chain (Linklater 2002). The precise measurement of land permitted a market by which land became yet another commodity. According to the land ordinance of 1785, "surveyors shall proceed to divide the said territory into townships of 6 miles square, by lines running due north and south, and others crossing these at right angles" (Linklater 2002, 73). The subsequent grid pattern marched westward from the Appalachian Mountains across prairies to the Pacific regardless of variations of topography, ecology, or the long-established use patterns of the natives, who did not believe in owning land, or in right angles for that matter. In the twentieth century, the drive for standardization was applied to the workplace (Taylorism), to public economics as cost-benefit analysis, and to business operations, education, agriculture, forestry, urban planning, and governance. The culmination was what James C. Scott calls "a high modernist ideology . . . a muscle-bound, version of the self-confidence about scientific and technical progress, the expansion of production, the growing satisfaction of human needs, the mastery of nature (including human nature), and above all, the rational design of social order commensurate with the scientific understanding of natural laws" (Scott 1998, 4).

The juggernaut of standardization, uniformity, and legibility is a product of large forces, including technological dynamism, the triumph of quantification, the extension of state control over its land and peoples, and the imperative of capitalism to grow without limit. Whatever their other differences, all modern societies converged around the goals of precise measurement, increasingly pervasive control, and economic growth. The inevitable result was to increase scale, complexity, velocity, profitability, pollution, lethality, and centralization. But the choices were not inevitable or perhaps even probable at the time they were made. The important point is that the particular form modernization took was not foreordained. It required choices and sometimes better choices could have been made and some might still be made.

The well-advertised gains of modernization over the past two centuries are obvious, including greater material comfort, longer lives, increased mobility, and a huge gain in material wealth. On the other hand, any honest reckoning of the price we have paid and continue to pay for standardization must include the unaccounted costs such as the destruction of natural systems, climate destabilization, social regimentation, militarization, extermination of indigenous peoples, and more. Still other costs are evident only in hindsight. In the words of one observer, part of the cost (of quantification) is that "inevitably meanings are lost . . . because it imposes order on hazy thinking, but this depends on the license it provides to ignore or reconfigure much of what is difficult or obscure" (Porter 1995, 85). And much is difficult and obscure, including the value of diversity itself in all of its manifestations and even the capacity to think diversely. It is logical that the drive for standardization and uniformity might someday impose a gridlike pattern to the ecology of our minds until we are permitted to have no thoughts without right angles.

On the other hand, the world is full of surprise and paradox and mocks the human pretension to mastery. As management, standardization, and uniformity have increased without limit, so have the number of unanticipated effects of all kinds. Often called "side effects," they are more accurately said to be logical outcomes that we weren't smart enough to foresee (like the market collapse of 2008). Predictably, our attempts to render nature more orderly often backfire. Large dams, pesticides, wonder drugs, improved forests, highways, industrial agriculture, and factory farms run afoul of larger forces and limitations. But the same applies everywhere, relentlessly. Human perversity or creativity, sometimes only a difference of perspective, will find ways to subvert clocks, walls, bureaucracy, straight lines, barriers, and regimentation. There are, in other words, limits to what we can control at any scale over any length of time and remorseless forces that care not one iota for capitalism, reasons of state, or big science and will, in due time, sweep all off the stage. Those limits—ecological, economic, political, organizational, and human—were evident in the collapse of the Soviet Union. But it is no less likely that a capitalist society organized around a diminishing number of ever-larger corporations will also self-destruct. The belief that we might organize the entire planet for the convenience of capital and a diminishing number of capitalists will one day be seen as a kind of perverse and self-defeating foolishness. By the same token the belief that we can render ourselves safe by building higher and thicker walls or imposing tighter surveillance or find safety in ever more heroic, elaborate, and expensive technology will one day be seen as about as effective as a child hiding under bedcovers in a thunderstorm.

We struggle to understand diversity and ecological complexity in part because our language, minds, categories of thought, perceptions, values, and intellectual tools have been honed to control. But it is apparent that the logic of human control and mastery runs counter to the 3.8 billion years of evolution and that discrepancy is not just a flaw but a fault line. The record of evolution is one of surprise, flow, networks, unpredictability, nonlinearity, collapse, and creativity. But the logic of the modern economy and state runs against the flow of evolution, and something will have to give.

Climate stability and the preservation of diversity will require a manner of thinking grounded in ecology and systems. We need to know why diversity is important, and the strongest arguments are not first and foremost economic. To arguments of self-interest, prudence, and intrinsic rights—all true enough—we should add the idea of celebration of beauty that is inherent in the diversity of life and human culture. Arguments from self-interest or duty often sound like a Puritanical sermon running on too long, the point of which is that guilt will move us to better behavior. Sometimes it does, but it is more likely that we will be moved farther and to better ends by shorter sermons and the power of wonder, joy, and celebration.

The protection of diversity will require that we confront what E. O. Wilson calls "the juggernaut of technology-based capitalism." (Wilson 2002, 156). But there's the rub. Whether capitalism transforms itself into a different order of "natural capitalism" that protects wildlife, ecosystems, oceans, climate stability, soils, and forests is unlikely without the leadership necessary to change the rules of a system of short-term shareholder value. And there are bigger questions ahead having to do with the scale of the economy; the point at which further growth becomes, not just superfluous, but destructive; and the growing gap between the rich and the poor and between present and future generations.

The protection of diversity, implying restraint, also runs counter to the libertarian idea of individual freedom. The original idea helped humanity surmount arbitrary authority of church, monarchy, and rigid hierarchy. But if we are to preserve diversity and a habitable Earth, we will need a more inclusive idea that does not confuse freedom with license. Edmund Burke, the founder of modern conservatism, put it this way:

> Men are qualified for civil liberty in exact proportion to their disposition to put moral chains on their own appetites . . . society cannot exist unless a controlling power upon will and appetite be placed somewhere, and the less of it there is within the more there must be without. It is ordained in the eternal constitution of things, that men of intemperate minds cannot be free. Their passions forge their fetters. (Ophuls 1977, quoted in frontispiece)

Burke understood there can be no freedom amidst social chaos, nor can freedom exist in a state of ecological ruin. This level of sophistication requires that people understand the linkages between the limits to human actions and ecological health.

Finally, the protection of diversity will require a larger and yet more limited view of science and what it means to know. It is assumed, wrongly I think, that knowing is equivalent to measuring, explaining, and controlling. The protection of diversity will require, to the contrary, that we recognize reality and value that exist beyond our limited ability to measure and control. The fact is that biological diversity can be measured and described at a superficial level but can never be fully explained or known. The scientific impulse is to add something like "not yet," in the faith that we will, given time, figure it all out. I think it more likely that the right word is *never*, in the recognition of the limits of human knowledge and the many ways that knowledge can be corrupted, co-opted, and misused. This is the kind of mature knowledge, once proposed by Aldo Leopold and Rachel Carson, rooted in the recognition of the kinship of all life and the limits of human knowledge. It is a science driven by wonder and disciplined by humility in the recognition that there are mysteries that we are powerless to name.

Chapter 16

All Sustainability Is Local: New Wilmington, Pennsylvania

(1994)

AUTHOR'S NOTE 2010: *I grew up in New Wilmington, Pennsylvania. In 1950, like many small towns, it was a great deal more sustainable than it was a half century later. It wasn't Nirvana in 1950, but neither was it as vulnerable to outside forces as it is now. There was a local economy that is now a shadow of what it once was. The principal reasons have to do with the practical effects of bad economic ideas that gave little thought for the morrow or consideration for real places and flesh-and-blood people. Looking back, however, is not just an exercise in nostalgia. On the contrary, it offers some standard by which to judge what we've lost and what we might relearn about practical sustainability and a workable future in particular places on what is becoming a different Earth.*

I GREW UP IN A SMALL TOWN amidst the rolling hills and farms of western Pennsylvania. As towns go, it wasn't much different from thousands of others throughout the United States. There were four churches and a small liberal arts college. It was a "dry" town filled with serious and hard-working Protestants and a disconcertingly large number of retired preachers and missionaries. It was not the kind of

This article was originally published in 1994.

place that greeted Elvis and rock and roll with open arms. The prevailing political sensibilities were sober and overwhelmingly Republican of the Eisenhower sort. The town would have seemed stuffy and parochial to a Sherwood Anderson or a Theodore Dreiser, and it probably was. But for a little boy on a bicycle it was a paradise. By the standards of the 1990s, the town, the college, and its residents would have failed even the most lax certification for political correctness. It was a man's world, neither multicultural nor multiracial. The sexual revolution lay ahead. And almost everyone who was anyone in town bought without question the assumptions of midcentury America about our inherent virtue, the certainty of economic progress, the evils of Communism, and the beneficence of technology. J. Edgar Hoover was a hero. Boys were measured for manhood on the baseball diamond or the basketball court. It was also a place, like most others, in transition from one kind of economy to another.

Typical of most small towns, the main street of New Wilmington, Pennsylvania, still reflected bits of the nineteenth-century agrarian economy. There was, for example, a dilapidated and unused livery stable behind the main street, where a funeral parlor parked a hearse. On Main Street, Mr. Meeks operated his watch repair shop and Mr. Fusco had his shoe repair shop. There were locally owned and operated businesses, including two grocery stores, a hardware and plumbing store, a five and dime, a good bakery, an electronics/appliance store, a dairy store, a bank, a dry goods store, a magazine and tobacco shop, a movie theater, a building supply store, and a butcher shop. The train station was located two blocks from Main Street. A half mile to the east a local entrepreneur operated a tool-making plant. A quarter mile beyond, the town dump festered on the banks of Neshannock Creek.

The small-town, repair-and-reuse economy was predominantly locally owned and operated. My mother bought groceries from the store on Main Street. She bought vegetables from local farmers, including the Amish who went door to door selling everything from farm fresh eggs to maple syrup. Milk was delivered daily in returnable glass bottles by a locally owned dairy company. Soda pop also came in returnable glass bottles from a bottling plant 8 miles distant. Broken machinery could be repaired in town. Mr. Hoover sharpened dull saws for a dime, but his tales of local history were told for free. Grown men played summer "town ball" on baseball diamonds that sometimes doubled as pastures. Hand-me-down clothing was standard, and as the youngest of three I was the last stop for lots of items. And some of the best Christmas presents I ever received were made by hand.

The forces that would undermine that sheltered world of small-town, midcentury America were on the march. But I knew nothing of these as I joined the great exodus of self-assured and expectant young people leaving their hometowns for some other place thought to offer greater opportunity and more excitement. Few of us could say with certainty why we were going or where we were headed other than that it was somewhere else and presumably upward. Nor could we have said what we were leaving behind.

Looking back, I can see that even then, things were changing as the larger industrial economy began to undermine local economies nearly everywhere. We bought our first television set the same year that Congress passed the National Interstate and Defense Highway Act. I recall the lights on the big shovel at the strip mine across the valley burning into the night. The contractor for whom I worked in the summer went out of business shortly after I graduated from college. The farmer who gave me part-time employment, and was thought to be the most progressive in the county, went bankrupt in 1975. He was not alone. People in New Wilmington now buy their milk in plastic jugs from interstate dairy cooperatives. The local bottling plant disappeared and with it the practice of returning bottles to the store. The nearby industrial cities of New Castle and Youngstown, Ohio, which I knew as busy and thriving manufacturing places, are now mostly derelict and abandoned, as are other cities in what was once a blue-collar industrial corridor stretching from Pittsburgh to Cleveland and on to Detroit. Interstate highways to the north and east of town now slash across what was once farm country. Tourism is the main economic hope. Drugs and interstate crime are growing problems.

In the years since the class of 1961 set out to find its way, world population has grown from 3.2 billion to 7 billion; carbon dioxide in the atmosphere went from about 318 parts per million to 392 and is still rising; perhaps a tenth of the life-forms on the Earth disappeared in that time, but no one knows for sure; a quarter of the world's rain forests were cut down; half or more of the forests in Europe were damaged by acid rain; careless farming and development caused the erosion of some 600 billion tons of topsoil worldwide; and the ozone shield was severely damaged. Before the class of 1961 is just a faint memory, the Earth may be about 2°C hotter, with consequences we can barely imagine; world population will be pushing on to 9 billion; perhaps 25 percent of the Earth's species will have disappeared; and humans will have turned an area roughly equivalent to the size of the United States into desert. Something of earth-shattering importance went wrong in our lifetime, and we were not prepared to see it and so we became unwitting accomplices in the undoing of lots of things.

Looking back with more or less 20/20 hindsight, I believe that despite all of the many good things in my town, there were blind spots. First, and most obvious, we were taught virtually nothing of ecology, systems, and interrelatedness. But neither were many others. This was a blind spot for a country determined to grow and armed with the philosophy of economic improvement. As a consequence we knew little of our ecological dependencies or, for that matter, our own vulnerabilities. The orchard beside our house was drenched with pesticides every spring and summer, and we never objected. The blight of nearby strip mines grew year by year, and we saw little wrong with that either.

We grew up in a bountiful region, which was virtually opaque to us. In school I learned about lots of other places, but I did not learn much about my own. We were not taught to think about how we lived in relation to where we lived. The Amish farms nearby, arguably the best example we have of a culture that fits sustainably in its locality, were regarded as a quaint relic of a bygone world that had nothing to offer us. There was no course in high school or the local college on the natural history of the area. To this day, little has been written about the area as a bioregion. So we grew up mostly ignorant of the biological and ecological conditions in which we lived and what these required of us.

I finished high school the year before publication of Rachel Carson's (1962) *Silent Spring* but not before M. King Hubbert's 1957 projections of peak U.S. oil production and some of the best writings of Lewis Mumford, Paul Sears, Fairfield Osborn, and William Vogt and earlier writings of John Muir, John Burroughs, George Perkins Marsh, and Henry David Thoreau. Our teachers and mentors had been through the Dust Bowl years, the Great Depression, and World War II, but it was the Depression that seemed to have affected them most, and that fact could not help but affect us. Almost by osmosis we absorbed the purported lessons of economic hardship, but not those of ecological collapse, which can also lead to privation and economic failure. When it came time to rebel, we did so over such things as "lifestyle" and music. But we in the class of 1961 had little concept of *enough* or any reason to think that limits of any sort were very important. Inadequate though it was, we did have an economic philosophy, but we had no articulate or ecologically solvent view of nature. We were sent out into the world armed with a creed of progress but had scarcely a clue about our starting point or how to "find our place and dig in," as poet Gary Snyder said we should. And none of us in 1961 would have had any idea of what those words meant.

Looking back, I can see a second missing element. On one hand I recall no skepticism or even serious discussion about technology. On the other, the college-bound students were steered into academic courses and away from vocational courses. As a result the upwardly mobile became both technologically illiterate and technologically incompetent. All the while, there was a gee-whiz kind of naïveté reinforced by advertisers hawking messages from a cartoon character "Reddy Kilowatt" about "living better electrically" and those from General Electric saying that progress was their "most important product." We bought it all without much thought. We were good at detecting the benefits of technology in parts per billion and did not suspect what it would someday cost us. Nor could we see the web of dependencies that was beginning to entrap us. The same "they" who would somehow figure it all out were taking the things that Americans once did for themselves as competent people, citizens, and neighbors and selling them back at a good markup. We were turned out into the world with the intellectual equivalent of a malfunctioning immune system, unable to think critically about technology. If we read Marlowe's *Faust* at all, we read it as a fable, not as a prophecy.

Had we known our place better, and had we been ecologically literate and technologically savvy, we still would have lacked the political wherewithal to be better stewards of our land and heritage. Our version of small-town, flag-waving patriotism was disconnected from the tangible things of livelihood and location, soils and stewardship. We mistook the large abstractions of nationalism, flag, and presidential authority for patriotism. Accordingly, we were vulnerable to the chicanery of Joe McCarthy and J. Edgar Hoover, and later to Lyndon Johnson's lies about Vietnam, Richard Nixon's lies about nearly everything, and Ronald Reagan's fantasies about "morning in America."

My classmates and I are, I think, typical of most Americans born and raised in the middle decades of the twentieth century. Ours has been a time of cheap energy, economic and technological optimism, lots of patriotic self-righteous huffing and puffing, and "auto-mobility." We are movers and we move on average 8 to 10 times in a lifetime. We were educated to be competent in an industrial world and nigh onto useless in any other. We did not much question the values and assumptions of the industrial "paradigm" or those underlying notions of progress. Those were givens. We were turned out into the world, vulnerable to whatever economic, technological, or even political changes would be thrust upon us, as long as they were said to be economically necessary or simply inevitable. We were not taught to question the physical, biological, and psychological reordering of the world going on all around us. Nor were we enabled to see it for what it was—a kind of large-scale vandalism.

New Wilmington, Pennsylvania, is still a nice town. Having little industry, it has not suffered the rusting fate of the nearby industrial cities. It has also been spared some of the uncontrolled growth that has desecrated many other regions. Housing developments outside town, though, are now filling up what was once good farmland. Aside from the Amish, the local farm economy is a shadow of what it once was. The effects of acid rain are beginning to show on trees. To make ends meet, the region is increasingly dependent on tourism. New Wilmington, like most small towns, is an island at the mercy of decisions made elsewhere. It has been spared mostly because no one noticed it or thought it a place likely to be profitable enough for an interstate mall, mine, regional airport, a Disney World, or a new industrial "park." Not yet anyway. In the meantime, it too has become a full-fledged member of the throwaway economy, and its young people still depart in large numbers for careers elsewhere. You can't buy much on Market Street anymore. There is no one to repair watches or shoes, or sharpen saws, or sell appliances or dry goods or hardware or baked goods. People now drive to nearby shopping centers and distant malls. The trains don't run anymore.

Still, New Wilmington has gotten off lightly so far, but other towns and regions have not. Within a few miles, New Castle and Youngstown are industrial disaster areas. The landfill on the outskirts of my present hometown sells space for garbage from as far away as New York City. In southern Ohio, the nuclear processing plant at Fernald has spread radioactive waste over several hundred square miles. The same is true of Maxey Flats, Kentucky; Rocky Flats, Colorado; and Hanford, Washington, all sacrificed in the name of "national security." Urban sprawl and decaying downtowns afflict hundreds of other towns and cities throughout the United States. Mobile capital and a large dose of economic idiocy did what no invading army could have done to Cleveland and Detroit. Large chunks of footloose capital ravage other places. In northern Alberta,

Canada, Mitsubishi Corporation has invested over $1 billion to build a pulp mill that will impair or destroy an ecosystem along with the indigenous culture. Others are wreaking havoc on a still larger scale to convert tar sands to fuels. Thousands of square kilometers of rain forest will be destroyed to supply Europe with cheap minerals and soybeans. The resulting devastation will not show up in the prices paid by Europeans. Nor will the devastation from the other mines, wells, clear-cuts, or feedlots around the world, which supply the insatiable appetite of the industrial economy, be subtracted from calculations of wealth. We are told that the gross world economy must increase fivefold by the middle of the century. That same global economy now uses, directly or indirectly, 25 percent of the Earth's net primary productivity. Can that increase fivefold as well?

Custodians of the conventional wisdom believe that economic growth is a good and necessary thing. Growth, in turn, requires capital mobility, free trade, and the willingness to take risks and make sacrifices. For the sake of growth, whole regions and entire industries may have to be sacrificed, as production and employment go elsewhere in search of cheaper labor and easier access to materials and markets. Such sacrifices are necessary, they say, so that "we" can remain competitive in the global economy and so that the things we buy can be as cheap as profit-maximizing corporations can make them.

Conventional wisdom denies the importance of place and environment in favor of global vandalism masquerading as progress. Its more progressive adherents believe that environmental improvement itself requires further expansion of the very activities that wreck environments. Devotees of the second piece of conventional wisdom ignore the political and ecological creativity of place-centered people, wishing us to believe that the same organizations that have ruined places around the world can be trusted to save the global environment.

On the contrary, a world that takes both its environment and prosperity seriously over the long run must pay careful attention to the patterns that connect the local and the regional with the global. I do not believe that global action is unnecessary or unimportant. It is, however, insufficient and inadequate. Taking places seriously would change what we think needs to happen at the global level. It does not imply parochialism or narrowness. It does not mean crawling into a hole and pulling the ground over our heads, or what economists call autarky. While we have heard for years that we should "think globally and act locally," these words are still more a slogan than a program. The national and the international are still given a disproportionate share of our attention, and the local not nearly enough. But places, localities, and what William Blake once called "minute particulars" matter for many reasons.

First, we are inescapably place-centric creatures shaped in important ways by the localities of our birth and upbringing (Gallagher 1993; Tuan 1977). We learn first those things in our immediate surroundings, and these we soak in consciously and subconsciously through sight, smell, feel, sound, taste, and perhaps other senses we do not yet understand. Our preferences, phobias, and behaviors begin in the experience of a place. If those places are ugly and violent, the behavior of many raised in them will also be ugly and violent. Children raised in ecologically barren settings, however affluent, are deprived of the sensory stimuli and the kind of imaginative experience that can only come from biological richness. Our preferences for landscapes are often shaped by what was familiar to us early on. There is, in other words, an inescapable correspondence between landscape and "mindscape" and between the quality of our places and the quality of the lives lived in them. In short, we need stable, safe, interesting settings, both rural and urban, in which to flourish as fully human creatures.

Second, the environmental movement has grown out of the efforts of courageous people to preserve and protect particular places: John Muir and Hetch Hetchy, Marjory Stoneman Douglas and the Everglades, Horace Kephart and the establishment of the Great Smoky Mountains National Park. Virtually all environmental activists, even those whose work is focused on global issues, were shaped early on by a relation to a specific place. What Rachel Carson (1984) once called the "sense of wonder" begins in the childhood response to a place that exerts a magical effect on the ecological imagination. And without such experiences, few have ever become ardent and articulate defenders of nature.

Third, as Garrett Hardin argues, problems that occur all over the world are not necessarily global problems, and some truly global problems may be solvable only by lots of local solutions. Potholes in roads, according to Hardin, are a big worldwide problem, but they are not a "global" problem that has a uniform cause and a single solution applicable everywhere (Hardin 1993, 278; Hardin 1986, 145–63). Any community with the will to do so can solve its pothole problem by itself. This is not true of climate change, which can be averted or minimized only by enforceable international agreements. No community or nation acting alone can avoid climate change. Even so, a great deal of the work necessary to make the transition to a solar-powered world that does not emit heat-trapping gases must be done at the level of households, neighborhoods, and communities.

Fourth, a purely global focus tends to reduce the Earth to a set of abstractions that blur what happens to real people in specific settings. An exclusively global focus risks what Alfred North Whitehead once called the "fallacy of misplaced concreteness" in which we mistake our models of reality for reality itself—equivalent, as someone put it, to eating the menu, not the meal. It is a short step from there to ideas of planetary management, which appeals to the industrial urge to control. Indeed, it is aimed mostly at the preservation of industrial economies, albeit with greater efficiency. Planetary managers seek homogenized solutions that work against cultural and ecological diversity. They talk about efficiency but not about sufficiency and the idea of self-limitation (Sachs 1992, 111). When the world and its problems are taken to be abstractions, it becomes easier to overlook the fine grain of social and ecological details for the "big picture," and it becomes easier for ecology to become just another science in service to planet managers and corporations.

A final reason why the preservation of places is essential to the preservation of the world has to do with the fact that we have not succeeded in making a global economy ecologically sustainable, and I doubt that we will ever be smart enough or wise enough to do it on a global scale. All of the fashionable talk about sustainable development is mostly about how to do more of the same, but with greater efficiency. The most prosperous economies still depend a great deal on the ruination of distant places, peoples, and ecologies. The imbalances of power between large wealthy economies and poor economies virtually assure that the extraction, processing, and trade in primary products and the disposal of industrial wastes rarely will be done sustainably. Having entered the global economy, the poor need cash at any ecological cost, and the buyers will deny responsibility for the long-term results, which are mostly out of sight. As a result, consumers have little or no idea of the full costs of their consumption. Even if the sale of timber, minerals, and food were not ruinous to their places of origin, moving them long distances would still be. The fossil fuels burned to move goods around the world add to pollution and global warming. The extraction, processing, and transport of fossil fuels is inevitably polluting. And the human results of the global trading economy include the effects of making people dependent on the global cash economy with all that it portends for those formerly operating as self-reliant, subsistence economies. Often it means leaving villages for overcrowded shantytowns on the outskirts of cities. It means growing for export markets while people nearby go hungry. It means undermining economic and ecological arrangements that worked well enough over long periods of time to join the world economy. It means Coca-Cola, automobiles, cigarettes, television, and the decay of old and venerable ways. The rush to join the industrial economy in the late years of the twentieth century is a little like coming on board the *Titanic* just after icebergs are spotted dead ahead. In both instances, celebrations should be somewhat muted.

The idea that place is important to our larger prospects comes as good news and bad news. On the positive side, it means that some problems that appear to be unsolvable in a global context may be solvable on a local scale if we are prepared to do so. The bad news is that much of Western history has conspired to make our places invisible and therefore inaccessible to us. In contrast to "dis-placed" people who are physically removed from their homes but who retain the idea of place and home, we have become "de-placed" people, mental refugees, homeless wherever we are. We no longer have a deep concept of place as a repository of meaning, history, livelihood, healing, recreation, and sacred memory and as a source of materials, energy, food, and collective action. For our economics, history, politics, and sciences, places have become just the intersection of two lines on a map, suitable for speculation, profiteering, another mall, another factory. So many of the abstract concepts that have shaped the modern world, such as economies of scale, invisible hands, the commodification of land and labor, the conquest of nature, quantification of virtually everything, and the search for general laws, have rendered the idea of place impotent and the idea of people being competent in their places an anachronism. This, in turn, is reinforced by our experience of the world. The velocity of modern travel has damaged our ability to be at home anywhere. We are increasingly indoor people whose sense of place is indoor space and whose minds are increasingly shaped by electronic stimuli. But what would it mean to take our places seriously?

It would mean restoring the idea of place in our minds by reordering educational priorities. It is commonly believed, however, that the role of education is only to equip young people for work in the new global economy in which trillions of dollars of capital roam the Earth in search of the highest rate of return. Those equipped to serve this economy, whom Robert Reich calls "symbolic analysts," earn their keep by "simplifying reality into abstract images that can be rearranged, juggled, experimented with, communicated to other specialists, and then, eventually, transformed back into reality" (Reich 1991, 177–9). Symbolic analysts "rarely come into direct contact with the ultimate beneficiaries of their work"; rather, they mostly

> sit before computer terminals—examining words and numbers, moving them, altering them, trying out new words and numbers, formulating and testing hypotheses, designing or strategizing, They also spend long hours in meetings or on the telephone, and even longer hours in jet planes and hotels—advising, making presentations, giving briefings, doing deals. (Reich 1991, 179)

Symbolic analysts seem to be a morally anemic bunch whose services "do not necessarily improve society," a fact that does not seem to matter to them, perhaps because they are too busy "mov[ing] from project to project . . . from one software problem to another, to another movie script, another advertising campaign, another financial restructuring" (Reich 1991, 185, 237). They are, in Reich's words, "America's fortunate citizens," perhaps 20 percent of the total population, but they are increasingly disconnected from any interaction with or sense of responsibility for the other four-fifths (Reich 1991, 250). People educated to be symbolic analysts neither have loyalty to the long-term human prospect nor are prepared by intellect or affection to improve any place. And they are sure signs of the failure of the schools and colleges that presumed to educate them but failed to tell them what an education is for on a planet with a biosphere.

The world does not need more rootless symbolic analysts or rootless people of any kind. It needs instead millions of young people equipped with the vision, moral stamina, and intellectual depth necessary to rebuild neighborhoods, towns, and communities around the planet. The kind of education presently available will not help them much. They will need to be students of their places and competent to become, as Wes Jackson puts it, native to their places. They will need to know a great deal about new fields of knowledge, such as restoration ecology, conservation biology, ecological engineering, and sustainable forestry and agriculture. They will need a more honest economics that enables them to account for all of the costs of economic-ecological transactions. They will need to master the skills necessary to make the transition to a solar-powered economy. But who will teach them these things?

Taking places seriously means learning how to build local prosperity without ruining some other place. It will require a revolution in economic thinking that challenges long-held dogmas about growth, capital mobility, the global economy, the nature of wealth, and the wealth of nature. My views about capital mobility and related subjects were influenced, no doubt, by growing up near a now derelict industrial city, a monument of sorts to mobile capital and failed ideas. Even the prosperous city of my memory, however, was an ecological disaster. On both counts, could it have been otherwise? What would "place-focused economies" look like (Kemmis 1990, 107)?

Historian Calvin Martin argued that the root of the problem dates back to the dawn of the Neolithic age and to the "gnawing fear that the earth does not truly take care of us, of our kind . . . that the world is not truly congenial to sapient Homo" (Martin 1992, 123). Perhaps this is why most indigenous cultures had no word for scarcity and why we, on the other hand, are so haunted by it. Long ago, out of fear and faithlessness, we broke our ancient covenant with the Earth. I believe that this is profoundly true. But we need not go so far back in time for workable ideas. Political scientist John Friedmann argued that in more recent times,

> we have been seduced into becoming secret accomplices in our own evisceration as active citizens. Two centuries after the battle cries of Liberty, Fraternity, and Justice, we remain as obedient as ever to a corporate state that is largely deaf to the genuine needs of people. And we have forfeited our identity as "producers" who are collectively responsible for our lives. (Friedmann 1987, 347)

What can be done? While believing that "the general movement of the last six hundred years toward greater global interdependency is not likely to be reversed," Friedmann argued for "the selective de-linking of territorial communities from the market economy" and "the recovery of political community." This work can only be done, as he put it, "within local communities, neighborhoods, and the household" (Friedmann 1987, 385–7).

But communities everywhere are now vulnerable to the migration of capital in search of higher rates of return. In the case of Youngstown, Ohio, after the purchase of Youngstown Sheet and Tube by the Lykes Corporation and eventually the LTV Corporation, its profits were used to support corporate investments elsewhere (Lynd 1982). This money should have been used for maintenance and reinvestment in plant and equipment. Eventually the business failed, taking with it many other businesses. The decision to divert profits out of the community was made by people who did not live in Youngstown and had no stake or interest in it as anything other than an abstraction on a balance sheet. Their decision had little to do with the productivity of the business and everything to do with shortsightedness and greed.

From this and all too many other cases like it, we can conclude that one requisite of resilient local economies is, as Daniel Kemmis states, "the capacity and the will to keep some locally generated capital from leaving the region and to invest that capital creatively and effectively in the regional economy" (Kemmis 1990, 103). This in turn means selectively challenging the "supremacy of the national market" where that restricts the capacity to build strong regional economies. It also means confronting what economist Thomas Michael Power calls a "narrow, market-oriented, quantitative definition of economics" in favor of one that gives priority to cultural, esthetic, and ecological quality (Power 1988, 3). Economic quality, according to Power, is not synonymous with economic growth. The choice between growth or stagnation is a false one that "leaves communities to choose between a disruptive explosion of commercial activity, which primarily benefits outsiders, while degrading values very important to residents and being left in the dust and decay of economic decline" (Power 1988, 114). There are alternative ways to develop that do not sell off the qualities that make particular communities desirable in the first place. Among these, Power proposed "import substitution" whereby local needs are increasingly met by local resources, not by imported goods and services. Energy efficiency, for example, can displace expensive imports of petroleum, fuel oil, electricity, and natural gas. Dollars not exported out of the community then circulate within the local economy, creating a "multiplier effect" by stimulating local jobs and investment.

Power, like Jane Jacobs in her 1984 book *Cities and the Wealth of Nations*, argues for development

> built around enterprising individuals and groups seeing a local opportunity and improvising, adapting, and substituting. Initially, these efforts start on a small scale and usually aim to serve a local market. (Power 1988, 186)

This approach stands in clear contrast to the standard model of economic development whereby communities attempt to lure outside industry and capital by lowering local taxes and regulations and providing free services, all of which lower the quality of the community.

The development of place-focused economies requires questioning old economic dogmas. The theory of free trade, for example, originated in an agrarian world in which state boundaries were relatively impermeable and capital flows stopped at national frontiers (Daly 1993; Daly and Cobb 1989, 209–35; Morris 1990, 190–95). These conditions no longer hold. Goods, services, and capital now wash around the world, dissolving national boundaries and sovereignty. Labor (i.e., people) and communities, however, are not so mobile. Workers in the developed world are forced to compete with cheap labor elsewhere, with the result of a sharp decline in workers' income (Batra 1993). For previously prosperous communities, free trade means economic decline and the accompanying social decay now evident throughout much of the United States.

In place of free trade, World Bank economist Herman Daly and theologian John Cobb recommend "balanced trade" that limits capital mobility and restricts the amount that a nation can borrow by importing more than it exports (Daly and Cobb 1989, 231). To restore competitiveness where it has been lost, they recommend enforcing national laws designed to prevent economic concentration (Daly and Cobb 1989, 291). To build resilient regional economies, they recommend enabling communities to bid for the purchase of local industries against outside buyers. To the argument that international capital is necessary for the development of third and fourth world economies, they respond that

> we have come, as have many others, to the painful conclusion that very little of First World development effort in the Third World, and even less of business investment, has been actually beneficial to the majority of the Third World's people. . . . For the most part the Third World would have been better off without international investment and aid [which] destroyed the self-sufficiency of nations and rendered masses of their formerly self-reliant people unable to care for themselves. (Daly and Cobb 1989, 289–90)

Daly and Cobb believe that economies should serve communities rather than elusive and mythical goals of economic growth.

Why does the idea that economies ought to support communities sound so utopian? The answer, I think, has to do with how fully we have accepted the radical inversion of purposes by which society is shaped to fit the economy instead of the economy being tailored to fit the society. Human needs are increasingly secondary to those of the abstractions of markets and growth. People need, among other things, healthy food, shelter, clothing, good work to do, friends, music, poetry, good books, a vital civic culture, animals, and wildness. But we are increasingly offered fantasy for reality, junk for quality, convenience for self-reliance, consumption for community, and stuff rather than spirit. Business spends hundreds of billions of dollars each year to convince us that this is good. But virtually nothing is spent to inform us of other alternatives that are better, cheaper, and satisfying. Our economy has not, on the whole, fostered largeness of heart or spirit. It has not satisfied the human need for meaning or roots. It is neither sustainable nor sustaining.

Taking the environment seriously means rethinking how our politics and civic life fit the places we inhabit. It makes sense, in Daniel Kemmis's words, "to begin with the place, with a sense of what it is, and then try to imagine a way of being public which would fit the place" (Kemmis 1990, 41). I do not think it is a coincidence that voter apathy has reached near epidemic proportions at the same time that our sense of place has withered and community-scaled economies have disintegrated. As with the economy, we have surrendered control of large parts of our lives to distant powers.

Rebuilding place-focused politics will require revitalizing the idea of citizenship rooted in the local community. Democracy, as John Dewey observed, "must begin at home, and its home is the neighborly community" (Dewey 1954, 213). But neighborly communities have been eviscerated by the physical imposition of freeways, shopping malls, the commercial strip, and mind-numbing sprawl. The idea of the neighborly community has receded from our minds as the centralization of power and wealth has advanced. But neither vital communities nor democracy is compatible with economic and political centralization, from either the right or the left.

We need an ecological concept of citizenship rooted in the understanding that activities that erode soils, waste resources, pollute, destroy biological diversity, and degrade the beauty and integrity of landscapes are forms of theft from the commonwealth as surely as is bank robbery. Ecological vandalism undermines future prosperity and democracy alike. For too long we have tried to deal with resource abuse from the top down and have pitifully little to show for our efforts and money. The problem, as Aldo Leopold noted, is that for conservation to become "real and important" it must "grow from the bottom up" (Leopold 1991, 300). It must, in other words, become fundamental to the day-to-day lives of millions of people, not just to those few professional resource managers working in public agencies.

Ecologically literate people, engaged in and by their place, will discover ways to conserve resources, to implement energy-efficiency programs that save thousands of dollars per household. They will discover ways to save farms through "community supported agriculture," where people pay farmers directly for a portion of their produce. They will limit absentee ownership of farmland and enable young farmers to buy farms. They will find the means to save historic and ecologically important landscapes. They will develop procedures to accommodate environmentalists and loggers, as did the residents of Missoula, Montana. They may even discover, as did residents of the Mondragon area of Spain or the state of Kerala in India, how to successfully address larger issues of equitable development (Whyte and Whyte 1988; Franke and Chasin 1991).

We are not without models and ideas, but we lack the vision of politics as something other than a game of winners and losers fought out by factions with irreconcilable private interests. The idea that politics is little more than the pursuit of self-interest is embedded in American political tradition, at least from the time James Madison wrote the 10th Federalist Paper. It is an idea, however, that tends to breed the very behavior it purports only to describe. In the words of political scientist Steven Kelman, "design your institution to assume self-interest, then and you may get more self-interest. And the more self-interest you get, the more draconian the institutions must become to prevent the generation of bad policies" (Kelman 1988, 51). Kelman proposed that institutions be designed not merely to restrain the unbridled pursuit of self-interest but to promote "public spirited behavior" in which "people see government as an appropriate forum for the display of the concern for others" (Kelman 1988). The norm of public spiritedness also changes how people define their self-interest. This is, I believe, what Vaclav Havel meant when he described "genuine politics" as "a matter of serving those around us: serving the community, and serving those who will come after us" (Havel 1992, 6). The roots of genuine politics are moral, originating in the belief that what we do matters deeply and is recorded "somewhere above us."

Is it utopian to believe that our politics can rise to public spiritedness and genuine service? I think not. Evidence shows that we are in fact considerably more public spirited than we have been led to believe, not always and everywhere to be sure, but more often than a cynical reading of human behavior would show (Kelman 1988, 43, notes 38–41). On the other hand, it is utopian to believe that the politics of narrow self-interest will enable us to avert the catastrophes on the horizon that can be forestalled only by foresight and collective action.

Conclusion

Western civilization irrupted on the Earth like a fever, causing, in historian Frederick Turner's words, "a crucial, profound estrangement of the inhabitants from their habitat." We have become, Turner continued, "a rootless, restless people with a culture of superhighways precluding rest and a furious penchant for tearing up last year's improvements in a ceaseless search for some gaudy ultimate" (Turner 1980, 5). European explorers arrived in the "new world" spiritually unprepared for the encounter with the place, its animals, and its peoples. American settlers' discontent spread to native peoples who were caught in the way. None were able to resist either the firepower or the seductions of technology.

More than just a symbol of a diseased spiritual state, that fever is now palpably evident in the rising temperature of the Earth itself. A world that takes its environment seriously must come to terms with the roots of its problems, beginning with the place called home. This is not a simple-minded return to a mythical past but a patient and disciplined effort to learn and, in some ways, to relearn the arts of inhabitation. These will differ from place to place, reflecting various cultures, values, and ecologies. They will, however, share a common sense of rootedness in a particular locality.

We are caught in the paradox that we cannot save the world without saving particular places. But neither can we save our places without national and global policies that limit predatory capital and that allow people to build resilient economies, to conserve cultural and biological diversity, and to preserve ecological integrities. Without waiting for national governments to act, there is a lot that can be done to equip people to find their place and dig in.

PART 3

On Ecological Design

AUTHOR'S NOTE 2010

The "ecological crisis" is the sum total of bad design with a tincture of bad intent, but the latter is not as easily solvable as the former. The emerging field of ecological design is the effort to recalibrate how we build, grow, make, power, move, live, and earn our keep so that they fit how the Earth works as a physical system. One day, that knowledge will help reshape and discipline human intentions as well. I intend the term design *broadly. The U.S. Constitution and the Federalist Papers, for example, are design blueprints for the conduct of the public business. The term applies more obviously to architecture, engineering, economics, finance, urban planning, manufacturing, and education. In all of its manifestations, ecological design is, in short, the harmonious integration of systems and functions within specific ecologies and places. At its most direct and tangible, good design requires local knowledge of soils, waters, topography, biota, animals, culture, history, and much more. The result of good design is, in a word, health—both human and ecological. Practically, good design means farms, buildings, neighborhoods, cities, and entire industries powered by renewable energy and discharging no waste and integrated into wholes in which the parts reinforce a larger emergent harmony. It is, in short, the art and science of applied resilience.*

Chapter 17

Designing Minds

(1992)

As THE ENTRY FROM *Homo sapiens* in any intergalactic design competition, industrial civilization would be tossed out at the qualifying round. It doesn't fit. It won't last. The scale is wrong. And even its apologists admit that it is not very pretty. The design failures of industrial technologically driven societies are manifest in the loss of diversity of all kinds, destabilization of the Earth's biogeochemical cycles, pollution, soil erosion, ugliness, poverty, injustice, social decay, violence, and economic instability.

Industrial civilization, of course, was not designed at all; mostly it just happened. Those who made it happen were mostly single-minded men and women innocent of any knowledge of what can be called the "ecological design arts," by which I mean the set of perceptual and analytical abilities, ecological wisdom, and practical wherewithal essential to making things that "fit" in a world of trees, microbes, rivers, animals, bugs, and small children. In other words, ecological design is the careful meshing of human purposes with the larger patterns and flows of the natural world and the study of those patterns and flows to inform human purposes.

Ecological designers aim to maximize resource and energy efficiency, take advantage of the free services of nature, eliminate waste, make ecologically smarter things, and educate ecologically smarter people. This means incorporating intelligence about how nature works, what David Wann (1990) called "biologic," into the way we think, design, build, and live. Design applies to the making of nearly everything that directly or indirectly requires energy and materials, or governs their use, including

---

This article was originally published in 1992.

farms, houses, communities, neighborhoods, cities, transportation systems, technologies, economies, and energy policies. When human artifacts and systems are well designed, they are in harmony with the larger patterns in which they are embedded. When poorly designed, they undermine those larger patterns, creating pollution, higher costs, and social stress in the name of spurious and short-run economizing. Bad design is not simply an engineering problem, although better engineering would often help. Its roots go deeper.

Good design begins, as Wendell Berry puts it, by asking, "What is here? What will nature permit us to do here? What will nature help us to do here?" (Berry 1987, 146). Good design everywhere has certain common characteristics, including right scale, simplicity, efficiency, a close fit between means and ends, durability, redundance, and resilience. Good designs also solve more than one problem at a time. They are often place specific or, in John Todd's words, "elegant solutions predicated on the uniqueness of place." Good design promotes

- human competence instead of addiction and dependence;
- efficient and frugal use of resources;
- sound regional economies;
- social resilience.

Where good design becomes part of the social fabric at all levels, unanticipated positive side effects (synergies) multiply. When people fail to design carefully and competently, unwanted side effects and disasters multiply.

As evidenced by the pollution, violence, social decay, and waste all around us, we have designed things badly. Why? There are, I think, three fundamental reasons. The first is that while energy and land were cheap and the world relatively "empty," we simply did not have to master the discipline of good design. We developed extensive rather than intensive economies. Accordingly, cities sprawled, wastes were dumped into rivers or landfills, farmers wore out one farm and moved on to another, houses and automobiles got bigger and less efficient, and whole forests were converted into junk mail and Kleenex. Meanwhile, the know-how necessary to a frugal, well-designed, intensive economy declined, and words like *realistic* or *convenience* became synonymous with habits of waste.

Second, design intelligence fails when greed, narrow self-interest, and individualism take over. Good design is a community process requiring people who know and value the positive things that bring them together and hold them together. Old-order Amish farmers, for example, refuse to buy combines, not because they would not make things easier or more profitable, but because they would undermine community by depriving people of the opportunity to help their neighbors. This is pound-wise and penny-foolish, the way intelligent design should be. In contrast, American cities, with their extremes of poverty and opulence, are products of people who believe that they have little in common with other people. Suspicion, greed, and fear undermine good community and good design alike. Gun sales soar.

Third, poor design results from poorly equipped minds. Good design can be done only by people who understand harmony, patterns, and systems. Good design requires a breadth of view that leads people to ask how human artifacts and purposes "fit" within the immediate locality and within the region. Industrial cleverness, however, is mostly evident in the minutiae of things, not in their totality or in their overall harmony. Moreover, good design uses nature as a standard and so requires ecological intelligence, by which I mean a broad and intimate familiarity with how nature works. For all of the recent interest in environment and ecology, this kind of knowledge, which is a product of both local experience and stable culture, is fast disappearing.

George Sturt, one of the last wheelwrights in England, in *The Wheelwright's Shop* describes "the age-long effort of Englishmen to fit themselves close and ever closer into England" (Sturt 1984, 66). Sturt built wagons crafted to fit the buyers' particular habits, fields, and topography. To do so, he needed to know a great deal about how his customers used a wagon, whether they drove fast or slow, whether their land was rocky or wet, and what they hauled. As a result, "we got curiously intimate with the peculiar needs of the neighborhood. In farm-wagon or dung-cart, barley-roller, plough, water barrel, or what not, the dimensions we chose, the curves we followed, were imposed upon us by the nature of the soil in this or that farm, the gradient of this or that hill, the temper of this or that customer or his choice perhaps in horseflesh" (Sturt 1984, 18).

Furthermore, a good wheelwright needed to know what kinds of trees gave particular parts extra strength, or flexibility, or weight, where these trees grew, and when they were ready to harvest. And finally he needed to know the traditions and skills unique to his craft that were passed down as folk knowledge:

> What we had to do was to live up to the local wisdom of our kind to follow the customs, and work to the measurements, which had been tested and corrected long before our time in every village shop all across the country. (Sturt 1984, 19)

The kind of mind that could design and build a good wagon depended a great deal on time-tested knowledge and intimate familiarity with place. The results were wagons that fit particular people and a particular landscape.

A contemporary example of ecological design can be found in John Todd's "living machines," which are carefully orchestrated ensembles of plants, aquatic animals, technology, solar energy, and high-tech materials to purify wastewater but without the expense, energy use, and chemical hazards of conventional sewage treatment technology. According to Todd,

> people accustomed to seeing mechanical moving parts, to experiencing the noise or exhaust of internal combustion engines or the silent geometry of electronic devices, often have difficulty imagining living machines. Complex life forms, housed within strange light-receptive structures, are at once familiar and bizarre. They are both garden and machine. They are alive yet framed and contained in vessels built of novel materials.... Living machines bring people and nature together in a fundamentally radical and transformative way. (Todd 1991, 335–43)

Todd has created several working examples of living machines, each resembling a greenhouse filled with exotic plants and aquatic animals. Wastewater enters at one end; purified water leaves at the other. In between, the work of sequestering heavy metals in plant tissues detoxifying toxics, and removing nutrients is done by plants and animals in an ecosystem driven by sunlight. A decade earlier he designed and built structures that similarly used aquatic systems to process waste, grow food, and store heat. Living machines and the logic of ecology imply changes in the way we process wastewater, grow food, and build houses and in the ways we integrate these and other functions into systems patterned after natural processes to do what industrial technology can only do expensively and destructively.

Ecological design also applies to the design of governments and public policies. Governmental planning and regulation require large and often ineffective or counterproductive bureaucracies. Design, in contrast, means

> the attempt to produce the outcome by establishing the criteria to govern the operations of the process so that the desired result will occur more or less automatically without further human intervention. (Ophuls 1977, 228–29)

In other words, well-designed policies and laws get the macro things right, like prices, taxes, and incentives, while preserving a high degree of micro freedom in how people and institutions respond. Design focuses on the structure of problems as opposed to their coefficients. For example, the Clean Air Act of 1970 required car manufacturers to install catalytic converters to remove air pollutants. Decades later, emissions per vehicle are down substantially, but with more cars on the road, air quality is about the same. A design approach to transportation would lead us to think more about creating access between housing, schools, jobs, and recreation that eliminate the need to move lots of people and materials over long distances. A design approach would have led us to reduce dependence on automobiles by building better public transit systems, restoring railroads, and creating bike trails and walkways. A design approach would also lead us to rethink the use of urban land and to reintegrate agriculture and wilderness into urban areas.

Ecological design requires the ability to comprehend patterns that connect, which means getting beyond the boxes we call disciplines to see things in their ecological context. It requires, in other words, a liberal education, but nearly everywhere the liberal arts have tended to become more specialized and narrow. Design competence requires the integration of firsthand experience and practical competence with theoretical knowledge, but the liberal arts have become more abstract, fragmented, and remote from lived reality. Design competence requires us to be students of the natural world, but the study of nature is being displaced by the effort to engineer nature to fit the economy instead of the other way around. Finally, design competence requires the ability to inquire deeply into the purposes and consequences of things, to know what is worth doing and what should not be done at all. But the ethical foundations of education have been diluted by the belief that values are relative. All of this is to say that from an ecological perspective, the "liberal arts" have not been liberal enough. I think this is evident in three respects.

First, the liberal arts have not been liberal enough in their response to the rapid decline in the habitability of the Earth. Changes in global and national policy are necessary but insufficient to reverse downward trends in the Earth's vital signs. It is also essential that we educate a citizen constituency that supports change and is competent to do the local work of rebuilding households, farms, institutions, communities, corporations, and economies that (1) do not emit carbon dioxide or other heat-trapping gases; (2) do not reduce biological diversity; (3) use energy, materials, and water with high efficiency; and (4) recycle wastes. In other words, a constituency that is capable of building economies that can be sustained without further reducing the Earth's potential to sustain life. At a minimum this will require a modification of the skills, aptitudes, abilities, and curriculum by which we learned how to industrialize the Earth.

Second, the liberal arts have come to mean an education largely divorced from practical competence. Inclusion of the ecological design arts in the liberal arts means bringing practical experience back into the curriculum in carefully conceived ways. The reasons, in Alfred North Whitehead's words, are straightforward: "First-hand knowledge is the ultimate basis of intellectual life . . . the second-handedness of the learned world is the secret of its mediocrity" (Whitehead 1967, 51). In contrast to the distinction that John Henry Newman once drew between desirable and useful knowledge (Newman 1982, 84–88), Whitehead argued that there is a "reciprocal influence between brain activity and material creative activity" essential for good thinking. In other words, good thinking and practical experience are mutually necessary. Accordingly, he thought, "the disuse of hand-craft is a contributory cause to the brain-lethargy of aristocracies" (Whitehead 1967). J. Glenn Gray has argued similarly that the exclusion of manual skills from the liberal arts is dangerous "because it first divorces us from our own dispositions at the level where intellect and emotions fuse." Purely analytical and abstract thinking "separates us from our natural and human environment" (Gray 1984, 85). Genuinely liberal education, in contrast, cultivates the full person, including manual competence and feeling as well as intellect.

Third, the liberal arts have come to include any number of fields, subfields, issues, and problems, excepting those that are closest at hand in the local community. Inclusion of the ecological design arts suggests a symbiotic relation between learning and locality. Here, too, the reasons are part of an older tradition going back to John Dewey. In 1899 John Dewey wrote that "the school has been so set apart, so isolated from the ordinary conditions and motives of life" that children cannot "get experience—the mother of all discipline" (Dewey 1990, 17). His solution required integrating opportunities for students "to make, to do, to create, to produce" and ending the separation of theory and practice. Dewey proposed that the immediate vicinity of the school be a focus of education, including the study of food, clothing, shelter, and nature. Through the study of these things, students might learn "the measure of the beauty and order about him, and respect for real achievement" (Dewey 1990). Gray has likewise argued that liberal education is "least dependent on formal instruction. It can be pursued in the kitchen, the workshop, on the ranch or farm" (Gray 1984, 81). It can also be pursued through the study of energy, water, materials, food, and waste flows on the campus.

How can competence in the ecological design arts be taught within the conventional curriculum? There are at least two broad possibilities. The best, but most difficult, approach is to make over entire institutions so that their operations and resource flows (food, energy, water, materials, waste, and investments) become a laboratory for the study of ecological design. There is a strong case for doing this for economic as well as educational reasons. A second possibility follows the suggestion of Herman Daly and John Cobb to establish separate centers or institutes within colleges and universities with the mission of fostering ecological design intelligence (Daly and Cobb 1989, 357–60). Ecological design arts centers would aim to (1) develop a series of ecological design projects that involve students, faculty, and staff; (2) study institutional resource flows; (3) develop curriculum; and (4) carry out studies on environmental trends throughout the region. Ecological design projects could include, for example,

- design and construction of zero discharge buildings using no fossil fuels, constructed with local materials;
- development of a bioregional directory of building materials;
- inventory and model campus resource flows;
- restoration of degraded ecosystems;
- design and development of a sustainable farm system;
- survey of resource and dollar flows in the local economy.

The list could be easily extended, but the point is clear. The functions of ecological design institutes would be to equip young people with a basic understanding of systems; develop habits of mind that seek out "patterns that connect" human and natural systems; teach young people the analytical skills necessary for thinking accurately about cause and effect; give students the practical competence necessary to solve local problems; and teach young people the habit of rolling up their sleeves and getting down to work.

Chapter 18

Loving Children: A Design Problem

(2002)

THE SKYMALL CATALOGUE, conveniently available as an anesthetic for irritated airplane passengers, recently offered an item that spoke volumes about our approach to raising children. For a price of several hundred dollars, parents could order a device that could be attached to a television set that would control access to the television. Each child would be given a kind of credit card, programmed to limit the hours he or she could watch TV. The child so disciplined would presumably benefit by imbibing fewer hours of mind-numbing junk. He or she might also benefit from the perverse challenge to discover the many exciting and ingenious ways to subvert the technology and the intention behind it, including a flank attack on parental rules and public decency via the Internet.

My parents had a rather different approach to the problem. It was the judicious and authoritative use of the word *no*. It cost nothing. My brother, sister, and I knew what it meant and the consequences for ignoring it. Still, I sometimes acted otherwise. It was a way to test the boundaries of freedom and parental love and the relation between the two.

The SkyMall device and the authoritative use of the word *no* both represent concern for the welfare of the child, but they are fundamentally different design approaches to the problem of raising children, and they have very different effects on the child. The device approach to discipline

This article was originally published in 2002.

is driven by three factors that are new to parenting. It is a product of a commercial culture in which we've come to believe that high-tech gadgetry can fix human problems, including that of teaching discipline and self-control to children. Moreover, the device is intended mostly for parents who are absent from the home for much of the day because they must (or think they must) work to make an expanding number of ends meet. And, all of our verbal assurances of love notwithstanding, it is a product of a society that does not love its children competently enough to teach them self-discipline. The device approach to parenting is merely emblematic of a larger problem that has to do with the situation of childhood within an increasingly dysfunctional society absorbed with things, economic growth, and self.

We claim to love our children, and I believe that most of us do. But like sheep, we have acquiesced in the design of a society that corrupts genuine love. One result is a growing mistrust of our children that easily turns to fear and dislike. In a recent survey, for example, only one-third of adults believed that today's young people "will eventually make this country a better place" (Applebome 1997). Instead, we find them "rude" and "irresponsible." And often they are. We find them overly materialistic and unconcerned about politics, values, and improving society. And many are too materialistic and detached from important issues (Bronner 1998). Not infrequently they are verbally and physically violent, mimicking a society saturated with drugs and violence. A few kill and rape other children. Why are the very children that we profess to cherish becoming less than likable and sometimes less than human?

Some will argue that nothing of the sort is happening and that every generation believes that its children are going to hell. Eventually, however, things work out. Such views are, I think, questionable because they ignore the sharp divide imposed between the hyperconsumerism and the needs of children for extended nurturing, mentoring, and imagining. The evidence indicates that it's the economy that we love most, not our children. The symptoms are all around us. We spend 40 percent less time with our children than we did in 1965. We spend, on average, 6 hours per week shopping, but only 40 minutes playing with our children (Suzuki 1997, 23). It can no longer be taken for granted that this civilization can pass on its highest values to enough of its children to survive. Without intending to do so, we have created a society that cannot love its children, indeed one in which the expression of real love is increasingly difficult.

No society that loved children would consign nearly one in five to poverty and leave them without adequate health care. No society that loved its children would put them in front of television for 4 hours or more every day. No society that loved its children would lace their food, air, water, and soil with hundreds of chemicals known or suspected of being carcinogenic, endocrine disrupters, or mutagenic. No society that loved its children would build so many prisons and so few parks and schools. No society that loved its children would teach them to recognize hundreds of corporate logos but hardly any plants and animals in their own regions. No society that loved its children would divorce them so completely from contact with soils, forests, streams, and wildlife. No society that loved its children would create places like the typical suburb or shopping mall. No society that loved its children would casually destroy real neighborhoods and communities in order to build even more highways. No society that loved its children would build so many glitzy sports stadiums and shopping malls while its public schools fall apart. No society that loved its children would pave over a million acres of prime farmland each year for even more shopping malls and parking lots. No society that loved its children would knowingly run even a small risk of future climatic disaster. No society that loved its children would use the practice of discounting in order to ignore its future problems. No society that loved its children would leave behind a legacy of ugliness and biotic impoverishment.

Of course we do all of these things in the belief that they are the price we must pay to create a better world for children. But at some level, I believe, many teenagers understand that such arguments are phony. That may explain some of their unfocused anger, which is no more than a reflection of the incivility and rudeness that we inflict on them. They mirror the larger self-indulgence of a society organized for machines, quick gratification, and excessive individualism. They know that the study of literature counts for considerably less in this society than making it big in sports or as another "American Idol," or dealing drugs. They understand intuitively that the real curriculum is not what's taught in schools but what's written on the face of the land. It is remarkable, in fact, that they are not angrier.

What would it mean to make a society that did in fact love all of its children? Properly understood, this is a design problem that calibrates what we intend as parents with how we earn our living, conduct our daily lives, build homes, create communities, manage landscapes, and provision ourselves with food, energy, and materials. The health and well-being of children, not the gross national product, is the best indicator of the health of our civilization. And I believe that it is the ultimate standard for ecological design. How do we design neighborhoods and cities that are good for children?

The starting point is the child's need for joy, safety, parental love, play, and the opportunity to safely explore the wider world. Such awareness must begin early in life with the development of what Edith Cobb once called "compassionate intelligence" rooted in "biological motivation deriving from nature's history" (Cobb 1977, 16). The child's "ecological sense of continuity with nature" is not mystical but is "basically aesthetic and infused with the joy in the power to know and to be" (Cobb 1977, 23). Childhood is the "point of intersection between biology and cosmology, where the structuring of our worldviews and our philosophies of human purpose takes place." In other words, our minds are rooted as much in the ecology in which our childhood is lived as in our ("over emphasized") animal instincts (Cobb 1977, 101). Similarly, Paul Shepard once argued that mind and body are imprinted in the most fundamental ways by the "pattern of place" experienced in childhood (Shepard 1996 93–108). For Shepard, the conclusion is that a child must have the opportunity to "soak in a place" and to "return to that place to ponder the visible substrate of his own personality" (Shepard1996, 106). Conversely, the child's sense of connection to the world can be damaged by ecologically impoverished surroundings. And it can be damaged as well by exposure to violence and poverty and even by too much affluence. It can be destroyed, in other words, when ugliness, both human and ecological, becomes the norm. Ecological design begins with the creation of places in which the ecology of imagination and ecological attachment can flourish. These would be safe urban and rural places that included biological diversity, wildness, flowing water, trees, animals, open fields, and room to roam—places where beauty is the standard.

At a larger scale the same logic applies to the ways children and adolescents are alienated from or bound to the surrounding region. Typical land-use patterns in recent decades taught young people that

- the highest and best use of land is for shopping malls, roads, and parking lots;
- land has little value beyond those of utility and economics;
- some land is expendable as landfills and waste dumps;
- the poor live on poor land, the well-to-do live on good land;

- roads to satisfy our cravings for mobility trump community needs;
- lawns are merely decoration maintained by use of chemicals and by fuels that will be exhausted in their lifetimes;
- prime farmland is far less important than development;
- biological diversity is less important than economic growth.

One consequence of the homogenized and utilitarian landscape is that most young people have learned little about how they are fed, clothed, and supplied with materials, and virtually nothing about better alternatives to meet their needs. By separating basic functions from daily lived experience, we have concealed a great deal of ecological reality from young people. Often this has come with a loss of real neighborhoods and real community. The things that we used to do for ourselves as competent citizens and neighbors we now purchase from one corporation or another at a considerable markup. It should astonish no one that civility, neighborliness, and communities have declined and that crime and anomie have risen. When living and livelihood become too widely separated, human bonds deteriorate because people no longer need each other as they once did. And when minds and landscapes are widely separated, whole categories of thought disappear, ecological competence declines, and awareness of our dependence on nature atrophies.

In an ecologically and esthetically impoverished landscape, it is harder for children and adolescents to find meaning and purpose for their lives. Consequently, many children grow up thinking themselves to be useless. In landscapes organized for convenience, commerce, mobility, and economic growth subsidized by cheap oil, in fact they are useless because we have little good work for them to do. Since we really do not need them to do real work, they learn few practical skills and little about responsibility. Their contacts with adults are frequently unsatisfactory. When they do work, it is all too often within a larger pattern of design failure. Flipping artery-clogging burgers made from chemically saturated feedlot cows, for example, is not good work, and neither is most of the other hourly work available to teenagers.

Over and over we profess love for our children, but the evidence says otherwise. We seldom work with them or mentor them or teach them practical skills. At an early age all too many are deposited in front of television and later in front of computers. And we are astonished to learn that in large numbers, neither do they respect adults nor are they equipped with many of the basic skills and aptitudes necessary to live responsible and productive lives. Increasingly, they imitate the values they perceive in us with characteristic juvenile exaggeration.

Assuming that we could muster the good sense to solve the problem, what would we do? Part of the solution, I believe, is to rejoin mind and habitat at the landscape level by reconnecting living with livelihood. This can only be done in places where a large part of our needs for shelter, warmth, energy, economic support, health, creativity, and conviviality are met locally in competently used and well-loved landscapes. To some this will sound either utopian or nostalgic for some mythical past. It is neither. In fact, it is an honest admission that we've tried utopia on industrial terms, and it did not work. It is merely to recognize the fact that, for better or worse, the organization of our landscapes arranges our possibilities, informs our minds, and directs our attention. A landscape organized for the convenience of the automobile and consumption tells young people more about our real values than anything taught in school. Worse, it deflects and distorts their intelligence at a critical point in life. It is possible, however, to organize landscapes to teach usefulness, practical competence, social responsibility, ecological skill, the values of good work, and the higher possibilities of adulthood. And it is possible for children and young adults to be instructed by birds, animals, soils, plants, water, seasons, and the ecology of their places.

Farms, feedlots, mines, wells, clear-cuts, waste dumps, and factories are mostly out of sight and so out of mind. As a result we do not know the full costs of what we consume. Ignorant of the damage we do, we leap to the conclusion that we are much richer than we really are. Ecological poverty and poverty of mind and spirit are reverse sides of the same coin. When we get the design right, however, the manner in which we provision ourselves becomes a reminder of our larger relationships and obligations. The true aim of ecological design, then, is not merely to improve the various technologies and techniques by which we meet our physical needs, but to improve the integration of the human mind with its habitat and to fit in a larger order of things. "To live," in Wendell Berry's words,

> we must daily break the body and shed the blood of Creation. When we do this knowingly, lovingly, skillfully, reverently, it is a sacrament. When we do it ignorantly, greedily, clumsily, destructively, it is a desecration. In such desecration we condemn ourselves to spiritual and moral loneliness, and others to want. (Berry 1981, 281)

Ecological design in its fullest measure is not just smarter management by technicians but, rather, a wider awareness and visible manifestation of our awareness that we are part of a larger pattern of order and obligation. Frank Lloyd Wright once commented that he could design a house that would cause a married couple to divorce within a matter of weeks. By the same logic it is possible to create buildings and cities so badly as to cause a culture to disintegrate socially and come unhinged from nature. Compare the architecture of the modern world with that of earlier civilizations. The ancient cities of India, Greece, and Rome, for example, were planned, in Peter Wilson's words, as "representations of microcosm and macrocosm, projections of the human body and distillations of the universe" (Wilson 1988, 75). The architecture of houses and public buildings were means to "portray to people their relation to one another as well as to important features of their environment," a kind of "diagram of how the system works" (Wilson 1988, 153). Buildings were not simply machines, as le Corbusier would have it, but a map showing "how the individual, the various orders of groups, and the cosmos are linked and related" (Wilson 1988, 75). For all of their imperfections as places and cultures, inhabitants in such cities were oriented to larger patterns.

Compare this with sprawling cities of the twentieth century that give no clue about any cosmology larger than the gross national product. They have become sprawling wastelands, islands of sybaritic affluence surrounded by a sea of necrotic urban tissue. For the most part, our buildings, in which we spend over 90 percent of our time, are poorly built. They are often made of materials that are toxic. They are often oversized and use energy and materials inefficiently. They are mostly disconnected from any discernible sense of community or any larger ecological or spiritual pattern. And what do such cities and buildings teach us? They teach us in exquisite detail that we are alone and powerless in the world, that energy and materials are cheap and can be consumed with impunity, that the highest purpose of life is consumption, and that the world is chaotic and dangerous.

Architecture, in other words, is also a form of instruction that works well or badly but never fails to instruct. When we get the design of buildings and communities right, they will help to inform us about our place within larger patterns of energy and materials flows and bind our affections and attention to the care of particular places. Architecture practiced as the art of ecological design promotes ecological competence and reflects larger patterns of order.

Conclusion

The goal of ecological design is not merely to meet our physical needs within the boundaries of ecological carrying capacity but, more importantly, to inform our desires. Good design would instruct us in what we need and the terms of our existence on Earth. In other words, the systems we devise to provision ourselves with food, energy, materials, shelter, and health constitute a larger form of education. But if these systems are designed to educate, they must give quick feedback about the consequences of our decisions, and they must work at a comprehensible scale. They must be devised in ways that create competence and practical understanding. They must be resonant with our deeper needs for meaning embedded in ritual and celebration. And design intelligence and the practical competence necessary to maintain it must be faithfully transferred from one generation to the next.

Good design must also meet other standards imposed by the way the physical world works. It must result in systems that are flexible and resilient in the face of changing circumstances. Given limits to our knowledge and foresight, good design would never lead us to bet it all, to risk the unforeseeable, or to commit acts that are irrevocable when the consequences are potentially large. And it would reorient our sense of time, giving greater weight to our future prospects and to long-term ecological processes as well. It would never cause us to discount the future.

Finally, designing ecologically begins in the belief that the world is not meaningless but coherent in ways that are often mysterious to us. Our task is to discern, as best we are able, the larger patterns and scales in which we live and act faithfully within those boundaries. Design, in this larger sense, is not simply the making of things but rather a striving for wholeness. At its best, ecological design is the ultimate manifestation of love—a gift of life, harmony, and beauty to our children.

∴ Chapter 19 ∴

Further Reflections on Architecture as Pedagogy

(1997)

AUTHOR'S NOTE 2010: *This is an amended version of a 1993 essay titled "Architecture as Pedagogy." It describes the origins of what became the Adam Joseph Lewis Center for Environmental Studies at Oberlin. The story is told in greater detail in* Design on the Edge *(Orr 2006).*

The worst thing we can do to our children is to convince them that ugliness is normal.
RENE DUBOS

THE CURRICULUM EMBEDDED in any building instructs as fully and as powerfully as any course taught in it. Most of my classes, for example, are taught in a building that I think Descartes would have liked. It is a building with lots of squareness and straight lines. There is nothing whatsoever that reflects its locality in northeast Ohio in what had once been a vast forested wetland (Sherman 1996). How it is cooled, heated, and lighted and at what true cost to the world is an utter mystery to its occupants. It offers no clue about the origins of the materials used to build it. It tells no story. With only minor modifications it could be converted to use as a factory or prison. When classes are over, students seldom linger for long. The building resonates with no part of

This article was originally published in 1997.

our biology, evolutionary experience, or esthetic sensibilities. It reflects no understanding of ecology or ecological processes. It is intended to be functional, efficient, minimally offensive, and little more. But what else does it do?

First, it tells its users that locality, knowing where you are, is unimportant. To be sure, this is not said in so many words anywhere in this or any other building. Rather, it is said tacitly throughout the entire building. Second, because it uses energy wastefully, the building tells its users that energy is cheap and abundant and can be squandered with no thought for the morrow. Third, nowhere in the building do students learn about the materials used in its construction or who was downwind or downstream from the wells, mines, forests, and manufacturing facilities where those materials originated or where they eventually will be discarded. And the lesson learned is mindlessness, which is to say it teaches that disconnectedness is normal. And try as one might to teach that we are implicated in the larger enterprise of life, standard architectural design mostly conveys other lessons. What is taught in classes and the way buildings actually work are often at cross-purposes. Buildings are provisioned with energy, materials, and water, and they dispose of their waste in ways that say to students that the world is linear and that we are no part of the larger web of life. Finally, there is no apparent connection in this or any other building on campus to the larger set of issues having to do with climatic change, biotic impoverishment, and the unraveling of the fabric of life on Earth. Students begin to suspect, I think, that those issues are unreal or that they are unsolvable in any practical way or that they occur somewhere else.

Is it possible to design buildings and entire campuses in ways that promote ecological competence and mindfulness (Lyle 1994)? Through better design is it possible to teach our students that our problems are solvable and that we are connected to the larger community of life? As an experiment, I organized a class of students in 1992–1993 to develop what architects call a preprogram for an environmental studies center at Oberlin College. Twenty-five students and a dozen architects met over two semesters to develop the core ideas for the project. The first order of business was to question why we ought to do anything at all. Once the need for facilities was established, the participants questioned whether we ought to build new facilities or renovate an existing building. Students and faculty examined possibilities to renovate an existing building but decided on new construction. The basic program that emerged from the yearlong class called for an approximately 14,000-square-foot building that

- discharged no wastewater (i.e., drinking water in, drinking water out);
- generated more electricity than it used;
- used no materials known to be carcinogenic, mutagenic, or endocrine disrupters;
- used energy and materials with great efficiency;
- promoted competence with environmental technologies;
- used products and materials grown or manufactured sustainably;
- was landscaped to promote biological diversity;
- promoted analytical skill in assessing full costs over the lifetime of the building;
- promoted ecological competence and mindfulness of place;
- became, in its design and operations, genuinely pedagogical;
- met rigorous requirements for full-cost accounting.

We intended, in other words, a building that did not impair human or ecological health somewhere else or at some later time.

Following approval by college trustees in June of 1995, I hired two graduates from the class of 1993 to help coordinate the design of the project and to enlist students, faculty, and the wider community in the design process. We also hired architect John Lyle to facilitate design charettes, or planning sessions, that began in the fall of 1995. Some 250 students, faculty, and community members eventually participated in the 13 charettes in which the goals for the environmental studies center were developed and refined. From 26 architectural firms, we selected William McDonough + Partners in Charlottesville, Virginia. In addition to hiring John Lyle and the McDonough firm, we assembled a design team that included Amory Lovins and Bill Browning from the Rocky Mountain Institute, scientists from NASA's Lewis Research Center, ecological engineers John Todd and Michael Shaw, the landscape design firm Andropogon, structural and mechanical engineers, and a contractor. During the programming and schematic design phase, this team and representatives from the college met by conference call weekly and in regular working sessions.

A "front-loaded" team approach to architectural design was new to the college. Typically, architects do the basic design, assign it to engineers to heat and cool it, and as a last step, hand it off to landscapers to make it look like it belongs. By engaging the full design team from the beginning, we intended to improve the integration of building systems and technologies and the relationship between the building and its site. Early on, we decided that the standard for technology in the building was to be "state of the shelf" but within state-of-the-art design. In other words, we did not want the risk of untried technologies, but we did want the overall product to be at the frontier of what it is now possible to do with ecologically smart design.

The building program called for major changes, not only in the design process, but also in the selection of materials, in the relationship to manufacturers, and in the way we counted the costs of the project. We intended to use materials that did not compromise human dignity or human health somewhere else. We also wanted to use materials that had as little embodied fossil energy as possible, hence giving preference to those locally manufactured or grown. In the process, we discovered how little is generally known about the ecological and human effects of building materials and how little the present tax and pricing system supports standards upholding ecological or human integrity. Unsurprisingly, we also discovered that building codes do little to encourage innovation and environmental quality.

Typically, buildings are a kind of snapshot of the state of technology at a given time. In this case, however, we intended for the building to remain technologically dynamic over a long period of time. In effect we proposed that the building adapt or learn as the state of technology changed and as our understanding of design became more sophisticated. We explored alternatives by which a third party would own, maintain, and operate the photovoltaic electric system, upgrading it as the technology improved. Unfortunately, in the late 1990s those possibilities were still undeveloped.

The same strategy applied to materials. McDonough + Partners regarded the building as a union of two different metabolisms, one industrial, the other ecological. Materials that might eventually decompose into soil were considered part of an ecological metabolism. Otherwise they were part of an industrial metabolism and might be leased from the manufacturer and eventually returned as a feedstock to be remanufactured into new product. That, too, proved to be way ahead of the times.

We intended, as well, to account for the life-cycle costs of the building, instead of following conventional practice, which accounts for only the "purchase price" of design and construction. In other words, we proposed to include all of those other costs to environment and human health not included in the prices of energy, materials, and waste disposal. The initial costs of the project, accordingly, came in at the high end of "average" costs for public buildings built in the late 1990s. The premium was slightly higher because we included the costs of

- student, faculty, and community participation in the design process;
- student research into materials and technologies to meet program goals;
- higher performance standards (e.g., zero discharge and on-site electricity production);
- more-sophisticated technologies;
- greater efforts to integrate technologies and systems;
- a building maintenance fund in the project budget.

In the longer term, we aimed as well to conduct an audit of the building, including an estimate of the amount of CO_2 released by the construction.

AUTHOR'S NOTE 2010: *Ground breaking for the Lewis Center occurred in October 1998, and the basic building was completed in January 2000. Design adjustments made in the first 18 months after occupancy allowed us to fulfill the goals of the project within reasonable costs. The building now generates all of its electricity from two photovoltaic arrays. It purifies wastewater on-site. It successfully minimized or eliminated the use of toxic materials. My colleague John Petersen and three students designed a real-time monitoring system to display energy use and other significant ecological data. Their ingenuity and diligence led to the creation of a highly successful company, Lucid Designs, Inc. The Lewis Center is landscaped to include a small restored wetland and forest, pond, and amphitheater, as well as gardens and orchards. In short, the Lewis Center became a laboratory for the study of applied sustainability and the arts of ecological design applied to buildings, energy systems, performance monitoring, wastewater purification, and landscape management.*

Buildings, however, are only means. The more important effects of the project have been its impact on the lives and careers of those who participated in the project. Some of the students who devoted time and energy to the project describe it now as their "legacy" to the college. Because of their work on the project, many of them learned about ecological design and how to solve real problems by working with some of the best practitioners in the world. Pessimists who thought change was impossible, perhaps, became somewhat more optimistic. And some of the trustees and administrators who initially saw this as a risky project, perhaps, came to regard risks incurred for the right goals as worthwhile.

The Adam Joseph Lewis Center is now the template for a larger project presently under way to (1) rebuild a 13-acre block in downtown Oberlin as a model of ecological design at the neighborhood scale and as a driver for post-carbon urban economic revitalization, (2) eliminate the use of fossil fuels in both the city and the college, (3) develop a 20,000-acre greenbelt for forestry and farming, and (4) engage students from the college, the public schools, a joint vocational school, and a community college in the transition.

Conclusion

By some estimates, humankind will build more in the next half century than it has built throughout all of recorded history. If we do this inefficiently and carelessly, the resulting ecological and human damage will be irreparable, and the dream of sustainability will have proved to be an unachievable fantasy. Ideas and ideals need to be rendered into working examples that make them visible, comprehensible, and compelling. But who will lead?

More than any other institutions in modern society, colleges and universities have a moral stake in the health, beauty, and integrity of the world our students will inherit. We have an obligation to provide them with tangible grounds for authentic hope and to equip them with the analytical skills and practical competence to lead in the transition to a sustainable future powered by sunlight. No generation ever faced a more daunting agenda. True. But none ever faced more exciting possibilities either.

Finally, the potential for ecologically smarter design in all of its manifestations in architecture, landscape design, community design, the management of agricultural and forest lands, manufacturing, and technology does not amount to a fix for all that ails us. Reducing the amount of damage we do to the world per capita will only buy us a few decades, perhaps a century if we are lucky. If we squander that reprieve, we will have succeeded only in delaying the eventual collision between unfettered human desires and the limits of the Earth. The default setting of our civilization needs to be reset to ensure that we build a sustainable world that is also humanly sustaining. This is not necessarily a battle between Left and Right or haves and have-nots, as it is often described. At a deeper level the issue has to do with art and beauty. In the largest sense, what we must do to ensure human tenure on the Earth is to cultivate a new standard that defines beauty as that which causes no ugliness somewhere else or at some later time.

Chapter 20

The Origins of Ecological Design

(2006)

THE ORIGINS OF ECOLOGICAL DESIGN can be traced back to our prehistoric ancestors' interest in natural regularities of seasons, sun, moon, and stars and later to the Greek conviction that humans, by the application of reason, could discern the laws of nature. Ecological design also rests on the theological conviction that we are obliged, not merely constrained, to respect larger harmonies and patterns. The Latin root word for the word *religion*—bind together—and the Greek root for *ecology*—household management—suggest a deeper compatibility and connection to order. Ecological design, further, builds on the science and technology of the industrial age but for the purpose of establishing a partnership with nature, not domination. The first models of ecological design can be found throughout the world in the vernacular architecture and the practical arts that are as old as recorded history. It is, accordingly, as much a recovery of old and established knowledge and practices as a discovery of anything new. The arts of building, agriculture, forestry, healing, and resilient economy were sometimes models of great ecological intelligence developed by cultures that we otherwise might dismiss as primitive. The art of applied wholeness was implicit in social customs such as the observance of the Sabbath and holy days, the Jubilee year, or the practice of potlatch, in which debts were forgiven and wealth was recirculated. It is evident still in all of those various ways by which com-

This article was originally published in 2006.

munities and societies gracefully cultivate the arts of generosity, kindness, prudence, love, humility, compassion, gentleness, forgiveness, gratitude, and ecological intelligence.

In its specifically modern form, ecological design has roots in the Romantic rebellion against the more extreme forms of modernism, particularly the belief that humans armed with science and a bit of technology were lords and masters of Creation. Francis Bacon, perhaps the most influential of the architects of modern science, proposed the kind of science that would reveal knowledge by putting nature on the rack and torturing her secrets from her, a view still congenial to some who have learned to say it more correctly. The science that grew from Bacon, Galileo, and Descartes overthrew older forms of knowing, which were based on the view that we are participants in the forming of knowledge and that nature is not inert. The result was a science based on the assumptions that we stand apart from nature, that knowledge is to be judged by its usefulness in extending human mastery over nature, and that nature is best understood by reducing it into its components. "The natural world," in the words of E. A. Burtt, "was portrayed as a vast, self-contained mathematical machine, consisting of motions of matter in space and time and man with his purposes, feelings, and secondary qualities was shoved apart as an unimportant spectator" (Burtt 1954, 104). Our minds are so completely stamped by that particular kind of science that it is difficult to imagine any other way to know, in which comparably valid knowledge might be derived from different assumptions and something akin to sympathy and a "feeling for the organism" (Keller 1983).

Among the dissidents of modern science, Goethe, best known as the author of *Faust*, stands out as one of the first theorists and practitioners of the science of wholeness. In contrast to a purely intellectual empiricism, what physicist and philosopher Henri Bortoft calls the "onlooker consciousness," Goethe stressed the importance of observation grounded in intuition so that objects under investigation could communicate to the observer (Goethe 1952). Descartes, in contrast, reportedly began his days in bed by withdrawing his attention from the contaminating influence of his own body and the cares of the world, in order to think deeply. He aimed, thereby, to establish the methodology for a science of quantity established by pure logic. Goethe, on the other hand, practiced an applied science of wholeness in which "the organizing idea in cognition comes from the phenomenon itself, instead of from the self-assertive thinking of the investigating scientist" (Bortoft 1996, 240). Instead of the intellectual inquisition proposed by Bacon and practiced subsequently, Goethe proposed something like a dialogue with nature by which scientists "offer their thinking to nature so that nature can think in them and the phenomenon disclose itself as idea" (Bortoft 1996, 242). Facilitation of that dialogue required "training new cognitive capacities" so that Goethean scientists, "far from being onlookers, detached from the phenomenon, or at most manipulating it externally . . . are engaged with it in a way which entails their own development," which requires overcoming a deeply ingrained habit of seeing things as only isolated parts, not in their wholeness (Bortoft 1996, 244). The mental leap, as Bortoft notes, is similar to that made by Helen Keller, who, blind and deaf, was nonetheless able to wake to what she called the "light of the world" without any preconceptions or prior metaphoric structure whatsoever. Goethe proposed, not to dispense with conventional science, but rather to find another, and complementary, doorway to the realm of knowledge in the belief that Truth is not to be had through any single method, nor by any one age or culture.

Implicit in Goethe's mode of science is the old view, still current among some native peoples, that the Earth and its creatures are kin and in some fashion sentient and that they communicate to us, that life comes to us as a gift, and that a spirit of trust, not fear, is essential to knowing anything worth knowing. That message, in Calvin Martin's words, "is riveting . . . offering a civilization strangled by fear, measuring everything in fear, the chance to love everything" and to rise above "the armored chauvinism" inherent in a kind of insane quantification (Martin 1992, 107, 113). It is, I think, what Albert Einstein meant in saying that

> a human being is part of a whole world, called by us the "Universe," a part limited in time and space. He experiences himself, his thoughts and feelings, as something separate from the rest—a kind of optical delusion of his consciousness. The striving to free oneself from this delusion is the one issue of true religion. Not to nourish it but to try to overcome it is the way to reach the attainable measure of peace of mind. (Calaprice 2005, 206)

Goethe earlier proposed a kind of jailbreak from the prison of Cartesian anthropocentricism and from beliefs that animals and natural systems were fit objects to be manipulated at will. His intellectual heirs include all of those who believe that the whole is more than the sum of its parts, including systems thinkers as diverse as mathematician and philosopher Alfred North Whitehead, politician and philosopher Jan Smuts, biologist Ludwig von Bertalanffy, economist Kenneth Boulding, and ecologist Eugene Odum. Goethe's approach continues in the study of nonlinear systems in places like the Santa Fe Institute. Biologist Brian Goodwin, for one, calls for a "science of qualities" that complements and extends existing science (Goodwin 1994, 198). Conventional science, in Goodwin's view, is incapable of describing "the rhythms and spatial patterns that emerge during the development of an organism and result in the morphology and behavior that identify it as a member of a particular species . . . or the emergent qualities [that] are expressed in biological form are directly linked to the nature of organisms as integrated wholes" (Goodwin 1994, 198–99). Goodwin, like Goethe, calls for a "new biology . . . with a new vision of our relationships with organisms and with nature in general . . . [one] that emphasizes the wholeness, health, and quality of life that emerge from a deep respect for other beings and their rights to full expression of their natures" (Goodwin 1994, 232). Goodwin, Goethe, and other systems scientists aim for a more scientific science, predicated on a rigor commensurate with the fullness of life in its lived context.

While Goethe's scientific work focused on the morphology of plants and the physics of light, D'Arcy Thompson, one of the most unusual polymaths of the twentieth century and one who "stands as the most influential biologist ever left on the fringes of legitimate science," approached design by studying how and why certain forms appeared in nature (Gleick 1987, 199). Sir Peter Medawar said of Thompson's 1917 magnum opus, *On Growth and Form*, that it was "beyond comparison the finest work of literature in all the annals of science that have been recorded in the English tongue" (Gleick 1987, 200). Thompson seems to have measured everything he encountered, notably natural forms and the structural features of plants and animals. In so doing he discovered the patterns by which form arises from physical forces, not just by evolutionary tinkering as proposed by Darwin. Why, for example, does the honeycomb of the bee consist of hexagonal chambers similar to soap bubbles compressed between two glass plates? The answer, Thompson discovered, was found in the response of materials to physical forces, applicable as well to "the cornea of the human eye, dry lake beds, and polygons of tundra and ice" (Willis 1995, 72). By showing the physical and mechanical forces behind life forms at all levels, Thompson challenged the Darwinian idea that heredity determined everything. His work inspired subsequent work in biomechanics, evolutionary biology, architecture, and biomimicry, including that by Paul Grillo, Karl von Frisch, and Steven Vogel.

Frisch, for example, explored the ingenuity of architecture evolved by birds, mammals, fishes, and insects. African termite mounds a dozen feet high, for example, maintain a constant temperature of about 78°F in tropical climates (Frisch 1974, 138–49). Nests are ventilated variously by permeable walls that exchange gases and by ventilation shafts opened and closed manually as needed with no other instructions than those given by instinct. Interior ducts move air and gases automatically by convection. The system is so ingeniously designed that chambers deep underground are fed a constant stream of cool, fresh air that rises as it warms before being ventilated to the outside. Termite nests are constructed of materials cemented together with the termites' own excretions, eliminating the problem of waste disposal. Desert termites, with no engineering degrees as far as we know, bore holes 40 meters below their nests to find water. Beavers construct dams 1000 feet or more in length; their houses are insulated to remain warm in subzero temperatures. Other animals, less studied, build with comparable skill (Tsui 1999, 86–131). Human ingenuity, considerable as it is, pales before that of many animals that design and build remarkably strong, adaptable, and resilient structures without toxic chemicals, machinery, hands with opposable thumbs, fossil fuels, and professional engineers.

The idea that nature is shaped by physical forces as much as by evolution is also evident in the work of Theodor Schwenk, who explored the role of water as a shaper of Earth's surfaces and biological systems. Of water Schwenk wrote:

> In the chemical realm, water lies exactly at the neutral point between acid and alkaline, and is therefore able to serve as the mediator of change in either direction. In fact, water is the instrument of chemical change wherever it occurs in life and nature. . . . In the light-realm, too, water occupies the middle ground between light and darkness. The rainbow, that primal phenomenon of color, makes its shining appearance in and through the agency of water. . . . In the realm of gravity, water counters heaviness with levity; thus, objects immersed in water take on buoyancy. . . . In the heat-realm water takes a middle position between radiation and conduction. It is the greatest heat conveyer in the earth's organism, transporting inconceivable amounts of warmth from hot regions to cooler ones by means of the process known as heat-convection. . . . In the morphological realm, water favors the spherical; we see this in the drop form. Pitting the round against the radial, it calls forth that primal form of life, the spiral. . . . In every area, water assumes the role of mediator. Encompassing both life and death, it constantly wrests the former from the latter. (Schwenk 1989, 24)

Moving water shapes landscapes. As ice it molds entire continents. At a micro scale, its movement shapes organs and the tiniest organisms. But at any scale it flows, dissolves, purifies, condenses, floats, washes, and conducts, and some believe that it even remembers. Our language is brim full of water metaphors, and we have streams of thought or dry spells. The brain literally floats on a water cushion. Water in its various metaphors is the heart of our language, religion, and philosophy. We are much given to the poetry of water as mists, rain, flows, springs, light reflected, waterfalls, tides, waves, storms. Some of us have been baptized in it. But all of us stand ignorant before the mystery that D. H. Lawrence called "the third thing," by which two atoms of hydrogen and one of oxygen become water, and no one knows what it is.

"Form patterns," Schwenk wrote, "such as those appearing in waves with new water constantly flowing through them, picture on the one hand the creation of form and on the other the constant exchange of material in the organic world" (Schwenk 1996, 34). Water is a shaper, but the physics of its movement is also the elementary pattern of larger systems "depicting in miniature the great starry universe" (Schwenk 1996, 45). Water is the medium by which and through which life is lived. Turbulence in air and water have the same forms and mechanics as vortices, whether in the ocean, the atmosphere, or space. Sound waves and waves in water operate similarly. Schwenk's great contribution to ecological design, in short, was to introduce water in its fullness as a geologic, biological, somatic, and spiritual force, a reminder that we are creatures of water, all of us eddies in one great watershed.

The profession of design as an ecological art probably begins with the great British and European landscapers such as Capability Brown (1716–1783), famous for developing pastoral vistas for the rich and famous of his day. Looking out from the massive ostentation of Blenheim Palace across the surrounding lakes, trees, and grazing sheep, you are witness not to the natural landscape but to Brown's version of the pastoral—an orderliness of considerable comfort to the creators of the British Empire. In American history the early beginnings of design as ecology are apparent in the work of the great landscape architect and creator of Central Park in New York, Frederick Law Olmsted, and, later, in that of Jens Jensen, who pioneered the use of native plants in designed landscapes of the Midwest. Ian McHarg, a brilliant revolutionary, merged the science of ecology with landscape architecture aiming to create human settlements in which "man and nature are indivisible, and . . . survival and health are contingent upon an understanding of nature and her processes" (McHarg 1969, 27). His students, including Frederick Steiner, Pliny Fisk, Carol Franklin, and Anne Whiston Spirn, continue that vision armed with sophisticated methodological tools of geographic information systems and ecological modeling applicable to broader problems of human ecology.

While the degree of influence varied, many early efforts toward ecological design were inspired by the arts and crafts movement in Britain, particularly the work of William Morris and John Ruskin. In U.S. architecture, for example, Frank Lloyd Wright's attempt to define an "organic architecture" has clear resonance with the work of Morris and Ruskin as well as the transcendentalism of Ralph Waldo Emerson. Speaking before the Royal Institute of British Architects in 1939, Wright described organic architecture as "architecture of nature, for nature . . . something more integral and consistent with the laws of nature" (Wright 1993, 302, 306). In words Morris and Ruskin would have applauded, Wright argued that a building "should love the ground on which it stands," reflecting the topography, materials, and life of the place (Wright 1993, 307). Organic architecture is "human scale in all proportions" but is a blending of nature with human-created space so that it would be difficult to "say where the garden ends and where the house begins . . . for we are by nature ground-loving animals . . . insofar as we court the ground, know the ground, and sympathize with what it has to give us" (Wright 1993, 309). Wright's vision extended beyond architecture to a vision of the larger settlement patterns that he called "Broadacre City," arguing that organic architecture had to be more than an island in a society with other values. Wright, with his attempts to harmonize building and ecology and in his pioneering efforts to use natural materials and solar energy, is a precursor to the green building movement. And in his often random musings about an "organic society," he foreshadowed the present dialogue about ecological design and the sustainability of modern society.

Ecological design, however, is not just about calibrating human activities with natural systems. It is also an inward search to find patterns and order of nature written in our senses, flesh, and human proclivities. There is no line dividing nature outside from inside; we are permeable creatures inseparable from nature and natural processes in which we live, move, and have our being. We are also sensual creatures with five senses that we know of and others that we only suspect. At its best, ecological design is a calibration, not just of our sense of proportion that the Greeks understood mathematically, but also a finer calibration of the full range of our sensuality with the built environment, landscapes, and natural systems. Our buildings are thoughts, words, theories, and entire philosophies crystallized for a brief time into physical form that reveals what's on our mind and what's not. When done right, they are a kind of dialogue with nature and our own deeper, sensual nature. The sights, smells, texture, and sounds of the built environment evoke memories, initiate streams of thought, engage, sooth, provoke, bind or block, open or close possibilities. When done badly, the result is spiritual emptiness characteristic of a great deal of modern design that reveals, in turn, a poverty of thought and perception and feeling manifest as ugliness.

We are creatures shaped inordinately by the faculty of sight, but seeing is anything but simple. Oliver Sacks once described a man blind since early childhood who, sight once restored, found it to be a terrible and confusing burden and preferred to return to blindness and his own inner world of touch. "When we open our eyes each morning," Sacks writes, "it is upon a world we have spent a lifetime learning to see" (Sacks 1993, 64). And we can lose not only the faculty of sight but the ability to see as well. Even with 20/20 vision, our perception is always selective because our eyes permit us to see only within certain ranges of the light spectrum and because personality, prejudice, interest, and culture further filter what we are able to see. Sacks notes that individual people can choose not to see, and I suspect the same is true for cultures as well. The affinity for nature, a kind of sight, is much diminished in modern cultures.

Collective vision cannot be easily restored by more clever thinking, but, as David Abram puts it, only "through a renewed attentiveness to this perceptual dimension that underlies all our logics, through a rejuvenation of our carnal, sensorial empathy with the living land that sustains us" (Abram 1996, 69). Abram describes perception as interactive and participatory, in which "perceived things are encountered by the perceiving body as animate, living powers that actively draw us into relation . . . both engender[ing] and support[ing] our more conscious, linguistic reciprocity with others" (Abram 1996, 90). Further, sight as well as language and thought are experienced bodily as colors, vibrations, sensations, and empathy, not simply as mental abstractions. The ideas that viewer and viewed are in a form of dialogue and that we experience perception bodily runs against the dominant strains of Western philosophy. For illustration, Plato's *Phaedrus* has Socrates say, "I'm a lover of learning, and trees and open country won't teach me anything whereas men in the town do." Plato's world of ideal forms existed only in the abstract. Similarly, the Christian heaven exists purely somewhere beyond earthly and bodily realities. Both reflected the shifting balance between the animated sacred, participatory world and the linear, abstract, intellectual world. Commenting on the rise of writing and the priority of the text, Abram says that "the voices of the forest, and of the river began to fade . . . language loosen[ed] its ancient association with the invisible breath, the spirit sever[ed] itself from the wind, and psyche dissociate[d] itself from the environing air" (Abram 1996, 254). As a result, "human awareness folds in upon itself and the senses—once the crucial site of our engagement with the wild and animate earth—become mere adjuncts of an isolate and abstract mind" (Abram 1996, 267).

Through the act of design we are invited to see larger realities. The creators of Stonehenge, I think, intended worshippers to see, not just circles of artfully arranged stone, but the cosmos above and perhaps within. The Parthenon is a temple to the goddess Athena but also a visible testimony to an ideal existing in mathematical harmonies, proportion, and symmetry discoverable by human reason. The builders of Gothic cathedrals intended not just monumental architecture but a glimpse of heaven and a home for sacred presence. For all of the crass, utilitarian ugliness of the factories, slums, and glittering office towers, the designers and builders of the industrial world intended to reveal possibilities for abundance and human improvement in a world they otherwise deemed uncertain and violent, ruled by the laws of the jungle.

Finally, the practice of ecological design is rooted in the emerging science of ecology and the natural characteristics of specific places. The ecological design revolution is, not merely a more efficient recalibration of energy, materials, and economy in accord with ecological realities, but a deeper and more coherent vision of the human place in nature. Ecological design is, in effect, the specific terms of a declaration of coevolution with nature that begins in the science of ecology and the recognition of our dependence on the web of life (Capra 1996; Capra 2002). In contrast to the belief that nature is little more than a machine and its parts merely resources, for ecological designers nature is, as Aldo Leopold put it,

> a fountain of energy flowing through a circuit of soils, plants, and animals. Food chains are the living channels which conduct energy upward; death and decay return it to the soil. The circuit is not closed; some energy is dissipated in decay, some is added by absorption from the air, some is stored in soils, peats, and long-lived forests; but it is a sustained circuit, like a slowly

> augmented revolving fund of life. There is always a net loss by downhill wash, but this is normally small and offset by the decay of rocks. (Leopold 1966, 216)

Energy flowing through the "biotic stream" moves "in long or short circuits, rapidly or slowly, uniformly or in spurts, in declining or ascending volume," through what ecologists call food chains. For designers, the important point is that the internal processes of the biotic community, the ecological books, in effect, must balance so that energy used or dissipated by various processes of growth is replenished (Leopold 1953, 162). Leopold proposed three basic ideas (Leopold 1987, 218):

- that land is not merely soil;
- that the native plants and animals keep the energy circuit open; others may or may not;
- that man-made changes are of a different order than evolutionary changes and have effects more comprehensive than is intended or foreseen.

Ecological design, as Leopold noted, begins in the recognition that nature is not simply dead material or simply a resource for the expression of human wants and needs but, rather, "a community of soils, waters, plants, and animals, or collectively: the land" of which we are a part (Leopold 1966, 204). But Leopold did not stop at the boundary of science and ethics; he went on to draw out the larger implications. For reasons that are both necessary and right, the recognition that we are members in the community of life "changes the role of *Homo sapiens* from conqueror of the land-community to plain member and citizen of it" (Leopold 1966, 204). The "upshot" is Leopold's classic statement that "a thing is right when it tends to preserve the integrity, stability, and beauty of the biotic community. It is wrong when it tends otherwise" (Leopold 1966, 224–25). We will be a long time understanding the full implications of that creed, but Leopold, late in his life, was beginning to ponder the larger social, political, and economic requisites of a fully functioning land ethic.

Like Leopold's land ethic, ecological design represents a practical marriage of ecologically enlightened self-interest with the recognition of the intrinsic values of natural systems. Once consummated, however, the marriage branches out into a myriad of possibilities. Economics rooted in the realities of ecology, for example, requires the preservation of natural capital of soils, forests, and biological diversity, which is to say economies that operate within the limits of the Earth's carrying capacity (Hawken et al. 1999; Daly 1996). An ecological politics requires the recalibration of the complexities and timescales of ecosystems with the conduct of the public business. An ecological view of health would begin with the recognition that the body exists within an environment, not as a kind of isolated machine (Kaptchuk 2000). Religion grounded in the operational realities of ecology would build on the human role as steward and the obligation to care for the Creation (Tucker 2003). An ecological view of agriculture would begin with the realities of natural systems, aiming to mimic the way nature "farms" (Jackson 1980). An ecological view of business and industry would aim to create solar-powered industrial and commercial ecologies so that every waste product cycles as an input in some other system (McDonough and Braungart 2002). And an ecological view of education would, among other things, foster the capacity to perceive systems and patterns and promote ecological competence.

Ecology, the "subversive science," is the recognition of our practical connections to the physical world, but it does not stop there. The awareness of the many ways by which we are connected to the web of life would lead intelligent and scientifically literate people to protect nature and the conditions necessary to it, for reasons of self-interest. But our knowledge, always incomplete and often dead wrong, is often inadequate to the task of knowing what's in our interest, let alone discerning exactly what parts of nature we must accordingly protect and how to do it. Science notwithstanding, often we do not know what we are doing or why. More subversive still are questions concerning the interests and rights of lives and life across the boundaries of species and time. Since they cannot speak for themselves, their only advocates are those willing to speak on their behalf. Many clever arguments purport to explain why we should or should not be concerned about those whose lives and circumstances would be affected by our action or inaction. Like so many tin soldiers, arrayed across the battlefield of abstract intellectual combat, they assault frontally or by flank, retreat only to regroup, and charge again, each battle giving rise to yet another. But in the end, I think, such questions will be decided, not by intellectual combat and argumentation, however smart, but rather more simply and profoundly by affection—all of those human emotions that we try to capture in words like *compassion*, *sympathy*, and *love*. Love, in other words, neither requires nor hinges on intellectual argument. It is a claim that we recognize as valid but for reasons we could never describe satisfactorily. In the end it is a nameless feeling that we accept as both a limitation on what we do and a gift we offer. Pascal's observation that the heart has reasons that reason does not know sums the matter. Love is a gift but the giver expects no return on the investment, and that defies logic, reason, and even arguments about selfish genes.

After all of the intellectualization is finished and all of the various arguments made, whether we choose to design with nature or not will come down to a profoundly simple matter of whether we love deeply enough, artfully enough, carefully enough to preserve the web of life. Ecological design is simply an informed love applied to the dialogue between humankind and natural systems. The origins of ecological design can be traced far back in time, but deeper origins are found in the recesses of the human heart.

Chapter 21

The Design Revolution: Notes for Practitioners

(2006)

When you build a thing you cannot merely build that thing in isolation, but must also repair the world around it, and within it so that the larger world at that one place becomes more coherent and more whole; and the thing which you make takes its place in the web of nature as you make it.
Christopher Alexander

The long-term goal of ecological design, in Aldo Leopold's words, is to go "from conqueror of the land-community to plain member and citizen of it." Drawing from Sim van der Ryn and Stuart Cowan (1996) and William McDonough and Michael Braungart (2002), the basic principles of ecological design are these:

- Use sunshine and wind.
- Preserve diversity.
- Account for all costs.
- Eliminate waste.
- Solve for pattern.
- Protect human dignity.
- Leave wide margins for error, malfeasance, and ignorance.

But there is no larger theory of ecological design, nor is there a textbook formula that works for practitioners across different fields and at varying

This article was originally published in 2006.

scales. And neither should we presume agreement on what it means for humankind to become a "plain member and citizen" of the biotic community. In other words, we have a compass but no map. Samuel Mockbee, founder of the Rural Studio, enjoined his architectural students working with the poor in Hale County, Alabama, simply to make their work "warm, dry, and noble." Warm and dry are easier for the most part because we feel them somatically, but noble is hard because it requires us to make judgments about what we ought to do relative to some standard higher than creature comfort. But in the best sense of the word it implies decent, worthy, generous, magnificent, proud, and resilient. And it ought to be synonymous with ecological design as well.

Having no theory to expound, I present what follows as notes for something like a bull session on ecological design.

1. Beginnings

The human sense of order and affinity for design, forged through our long evolutionary history, goes back to our dawning sensations and experiences of life. The first safe haven we sense is our mother's womb. Our first awareness of regularity is the rhythm of our mother's heartbeat. Our first passageway is her birth canal. Our first sign of benevolence is at her breast. Our first awareness of self and other comes from sounds made and reciprocated. Our first feelings of ecstasy come from bodily release. The first window through which we see is our eye. The first tool we master is our own hand. The world is first revealed to us through the senses of touch and taste. Our first worldview is formed within small places of childhood. Our ancestors' first inkling that they were not alone was the empathetic encounter with animals. The first music they heard was sounds made by birds, animals, wind, and water. Their first source of wonder, perhaps, was the undimmed night sky. Their first models of shelter were those created by birds and animals. The first materials humans used for building were mud, grass, stone, wood, and animal skins. Their first metaphors were likely formed from daily experiences of nature. The first models for worship were found in what early humans perceived as cosmic harmony, often replicated in the design of dwelling places.

We are creatures shaped by such experiences and by the interplay between our senses and the world around us. We know of five senses and have reason to believe that there are others. Some evidence suggests, for example, that we have a rudimentary awareness of being watched.

Aboriginal peoples can walk with unerring accuracy through trackless landscapes in the dark of night. Across all cultures and times, good design is a close calibration of our sensuality with inspiration, creativity, place, form, and materials. Good design feels right and is a pleasure to behold and experience for reasons that we understand at an intuitive level but have difficulty explaining (Alexander 2001; Kellert 1996).

2. *Evolution as Model/Nature as Standard*

The starting point for ecological design is the 3.8 billion years of evolving life on Earth. Nature, for ecological designers, is not something just to be mastered but a tutor and mentor for human actions. Janine Benyus, author of *Biomimicry*, points out, for example, that spiders make biodegradable materials stronger than steel and tougher than Kevlar without fossil fuels or toxic chemicals (Benyus 1997). From nothing more than substances in seawater, mollusks make ceramic-like materials that are stronger and more durable than anything we presently know how to make. These and thousands of other examples are models for manufacturing, the design of technologies, farming, machines, and architecture that are orders of magnitude more efficient and elegant than our best industrial capabilities.

Ecological design, however, is not simply a mimicking of nature toward a smarter kind of industrialization but, rather, a deeper revolution in the place of humans in nature. In Wendell Berry's words, design begins with the questions "What is here? What will nature permit us to do here? What will nature help us do here?" (Berry 1987, 146). The capacity to question presumes the humility to ask, the good sense to ask the right questions, and the wisdom to follow the answers to their logical conclusions. Ecological design is not a monologue of humans talking to nature but a dialogue that requires the capacity to listen, discern, and learn from nature. When we get it right, the results, in John Todd's words, are "elegant solutions predicated on the uniqueness of place." The industrial standard, in contrast, is based on the idea that nature can be tortured into revealing her secrets, as Francis Bacon so revealingly put it, and then by brute force and human cleverness coerced to do whatever those with power intend. One size fits all, so industrial design looks the same and operates by the same narrow logic everywhere. But this is no great victory for humankind, because the mastery of nature, in truth, represents the mastery of some men over other men using nature as the medium, as C. S. Lewis once put it (Lewis 1947).

3. All Design Is Political

Design inevitably involves decisions about how society provides food, energy, shelter, materials, water, and waste cycling and distributes risks, costs, and benefits. In other words, design affects who gets what, when, and how—a standard definition of politics. The environment, then, is a mirror reflecting decisions that we make about energy, forests, land, water, biological diversity, resources, and the distribution of wealth, risks, and benefits. Often cast as "liberal" or "conservative," such decisions are, in fact, often about how the present generation orients itself to the interests of its children and grandchildren. One can arrive at a decent regard for their prospects as either a conservative or a liberal. These are not opposing positions so much as they are different sides of a single coin. But neither conservatives nor liberals have yet invested much energy, time, or thought to the design requirements of the transition to sustainability. The point is that harmonizing social and economic life with ecological realities will require choices about energy technologies, agriculture, land use, settlement patterns, materials, the handling of wastes, and water that are inescapably political and will distribute risks and benefits in one way or another.

Further, as the Greeks understood, design entails choices that enhance or retard civic life and the prospects for citizenship. But in our time "we are witnessing the destruction of the very idea of the inclusive city" and with it the arts of civility, citizenship, and civilization (Rogers 1997; Rogers and Power 2000). By including or excluding possibilities to engage each other in convivial dialogue, the creators of urban spaces enhance or diminish civility, urbanity, and the civic prospect. It is no accident, I think, that crime, loneliness, and low participation became epidemic as spaces such as town squares, street markets, front porches, corner pubs, and parks were sacrificed to the automobile, parking lots, and urban sprawl. Better architecture and landscape architecture alone cannot cure these problems, but they can create convivial spaces where people talk, argue, reason together, play, bargain, and learn the art of being citizens.

4. Honest Accounting

In an age much devoted to the theology of the market, disciples of the conventional wisdom believe it imprudent to design ecologically if the costs are even marginally more than the costs of conventional design. Based on incomplete and highly selective accounting, that view is almost always wrong because it overlooks the fact that we—or someone—sooner or later will pay the full costs of bad design, one way or another. In other words, society pays for ecological design whether it gets the benefits of it or not. Honest accounting, accordingly, requires that we keep the boundaries of consideration as wide as possible over the long term and have the wit to deduct the collateral benefits that come from doing the right things in the right way. For example, ignoring the costs of wars fought for "cheap" oil, the costs of climate change and air pollution, and the health effects of urban sprawl, an SUV is cheap enough. But price and cost should not be confused. It is the height of folly to believe that we can eliminate forests, pollute, squander resources, erode soils, destroy biological diversity, remodel the biogeochemical cycles of the Earth, and create ugliness, human and ecological, without consequence. The truth is that, sooner or later, the full costs will be paid one way or another. The problem, however, is that the costs of environmental dereliction are diffuse and often can be deferred to some other persons and to some later time. But they do not thereby disappear. The upshot is that much of our apparent prosperity is phony and so too the intellectual and ideological justifications for it.

The standard of neoclassical economics applied to architecture, in particular, has been little short of disastrous. "The rich complexity of human motivation that generated architecture," in architect Richard Rogers and Anne Power's words, "is being stripped bare. Building is pursued almost exclusively for profit" (Rogers 1997, 67). By such logic we cannot afford to design well and build for the distant future. The results have been evident for a long time. In the mid-nineteenth century, John Ruskin noted, "Ours has the look of a lazy compliance with low conditions" (Ruskin 1989, 21). But even Ruskin could not have foreseen the blight of suburban sprawl, strip development, and urban decay driven by our near terminal love affair with the automobile and inability to plan sensibly. The true costs, however, are passed on to others as "externalities," thereby privatizing the gains while socializing the costs. The truth is, as it has always been, that phony prosperity is no good economy at all. False economic reckoning has caused us to lay waste to our countryside, abandon our inner cities and the poor, and build auto-dependent communities that are contributing mightily to destabilizing climate and rendering us dependent on politically volatile regions for oil.

An economy judged by the narrow industrial standards of efficiency will destroy values that it cannot embrace. Maximizing efficiency, measured as the output for a given level of input, creates disorder, that is to say, inefficiency at higher levels. The reasons are complex but have a great deal to do with our tendency to confuse means with ends. As a result efficiency often becomes an end in itself while the original purposes (prosperity, security, benevolence, reputation, etc.) are forgotten. The assembly line was efficient for the manufacturing firm, but its larger effects on workers, communities, and ecologies were often destructive, and the problems for which mass production was once a solution have been compounded many times over. Neighborliness is certainly an inefficient use of time on any given day, but not when considered as a design principle for communities assessed over months and years or generations. For engineers, freeways are efficient at moving people up to a point, but they destroy communities, promote pollution, lead to congestion, change foreign policies, and eliminate better alternatives, including design that eliminates some of the need for mobility. Walmart, similarly, is an efficient marketing enterprise, but it eliminates its competitors and many things that make for good communities, including jobs that pay decent wages that allow people to buy at any price. And, of course, nuclear weapons are wonderfully efficient devices as well. Ecological design, in contrast, implies a different standard of efficiency oriented toward ends, not means; the whole, not parts; and the long term, not the short term.

5. Design for Human Limitations

The limits of ecological design are those of nature and of human nature, including our incurable ignorance. The reasons for ignorance are many, as previously noted. Designers must also reckon with the uncomfortable probability that the amount of credulity in human societies remains constant. This is readily apparent by looking backward through the rearview mirror of history to see the foibles, fantasies, and follies of people in previous ages (Tuchman 1984). For all our pretensions to rationality, at some later time others will see us similarly. The fact is that humans presently are inclined to be as unskeptical and sometimes as gullible as those living in any other time—only the sources of our befuddlement change. People of previous ages read chicken entrails, relied on shamanism, consulted oracles. We, far more sophisticated but similarly limited, use computer models, believe experts, and exhibit a touching faith in technology to fix virtually everything. But who among us really understands how computers or computer models work? Who is aware of the many limits of expertise or the ironic ways in which technology "bites back"? Has gullibility declined as science has grown more powerful? No, if anything, it is growing because science and technology are increasingly esoteric and specialized, hence removed from daily experience. Understanding less and less of either, we will believe almost anything. Gullibility feeds on mental laziness and is enforced by social factors of ostracism, social pressures for conformity, and the pathologies of groupthink that penalize deviance.

This line of thought raises the related and equally unflattering possibility noted above that stupidity may be randomly distributed up and down the social, economic, and educational ladder. There are likely as many thoroughgoing, fully degreed fools as there are undegreed fools. In other words, intelligence and intellectual clarity can be focused and sharpened a bit but can be neither taught nor conjured. The numerous examples of the undereducated or those who were outright failures in the academic sense include Albert Einstein, Winston Churchill, and Frank Lloyd Wright. One should conclude, however, not that formal schooling is useless, but that its effectiveness, for all of the puffery that adorns college catalogues and educational magazines, is considerably less than advertised. And there are those made more errant by the belief that their ignorance has been erased by the possession of facts, theories, and the adornment of weighty learnedness.

Nor does the outlook for intelligence necessarily brighten when we consider the limitations of large organizations. These too are infected with our debilities. Most of us live out our professional lives in organizations or work for them as clients and discover to our dismay that the collective intelligence of organizations and bureaucracies is often considerably less than that of any one of its individual employees. We are baffled by the discrepancy between smart people and the organizations that employ them which exhibit a collective IQ of less than, say, Kitty Litter. We understand human stupidity and dysfunction because we encounter it at a scale commensurate with our own. But confronted with large organizations, whether corporations, governments, or colleges and universities, we tend to equate scale, prestige, and power with perspicacity and infallibility. Nothing could be further from the truth. The intelligence of big organizations (oxymoron?) is limited by the obligation to earn a profit, enlarge their domain, preserve entitlements, or maintain a suitable stockpile of prestige.

Our frailties infect design professions as well. Buildings and bridges sometimes fall down (Levy and Salvadori 1992). Clever designs can induce an astonishing level of illness and destruction. Beyond some limit, design becomes guesswork. British engineer A. R. Dykes puts it this way: "Engineering is the art of modeling materials we do not wholly understand, into shapes we cannot precisely analyze so as to withstand forces we cannot properly assess, in such a way that the public has no reason to suspect the extent of our ignorance." In various ways the same is true in other design professions and virtually every other field of human endeavor.

The point is simply to say that human limitations will dog designers at every turn. They will infect every design, every project, and the evolution of every system, however clever. From this there are, I think, two conclusions to be drawn. The first is simply that design, whether of bridges, buildings, communities, factories, or farms and food systems, ought to maximize the capacity of a system to withstand disturbance without impairment, which is to say its resilience. Ecological design does not assume human infallibility, or that technologies will work as intended, or that some deus ex machina will magically rescue us from our own folly. Rather it does things at a manageable scale aiming for flexibility, redundancy, and multiple checks and balances characteristic of healthy ecosystems, and in so doing, it avoids transgressing thresholds of the irreversible and irrevocable (Lovins and Lovins 1982, chapter 13; Lovins 2002).

Forewarned about human limitations, we might further conclude that a principal goal of designers ought to be the improvement of our collective intelligence by promoting mindfulness, transparency, and ecological competence. The public is less aware of how it is provisioned with food, energy, water, materials, security, and shelter and how its wastes are handled than people of any previous time. Industrial design cloaked the ecological fine print of what are often little better than Faustian bargains providing luxury and convenience now, while deferring ruin to some later time. Ecological design, on the contrary, ought to demystify the world, making us mindful of the ecological fine print by which we live, move, and have our being.

Design is always a powerful form of education. Only the terminally pedantic believe that learning happens just in schools and classrooms. The built environment in which we spend over 90 percent of our lives is at least as powerful in shaping our ideas and views of the world as anything learned in a classroom. Suburbs, shopping malls, freeways, parking lots, and derelict urban spaces have considerable impacts on how we think, what we think about, and what we can think about. The practice of design as a form of public instruction ought to free the ecological imagination from the tyranny of imposed forms and relationships characteristic of the fossil fuel–powered industrial age. Architecture, landscape architecture, and planning carried out as a form of pedagogy aim to instruct about energy, materials, history, rhythms of time and seasons, and the ecology of the places in which we live. It would help us become mindful of ecological relationships and engage our places creatively.

6. Vernacular

Many of the best examples of ecological design have been created by people at the periphery of power, money, and influence and living in out-of-the-way places. The truth is that practical adaptation to the ecologies of particular places over long periods of time has often resulted in spectacularly successful models of vernacular design (Rudofsky 1964). It may well be that the ecological design revolution will be driven, at least in part, by experience accumulated from the periphery, not from the center, and led by people skilled at solving the practical problems of living artfully by their wits and good sense in particular places. The success of vernacular design across all cultures and times underscores the possibility that design intelligence may be more accurately measured at the level of the community or culture, rather than at the individual level.

7. The Standard

The esthetic standard for ecological design is to work so artfully as to cause no ugliness, human or ecological, somewhere else or at some later time. The standard, in other words, requires a robust sense of esthetics that rises above the belief that beauty is wholly synonymous with form alone. Every great designer from Vitruvius (90–20 BC) through Frank Lloyd Wright demonstrated that beauty in the large sense had to do with the effects of buildings on the human spirit and our sense of humanity. But the standards for beauty must be measured on a global scale and longer time horizon so that beauty includes the upstream effects at wells, mines, and forests where materials originate as well as the downstream effects on climate, human health, and ecological resilience. Things judged truly beautiful will in time be regarded as those that raised the human spirit without compromising human dignity or ecological functions elsewhere. Architecture and landscape architecture, in other words, are a means to higher ends, not ends in themselves.

8. Education of Designers

As much art as science, the design professions are not simply technical disciplines, having to do with the intersection of form, materials, technology, and real estate. The design professions such as architecture, landscape architecture, and urban planning are first and foremost practical liberal arts with technical aspects. Long ago Vitruvius proposed that architects "be educated, skilful with the pencil, instructed in geometry, know much history, have followed the philosophers with attention, understand music, have some knowledge of medicine, know the opinions of the jurists, and be acquainted with astronomy and the theory of the heavens" (Vitruvius 1960, 5–6). That is a start of a liberal and liberating education. Design education, therefore, ought to be a part of a broad conversation that includes all of the liberal arts. This is what George Steiner means by saying that

> architecture takes us to the border. It has perennially busied the philosophic imagination, from Plato to Valery and Heidegger. More insistently than any other realization of form, architecture modifies the human environment, edifying alternative and counter-worlds in relationships at once concordant with and opposed to nature. (Steiner 2001, 251–52)

In countless ways all design, even the best, damages the natural world. Extraction and processing of materials depletes landscapes and pollutes. Building construction, operation, and demolition creates large amounts of debris. Agriculture inevitably simplifies ecosystems. A new breed of ecological designers, accordingly, must be even more intellectually agile and broader, capable of orchestrating the wide array of talents and fields of knowledge necessary to design outcomes that can be sustained within the ecological carrying capacity of particular places.

9. Design as a Healing Profession

The design professions are a form of the healing arts, an ideal with roots again in Vitruvius's advice that architects ought to pay close attention to sunlight, the purity of water, air movements, and the effects of the building site on human health. The word *healing* has a close affinity with other words such as *holy* and *wholeness*. A larger sense of the profession of architecture, which architect Thomas Fisher (2001) deems a "calling," would aim for the kind of wholeness that creates not just buildings but integral homes and communities. Compare, for example, the idea that "architecture applies only to buildings designed with a view to aesthetic appeal" (Pevsner 1990, 15) with architecture defined as "the art of place-making" and creation of "healing places" (Day 2002, 10, 5). In the former sense, design changes with trends in fashionable forms and materials. It is often indifferent to place, people, and time. The goal is to make monumental, novel, and photogenic buildings and landscapes that often express only the ego and power of the designer and owner. In contrast, the making of healing places signals a larger allegiance to place that means, in turn, a commitment to the health of other places. Place making is an art and science disciplined by locality, culture, and ecology requiring detailed knowledge of local materials, weather, topography, and the nature of particular places and a creative dialogue between past, present, and future possibilities. It is slow work in the same way that carefulness has a different clockspeed than carelessness. Place making uses local resources, thereby buffering local communities from the ups and downs of the global economy, unemployment, and resource shortages.

Practiced as a healing art, architecture, for example, would result in buildings and communities that would not compromise the health of people and places. Architects would aim to design buildings and neighborhoods in which community and conviviality could thrive. At larger scales the challenge is to extend healing to urban ecologies. Half of humankind now lives in urban areas, a percentage that will rise in coming decades to perhaps 80 percent. Cities built in the industrial model to accommodate the automobile are widely recognized as human, ecological, and, increasingly, economic disasters. Given a choice, people abandon such places in droves. But we have good examples of cities as diverse as Copenhagen, Chattanooga, and Curitiba that have taken charge of their futures to create livable, vital, and prosperous urban places—what Peter Hall and Colin Ward (1998) have called "sociable cities." In order to do that, however, designers must see their work as fitting in a larger human and ecological tapestry.

As a healing art, ecological design aims toward harmony, which is the proper relation of parts to the whole. Is there a design equivalent to the Hippocratic oath in medicine that has informed medical ethics for two millennia? Are there things that designers should not design? What would it mean for designers to "do no harm"?

Looking ahead, the challenge to the design professions is to join ecology and design in order to create buildings, communities, cities, landscapes, farms, industries, and entire economies that accrue natural capital and are powered by current sunlight—perhaps, one day, having no net ecological footprint. The standard is that of the healthy, regenerative ecosystem. In the years ahead we will discover a great deal that is new and rediscover the value of vernacular traditions such as front porches, village squares, urban parks, corner pubs, bicycles, pedestrian-scaled communities, small and winding streets, local stores, riparian corridors, urban farms and wild areas, and well-used landscapes.

Design practiced as a healing art is not a panacea for the failures of the industrial age. However well designed, a world of 7 to 10 billion human beings with unlimited material aspirations will sooner than later overwhelm the carrying capacity of natural systems as well as our own management abilities. There is considerable evidence that humans already exceed the limits of many natural systems. Further, ecological design does not require building; often the best design choices require adaptive reuse or more intense and creative uses of existing infrastructure. Sometimes it means doing nothing at all, a choice that requires a clearer and wiser distinction between our needs and wants.

What ecological designers can do, and all they can do, is to help reduce our ecological impacts and buy us time to reckon with the deeper sources of our problems, which have to do with age-old questions about how we relate to each other across the boundaries and sometimes chasms of gender, ethnicity, nationality, culture, and time and how we fit into the larger community of life. Ecological design, as a healing art, is a necessary but insufficient part of a larger strategy of healing, health, and wholeness, which brings me to the soul of the matter.

10. Design for Spirit

For designers it is significant that humans are inescapably spiritual beings, if only intermittently religious. Our choice is not whether we are spiritual or not but whether our spiritual energy is directed to authentic purposes or not. But much of the modern world, however, has been assembled as if people were machines without deeper needs for order, pattern, and roots. Modern designers filled the world with buildings and developments divorced from their context, existing as if in some alien realm disconnected from ecology, history, culture, people, and place. Ecological design, on the other hand, is a process by which we grow into a particular place, becoming citizens of the life-community in that place. It is a process by which dwellings and landscapes and the uses we make of them become part of a larger story. As a kind of storytelling, design is a celebration of the life that connects us with the nature of the places in which we live and work and grounds us in the still larger story of the human journey.

~

Ecological design is not a formula but rather a complex process of adapting human intentions to ecological realities. It is art as much as science, ethics as much as economics, ecology as much as engineering. And it is a messy, uncertain, difficult, sometimes contentious process demanding a high order of competence, creativity, and goodwill. Properly done, it changes routines of institutional decision making and management. Rules of finance and budgeting, for example, that worked in the industrial era, when the natural capital of soils, forests, water, and climate stability was assumed to be free, no longer do so. Designing ecologically requires the integration of expertise across many disciplines, perspectives, and professions, such as energy specialists, ecological engineers, materials scientists, lighting consultants, ecologically adept landscape architects, and engineers who understand buildings as whole systems, and those who will live and work there.

Finally, beyond performance of the obvious functions such as durable shelter, usefulness, and beauty, what more do we want from our buildings, landscapes, and communities? We should want our buildings, neighborhoods, communities, and cities to honor the ecologies and cultures of the places in which they are built. They should promote rootedness, not anomie. They ought to foster an awareness of connections and ecological competence. They ought to make us smarter and more competent people, not dumb us down. They ought to be designed to regenerate natural capital of soils, trees, and biological diversity. They ought to foster possibilities for real human engagement. They ought to be paid for fairly and not offload costs on others. But these, too, are means to still larger ends.

PART 4

On Education

∾ AUTHOR'S NOTE 2010 ∾

I had a high school diploma, a bachelor's degree, a master's degree, and a PhD and presumed myself to be educated. But 11 years of living in a small valley in the southern Ozarks in the 1980s showed me how little I knew and how little I could ever know but also the importance of striving to know. The first essay in this section is about the start of my own education; the rest of the essays are footnotes to a conversation with that particular place. We talk a great deal about training (what one does to a dog) and education as something that happens between the ages of 6 and 21 and aims mostly to improve career prospects. Learning, however, is a lifelong process, the end of which is a life well lived, which has little to do with careers, money, and success as we commonly measure it. I learned that when you don't have much to gain or lose, no vested interests, and no particular reputation at stake that something like learning can begin.

Chapter 22

Place as Teacher

(2006)

Once in his life a man . . . ought to give himself up to a particular landscape in his experience, to look at it from as many angles as he can, to wonder about it, to dwell upon it. He ought to imagine that he touches it with his hands at every season and listen to the sounds that are made upon it. He ought to imagine the creatures there and all the faintest motions of the wind. He ought to recollect the glare of noon and all the colors of the dawn and dusk.

N. Scott Momaday

I have lived in nine places in my life, but I dream about only one: a small valley in the southern Ozarks carved out over the last million years or so by a clear stream that the local people know as Meadowcreek. I lived in the Meadowcreek Valley for 11 years, and in some ways I still do and probably always will. As places go, it had a lot going against it. Meadowcreek was remote from some of the essential amenities of the good life. The nearest bank was 25 miles away. The nearest shopping mall was 100 miles to the south. The nearest town, Fox, was 3 miles distant by treacherous dirt roads. It has never made anyone's annual listing of the most desirable places to live. It had no Starbucks or fine restaurants. The general store on county highway 263 stocked mostly white bread, soft drinks, canned goods, cigarettes, and some hardware items. It functioned as the de facto town hall, where the conversation was slow but nonstop until a stranger wandered in to ask directions. The post office across the road was the only other establishment of note. There you could get

This article was originally published in 2006.

your mail, opinions about the weather, and a sympathetic hearing about what hurt. Within a quarter mile of the post office were four churches, all of the kind that connect Christianity with sin, tears, redemption in the blood, and *glory* spelled with a capital *G*, punctuated by hallelujahs. JD's garage was down the road a bit, along with most of the mechanical detritus he'd accumulated over a half century of repairing all manner of things. He would take nothing more than $2 for a tire change. The vacant building across an unpaved street housed any number of dreams. Donny Branscomb tried to make a go of a café there, but people in Fox don't eat out much and selling coffee and cigarettes didn't pay his bills. His next line of work was driving a tour bus out of Little Rock.

The surrounding Ozark hills have little of the grandeur of the Rockies, or of the Appalachian Mountains, for that matter, although they are scenic enough. Summers can be brutally hot and muggy. If the heat and humidity don't kill you, the ticks and saber-toothed chiggers might. The Ozark region looks something like a parallelogram stretching along an axis from east central Missouri southwest into Oklahoma. What are called mountains in the Ozarks are not particularly mountainous; the highest elevations seldom exceed 2000 feet. For all of their rural charm, the Ozarks remain an economic backwater roughly equidistant between St. Louis and Little Rock, and Memphis and Tulsa. To the south, I-40 runs east and west. To the northwest, I-44 runs between Joplin, Missouri, and St. Louis. In between, hardly a straight stretch of highway can be found. If you manage to get in, it's not easy to get out. Stay long enough and you may not want to.

The word *Ozarks* came from the French *Aux-Arcs*, which means "to the Arkansas post" (Rafferty 1980, 4). The French named them, but geology and water shaped them. Between the Precambrian and Pennsylvanian ages, most of the Ozarks were covered by an ancient sea. Sixty-five million years ago the first of a series of uplifts occurred, raising the Ozarks above the surrounding country. The resulting plateau is the highest ground between the Appalachians and the Rockies. The rugged landscape of the Ozarks, however, is the ongoing project of water working its will on land above an ancient seabed. The Ozarks are known for limestone caves, clear springs, and spectacular bluffs overlooking pristine, slightly bluish streams below.

The first human occupants of the region reportedly were Osage Indians. They were evicted in the early nineteenth century by land-hungry Scotch-Irish settlers spilling across Tennessee and Kentucky from the

Appalachians. These self-reliant settlers came armed with axes, biblical self-assurance fortified by homemade whiskey, and strong beliefs in the rights of property. This was not, however, as the Osage and later the Cherokee peoples surely noted, an equal opportunity belief. After the native people, the first thing to go was the virgin forests, which were cut over in less than 50 years. Prime Ozark white oak went to Memphis and St. Louis to make furniture, railroad ties, and barrel staves. Having sold their forests for a pittance, Ozark settlers turned to agriculture in earnest, but without much success. Their ideas about farming originated mostly in England, where, by comparison, soils were deep, topography was rolling, and rainfall tended to be gentle. In the Ozarks, however, thin rocky soils, steep hillsides, and summer drought punctuated by violent downpours typical of the southern midcontinent conspired against prosperity. Instead, the Ozark economy formed around subsistence farming with cattle, hogs, chickens, marginal timbering, and lots of doing without. All of this is to say that the nineteenth century settlers came with habits and expectations that did not fit well with the ecology and topography of the region. It is an old story.

If geology and water shaped the Ozark landscape, its mindscape was formed in the union of isolation with hardscrabble poverty. The difference between aspiration and situation was made up by evangelical religion, alcohol, resignation, folk music, and a love of the land. But the national stereotype of the Ozark personality created by Al Capp in his Dogpatch cartoons bears scant resemblance to the human reality. Ozark people, like rural people virtually everywhere, have learned to make do with what they have, which for the most part isn't much. On the whole, they do so without much self-consciousness of being victims of economic oppression or poverty. They'd much prefer being left alone to being helped. They are independent, self-reliant, often suspicious of outsiders, resistant to new ideas, and clannish, but not more parochial in their way than, say, cosmopolitan New Yorkers are in theirs. And if you have a choice of where to have your car break down at 2 AM some dark, rainy night, you'd be smart to arrange it in the Ozarks, where the word *neighbor* is still regarded as a verb.

Ecologically and culturally the Ozark region is a meeting ground. The oak, hickory, and ash forests are similar to those of the southern Appalachians, but I often found cactus on south-facing ridgetops, survivors of a hotter and drier age. Similarly, the humble armadillo, a native of the southeast, is migrating northward to take up residence in the Ozarks.

Not a few have become imbedded in the highway system. Culturally, too, the Ozarks are a meeting ground between the mountain culture of the Appalachians and the cowboy culture of the southwest. There were as many cowboy hats as baseball caps on the hat rack in the Rainbow Café in Mountain View. A rodeo came to town each fall but did not travel much farther east. There were a scattering of mailboxes with such and such "Ranch" posted on what otherwise looked a lot like a hill farm plus a few worn-out cows.

Split personality and all, few regions of the United States arouse such devotion and loyalty in their residents. Ozark people oftentimes think of themselves as part of the region first, before listing other and lesser loyalties to state, church, and nation. Many high school graduates stay put despite the lack of local opportunities for "upward mobility." Those who do leave rarely go far away, and they tend to return when they've saved enough money. There is a rich literature about the life and natural history of the region. The contrast between a region that seems to give so little yet arouse such a strong sense of place is striking. I grew up in western Pennsylvania, which, by comparison, is a lush land of milk and honey with rolling hills, fertile soils, and a temperate climate. Yet most of the people I knew while growing up had little sense of regional identity and only a superficial knowledge of the place. To this day I know of no significant book about the natural history of the region despite its apparent economic and ecological advantages. And once gone from Pennsylvania, few return.

Meadowcreek flows through the southwestern corner of Stone County in the Boston Mountains of the southern Ozarks toward the middle fork of the Little Red River. It is in Stone County, 110 miles due north of Little Rock. On government maps, Stone County ranks as the fifth poorest county in the 49th wealthiest state of the union. State bean counters were often moved to thank God for creating Mississippi—a thin statistical film between Arkansas and the bottom of the barrel.

The Meadowcreek Valley is 3 miles west of Fox, 3 miles southeast by jeep trail and deep faith from Flag, and about 5 miles north of the ghost town of Arlberg. Coming from any direction, however, you have to want to get there to get there. Few arrived by accident. It was a test of determination, nerves, tires, tie-rods, and brakes. Some found the precipitous descent into the valley on a rough, narrow, unpaved road with a sheer drop of 200 feet on one side something of a spiritual experience. I recall the driver of a semi truck who was delivering a load of concrete and forgot to gear down at the top of the hill. Halfway down he'd exhausted the reservoir of air for his air brakes but in that omission found an urgent need for Jesus. At the bottom, one could infer from his incoherent stammer and the color of his face that he had undergone a high-speed conversion. He swore he'd never do it again.

The valley is 3 miles long, running north-south, by a mile to a mile and a half wide. To the north the valley forks into Bear Pen Hollow and another, unnamed hollow leading to Flag. To the south, the valley opens into a U-shaped gorge through which the middle fork of the Little Red River flows on its way to the White River. On each side, the valley floor rises up to flat benches and then rises more steeply to the ridgetops above. Rock outcroppings at the same elevation all around make the valley look like a giant bathtub with a crusty ring. From valley floor to ridgetop the elevation averages 600 feet.

From the bluff known as Pinnacle Point at the southwest corner of the valley, you can see the length of the Meadowcreek Valley to the north and the gorge cut by the middle fork to the south. Below, on the east side of the valley, is what remains of the Bond family homestead, an Arkansas "dogtrot" house with two rooms on either side of an enclosed walkway. Most people in Stone County were reportedly born, courted, married, or shot there. It now sits abandoned and derelict. Southwest of Pinnacle Point is Bee Bluff with a sheer rock face on the south side that looks as if it had been cut with a knife. On the bench immediately below the eastern face of the bluff is a wooded cemetery containing a catalog of rural tragedy and hardship chiseled on primitive tombstones.

Angel sent from God 1-12-1901
Returned to her Savior 4-7-1903

At one time the valley reportedly had some 40 homesteads and the largest school in the county. Little remains other than the stones around an occasional well or door threshold and the daffodils that bloom each spring where once cabins stood. When we first arrived in the valley in the spring of 1979, the only human residents were a Baptist preacher and his sad-faced, heavily burdened wife, who rented a rundown house at the north end, and a couple the locals called hippies, who lived in what was left of an old homestead 2 miles to the south under the shadow of Pinnacle Point. Most of the valley was owned by a local doctor who used it for grazing cattle. Otherwise the land was becoming forest again. Fencerows were overgrown with cedar and greenbrier. Lichen-covered rock walls were falling down. Deer, raccoon, and stray hunting dogs had the run of the place.

I first saw the valley on a somber, cold, and blustery February day in 1979. The region had been through some of the worst freeze-thaw weather that anyone could remember. Creeks were swollen by heavy rains, and the roads were nearly impassable, even in a four-wheel drive vehicle. We hiked and drove around the valley until well after dark, comforted somehow that we had seen it at its worst. Later, we discovered how relative that word can be. On our way out, in the darkness of evening, the road bottomed out and we were stuck in mud that nearly covered the wheels. We had passed a house a quarter mile or so back and slogged through the dark and the mud to ask for help. Before we could knock on the door a voice inside boomed out, "I figured you'd be acoming back. I'll get the tractor." His name was Lonnie Lee, a bull of a man in his prime, and as famous for his hospitality as for his temper. A logger and woodsman by profession, but a musician and storyteller at heart, Lonnie had us on our way, or so we thought. Another mile and we heard the sound of metal on stone and discovered that we had lost a tire in the mud and were traveling on three tires and one bare wheel. Things are like that in the Ozarks. Easy becomes hard. Fast goes slow. Certainties are less certain. Tires fall off. A spare change and we were on our way again. We moved into the valley the following June.

We came as interlopers to a place to which we had neither attachments nor roots. What we had were ideas, energy, a bit of cash, and a belief that we might do great and good things in that place. Our intent was to create an educational center without the disciplinary blinders, shortsightedness, and bureaucracy of conventional educational institutions. We found this place quite by serendipity; it was a good choice for reasons that we could not have known in advance and a poor choice for obvious reasons we refused to see. We, of course, became the first students and the place itself became both our tutor and the curriculum.

Like most Americans, I had not thought much about the importance of place. By 1979 I had lived in seven other locations and could not tell you much about them that you could not discover for yourself with a map and a day's tour. I fancied myself an environmentalist, but I would have flunked the most basic test of bioregional knowledge about the seven previous places where I'd lived. In this regard I was typical. On average, Americans are increasingly ignorant about where they live and how they are provisioned with food, energy, water, material, and the services of nature. The reasons are not hard to find. We live like nomads, moving 8 to 10 times in a lifetime. Restlessness is part of the national psyche. America was discovered by tribes that walked east across the Bering Strait when it was above water, and later by Europeans who sailed from the opposite direction looking for India. Descendants of the latter included Daniel Boone, swarms of pioneers, armies of salesmen, herds of tourists, consultants by the thousands, and tribes of migrants in their fossil-fueled SUVs and mobile homes. Our cultural heroes have usually been one variation or another on the theme of lonely stranger who wanders into town and does some awesome and mostly violent thing, departs, and is never heard from again. The settlers who clean up the mess and get the kids back to school do not make such salable or salacious movie subjects. I know of no movie about, say, Henry David Thoreau, who said he did most of his traveling at home. What is the cause of our restlessness and our fascination with restlessness?

Perhaps it is hardwired into us; after all, many of our ancestral tribes migrated with the seasons and the food supply. That's true enough, but our mobility is driven by neither calories nor the calendar. It's a deeper kind of itch for opportunity, the chance to get rich, and the lure of excitement that infects bored people. With us, in other words, it's a mind thing, not a physical or even spiritual necessity. And movement can become addictive. A friend of mine drives an 18 wheeler for a living. He's tried to settle into a nine-to-five job at home but cannot do it for long to save himself or his family. A couple of weeks at home and he comes unglued and has got to get back on the road to preserve his sanity in an insane system. He just has it a bit worse than the rest of us.

We've made it easy to get up and go. First on post roads carved into the wilderness and then, in succession, canals, railroads, interstate highways, and airports: the great American motion sickness. We talk about colonizing space, and I suppose we may try that too. More likely, however, our restlessness will be met by purveyors of virtual reality who will sell us the simulated version of any fantasy or destination we—or they—can dream up. Want to go to the moon? Step into a virtual reality simulator and off you go! Reality, or their version of it, for a price.

This gets closer to the heart of the problem. Whatever our hardwiring, motion in service to fantasy is now the core of the national economy. Imagine for a moment what would happen if Americans one day decided to stay put. Car companies would go even more broke, along with all of the other companies that sell us roads, tires, gasoline, insurance, lodging, and hamburgers. The national economy would collapse, and I think "they" know that very well, which explains why a sizable part of the national advertising budget is spent to keep us restless and on the go. Whatever wanderlust exists in the human soul has been amplified into a positive feedback system that goes like this: more roads and airports → more oil wells, oil spills, oil refineries, oil wars, military spending, mines, malls, Disney Worlds, sprawl, ugliness, pollution, and noise → fewer neighbors, neighborhoods, livable communities, distinctive places, and solitude → more people trying to escape → more roads and airports . . . a cycle of futility, destruction, violence.

Of course the lack of a sense of place is not just a function of rootlessness. It also has to do with the way we are fed, clothed, supplied, and fueled. Modern technology has unhitched us from our places. We are no longer competent to do much for ourselves. Most of us are effortlessly provisioned from distant agribusiness, feedlots, wells, mines, and factories that we know nothing about. We consign our wastes to other, equally unknown places. All of this is said to be economically efficient, but for whom, how, and how long is never explained, because it cannot be both explained and justified.

Our relation to our places has been further weakened by the American tendency to commercialize land so that places come to be regarded solely as real estate. For many people, however, land is abstract because they neither own any nor have easy access to it. The experience of place as an enduring relationship with a landscape and all of its life forms is increasingly unlikely for the 80 percent of Americans who live in urban areas and for the growing number on the downhill side of the middle class.

The weakening sense of place and the competence necessary to live well in a particular place is now epidemic in our culture. It is, I think, at the heart of what is called the ecological crisis. All of the numbers foreshadowing one disaster or another, all of the sigmoid trend lines surging upward and others in freefall, represent the sum total of our collective disconnectedness to the places in which we live and in which we earn our livelihood. The reasons are straightforward.

The growing distance between consumers and producers creates innumerable possibilities for political and ecological mischief. An economy grown to a global scale not only invites irresponsibility, it cannot work otherwise and remain profitable for the few who run it. The global economy entices consumers to consume more than they need. To do so they must be largely ignorant about the ecological and human consequences of their consumption, including the effects of it on themselves. The global economy created the kind of dependence that breeds what Thomas Jefferson called "venality," which inevitably corrupts political life as thoroughly as it debases citizens. A global economy can only exist at a scale beyond the possibility of democratic control, and perhaps beyond control of any kind. It is defended, nonetheless, because of its supposed efficiency. But no estimate of its true efficiency can be made unless all of its costs can be known and compared with those of alternative ways to do the same or better things. Finally, by destroying all other economies and cultural possibilities, the global economy places the human future in extreme jeopardy. By homogenizing the human enterprise in the name of "development" or "progress," we are, in effect, betting it all on one roll of the dice.

⁓

In the late fall of 1983 we moved into a passive solar house that we built on the site of what had once been a steam-powered sawmill. Little of the mill remained but the rock pad where the boiler and steam engine once sat, along with rusted pipes, wrenches, axe heads, and bolts, all overgrown with greenbrier, cedar, and sweet gum. The place had become so overgrown that it was an eyesore to the few who traveled the dirt road that ran along the east edge of the site at the foot of a steep hill. The house was nestled in the arm of a steep hill to the east and a low boulder-strewn wooded hill to the north. Looking to the west through a patch of second-growth trees, across what local people called the "sand field," past Meadowcreek, the west ridge rose 600 feet to rock bluffs and chimney rocks at the top. To the south the house looked down the 3 miles of the Meadowcreek Valley to the gorge of the Middle Fork and the bluffs beyond. At night the only visible evidence of human occupation was a light at a Methodist church camp 7 miles distant.

I began to clear the site in spare time in the late fall of 1982, mostly because it offended my idea of what an edge ought to look like. Farm boundaries, fencerows, and the edges of fields, I'd learned, should be neat and manicured. And this was a conviction for which I was, then, prepared to shed blood. Those familiar with greenbrier may know how much blood can be shed in the clearing of roughly an acre of land overrun with it. As the brush, vines, briers, and small trees gave way, traces of the old sawmill became apparent. The owners of the mill had dug out a basin, long since overgrown, that collected water from a natural seep at the back of the site. This water was used to cool the boiler, which sat on a rock pad 15 feet long by 5 feet wide, which had become anchored at one end by a giant sycamore tree. Heat had made the upper layers of rock brittle, so they could be broken apart by hand. Still, most of the rock was useful for building retaining walls around the house.

Remnants of rusty hand-forged tools and metalware lay all about: head blocks from the sawmill, old buggy-wheel rims, pipes, and other things I could not identify. My collection, carefully cleaned and painted with rust-resistant paint, was eventually attached to the side of my woodshed. The collection testified to human ingenuity and perseverance in the face of necessity. Some nameless person, for example, had taken two pieces of strap metal, hot-welded them together, and beveled one edge to make a workable chisel. We discovered dozens of wrenches, perhaps made by the same person with similar homespun resourcefulness. I showed one piece of rusty pipe split at the seams to an itinerant philosopher with a keen sense of place and a compassionate heart. He uttered a low sigh and said he hoped that the child who had forgotten to drain the boiler some frosty night long ago was not rebuked too harshly. So did I.

While I cleared the site, the place was working on me in its own fashion. Often I would stop work to gaze down the valley or look up at the bluffs to the west. I wondered who had owned the mill. What were they like? What kind of life did they have in this place? Why did they leave? Several hundred yards to the south at the end of the sand field, where Meadowcreek had once run diagonally across the valley floor, was the site of an ancient Osage Indian village recently excavated by local archeologists. What were their lives like here? Were they, in some sense, still here? The place, I tell you, had voices.

It also had sounds. Across the sand field, Meadowcreek, on its way to the Gulf of Mexico via the Middle Fork, White River, and Mississippi, tumbles over and around boulders the size of cars. The first heavy rains in the late fall would raise the water level, and the sound of rushing water would again fill the valley. In the late evening, owls in the woods across the field would begin their nightly conversations. Occasionally I'd join in until they discovered that I had nothing sensible to say, at which point they would descend into a sullen silence so as not to encourage me further. In the spring and early summer the chuck-will's-widows and tree frogs would hold their evening serenades. Once a month or so, a pack of coyotes would interrupt their raids on the local chicken houses to hold a symposium in the valley. Unlike owls, who converse patiently throughout the night, coyotes handle their business quickly, seldom taking longer than 30 minutes, and then get back to work. By late fall the wind, which blows hot straight up the valley all summer long, shifts and comes cold down the valley out of the Bear Pen. Pieces of ancient seabed raised to bluff height would sometimes be heard breaking loose and crashing to the forest floor below. Except for an occasional pickup truck, however, few human-made sounds intruded on the symphony of wind and rushing water. And although humans in the past century had taken a terrible toll on the valley, the wounds were healing. One could imagine this as a wilderness in the remaking.

I do not recall when the thought of building a house in this place first came to us, but the logic of the location was clear. The site was sheltered from the north wind yet open to the summer sun and winds to the south. It was shaded from the blistering summer sun by woods on the west side, and daytime heat was tempered by cooler air descending in the night. Built in the valley, it was still high enough to be above the floodplain. And the view down Meadowcreek Valley framed by high ridges on either side was an endless and ever-changing delight. But logic was just a rationalization for holding a deeper conversation with a particular place and its nameless guardian spirits. We had to build there.

Once I invited a well-known cosmopolitan writer from San Francisco to give a talk at Meadowcreek to our students and staff on the theme of the importance of place. Her talk was sophisticated, smart, and full of allusions to great writers and big ideas. But she was honest enough to admit that she had no sense of place, only words and thoughts about it. By her own admission, place was only an alluring abstraction. In the back of the room, listening intently, were several Ozark women whose daily lives were lived to the rhythms and demands of place. They competently lived the reality, privations, pains, and joys the other woman for the most part could only talk about. They, however, could no more intellectualize about place and its importance than they could repeal the law of gravity or make their husbands give up tobacco and alcohol. Afterward, I asked several of them what they thought about the talk, to which they responded that they did not understand a word of it. "One who knows does not say and one who says does not know"—Lao-tzu.

Attachment to place grows by stealth, by which mere words and thoughts give way to something deeper. In time the boundary of the person and the place can become almost indistinguishable. There are people who die quickly when uprooted from their ancestral homes. I have come to believe that driving people from the places in which they are rooted is about the most cruel punishment that one human can inflict on another. But I do not believe that one can plan to become attached or centered in a place. It takes time, patience, perhaps poverty, but most certainly a great deal of necessity. It cannot happen during a vacation, although a kind of infatuation with a place can occur in that length of time. It will not happen without something akin perhaps to a marriage vow, a commitment to a particular location for better or for worse. Can it happen in a city? Not likely, at least not likely in the cities that we've built. My urban friends will protest that they too have a sense of place. By my reckoning, however, what they have is a sense of habitat shaped by familiarity. The sense of place is the affinity for what nature, not humans, has done in a particular location and the competence to live accordingly.

I doubt that we can ever come to love the planet as some claim to do, but I know that we can learn to love particular places, and that will require a great deal of competence and forbearance. I believe that the love of place and the acceptance of the discipline of place, far from being a quaint relic of a bygone age, will prove to be essential to anything like a fair, decent, and durable civilization.

The world is now engaged in the early stages of what will be a very long and contentious debate about the human prospect in a future without cheap oil and on the brink of nasty climate surprises. On one side are those who see problems but not dilemmas and certainly no cause for alarm. A bit of technology here, a policy change there, add a dash of luck, and we will arrive at the magic kingdom of sustainability. In other words, we don't have to prove ourselves worthy, just clever. On the other side are those who believe that we must first "become native" to our places before all of these other things can be added unto us—a more arduous route, with the aroma of brimstone and repentance to it. Advocates of the former often prefer to eat organically grown vegetables and vacation far from the ecological effects of their vocation. Advocates of the latter sometimes motor about in four-wheel drive trucks, use chainsaws, and communicate by e-mail. Meanwhile the bottleneck ahead comes closer.

~

We left the Meadowcreek Valley in June of 1990 after 11 years that changed us in more ways than I can say. We'd arrived in 1979 from one of the centers of wealth and power in American society: Chapel Hill, North Carolina. Fox, Arkansas, is by every measure at the periphery, and the world of power and wealth looks very different from the outside looking in. I'd arrived full of the self-assurance of thinking myself well educated, knowledgeable, and armed with a compelling point of view. Eleven years later I knew how phony that assurance can be. We set out to create an educational experiment, a cross between places like Black Mountain College, Deep Springs College, and a few others at the periphery of American education and imagination. I thought my own education and background in and around the academy would be adequate to the challenge. From the age of 5 onward I had been in or around higher education as the son of a college president, a student, and a faculty member. I soon discovered how irrelevant much of that experience was. In all of that time, I recall few serious conversations about the purposes and nature of education and none at all about the adequacy of formal education relative to our role as members in the community of life. It was assumed that mastery of a subject matter was sufficient in order to teach others and that those very subjects are properly conceived and important.

In the 1970s I had grown disillusioned by the rigid separation of disciplines in the academy, its complacency, and its indifference to big questions about the human future. I was disillusioned, too, about what I perceived to be the separation of head, hands, and heart in the learned world. Education, it was assumed, began at the neck and worked up, but it dealt with only half of what remained. The other half, that part of mind where feeling, humor, poetry, and integration reside, was considered lacking in rigor by people who were often, I thought, unable to distinguish between rigor and rigor mortis. The resulting wars among head, hands, and heart and between the world of theory and practical experience were fought, but without much awareness, in every classroom, school, and college in the land and in the minds and lives of every student. Problems we often diagnose as ones of bad behavior and low motivation among those to be educated more likely reflect the miscalibration between schooling and our full humanity trying to break free; they are made more difficult by bad parenting and too much television, affluence, sugar, caffeine, and drugs.

The idea behind the Meadowcreek experiment was that we would draw a line around the 1500 acres we'd bought and make everything that happened inside that line curriculum: how we farmed the 250 acres of farmland, how we built, how we managed the 1200 acres of forest, how we applied the ecological knowledge necessary to manage the place, how we supplied ourselves with energy. We intended this valley to be a laboratory to study some of the problems of sustainable living and livelihood.

Our curriculum coalesced around sustainable agriculture, forestry, applied ecology, rural economic development, and renewable energy technology delivered through internships with college graduates, January terms, conferences, seminars, and scholar-in-residence programs. Broadly, if it had to do with the subject of sustainability, it was fair game for us. Over a decade or so, the number of conference guests, students, and visitors rose to several thousand per year, and the list of attendees, visiting faculty, and conference participants included a roster of the most prominent thinkers and activists in the country.

The place itself became part of the curriculum in ways we did not anticipate. The land, as Thoreau noted, had its own expectations lurking below all of our confident talk about education and our clumsy efforts to render place into pedagogy. Places have a mind of their own that we aren't privy to. The curriculum of that place came to include particular events, such as a 500-year flood, the hottest and driest summer on record, and the coldest winter ever recorded, along with the mysterious events we sterilize and pigeonhole with academic words like *ecology*, *forestry*, *botany*, *soil science*, and *animal behavior*.

One moonlit night I decided to walk south down the valley toward the Middle Fork, about an hour-long walk. On my return through the tree breaks, the moon rising above the east ridge, I became aware that I was being followed. The safety of home was a long way off. Heart racing, I quickened my pace through a tree break dividing one field from another, went another 20 paces or so, and then turned around. Following me close behind was a lone coyote, perhaps crossed with a bit of red wolf, a formidably large animal. I had no weapon and wasn't nearly fast enough to outrun it. But when I stopped, it stopped and then did not budge. We were eye to eye in the awkward, wordless boundary between species. His intentions were unknown to me, and, I suppose, mine were to him. Not knowing what else to do, I spoke a few words, assuming we ought to talk this out and that language might be an advantage of sorts. The coyote cocked his head to one side, ears perked up. He would occasionally look away and then look back with what I interpreted hopefully as a quizzical but slightly interested look on his face. I was encouraged and greatly relieved. After a few minutes of monologue and perked ears, I decided to sit down; he reciprocated. I took this as a good sign and continued to talk softly, even tried to sing a bit, and from time to time our eyes met and I heard him make something like a low yip, yip that sounded friendly enough. Interspecies communication of sorts. By now the moon was nearly overhead and we were fully visible to each other. After what may have been 5 or 10 minutes, I stood up and he stood as well. I took one slow step forward; he responded by splaying out his feet, ready to bolt. Another step and he bounded off, turned and looked back, and then disappeared into the night. I stood and watched him fade into the trees along the creek and then walked home blessed in some nameless way.

I had ventured into the coyote's world of night foraging and mating, and I think he was simply curious about this lone, misplaced human. I had no weapon and no machine, which made me more approachable, and I think we did communicate in a fashion. Extending a bit further, he was both curious and courteous. And those who do not believe that animals think have never ventured alone and vulnerable into a conversation with one on its terms and in its native habitat. We still regard nature as a mere commodity and animals as abstractions, much as Descartes did. For the rising generation, the experience of nature, in any form, is rare, and it is increasingly alien to the enclosed curriculum of the academy where the matters of greatest consequence have to do with grade point averages, course units, careers, routines, tenure, and *US News & World Report*'s annual ranking. And I think this to be a serious loss to our ability to think and to our humanity.

I had a PhD but had not been educated to think much about *education*, the Latin root for which means to draw forth. Who is qualified, and by what standards, to midwife the birth of personhood in another, or to spark another's mind into the state of awareness, or to properly appraise the results? What does it mean to be educated, and by what standard is that mysterious process appraised? In some circles, great stock is placed in the mastery of routine knowledge, what Brazilian educator Paulo Freiere describes as the banking model of education. Others, deemed more progressive, emphasize the process of learning, which mostly means the cultivation of a kind of disciplined curiosity. Both, however, conceive education, in philosopher Mary Midgley's word, as a form of "anthropolatry," the worship of human accomplishments, history, and mastery over nature. As anthropolatry, the study of nature is mostly intended to fathom how the world works so as to permit a more complete human mastery and a finer level of manipulation extending down into genes and atoms. My experience at Meadowcreek opened the door to the different possibility that education ought, somehow, to be more of a dialogue requiring the capacity to listen in silence to wind, water, animals, the sky, nighttime sounds, and what a Native American once described as earthsong—the sort of things dismissed by anthropolators as romantic nonsense.

Confronted by the mysteries of a place I did not know, and slightly bookish by nature, I turned to all of those writers on education that I had avoided in my earlier years as a college teacher, including John Dewey, Albert Schweitzer, Maria Montessori, J. Glenn Gray, and Alfred North Whitehead. I discovered in their writings a useful criticism of the foundations of contemporary education that emphasizes the importance of place, individual creativity, our implicatedness in the world, reverence. From a variety of sources, we know that the things most deeply embedded in us are formed by the combination of experience and doing with the practice of reflection and articulation. And we know, too, that what Rachel Carson called "the sense of wonder" requires childhood experience in nature and constant practice as well as early validation by adults. The cultivation of the sense of wonder, however, takes us to the edge, where language loses its power to describe and where analysis, the taking apart of things, goes limp before the mystery of Creation, where the only appropriate response is prayerful silence.

Chapter 23

The Problem of Education

(1988)

AFTER DEEP REFLECTION, H. L. Mencken once proposed to improve education by burning down the schools and hanging the professoriate. For better or worse, the suggestion was ignored. Made today, however, it might find a more receptive public, ready to purchase the gasoline and rope. Americans, united on little else, are joined in the belief that the educational system is too expensive, too cumbersome, and not, on the whole, very effective. But reformers are deeply divided on how to improve it. All sides of the debate, nonetheless, agree on the basic aims and purposes of education, which are to equip our nation with a "world-class" labor force, first, to compete more favorably in the global economy and, second, to provide each individual with the means for maximum upward mobility. On purposes of education both higher and lower, would-be reformers seem to be of one mind.

There are, nonetheless, better reasons to rethink education that have to do with the issues of human survival, which will dominate the world of the twenty-first century and beyond. Those now being educated will have to do what we, the present generation, have been unable or unwilling to do: stabilize and then reduce the emission of greenhouse gases, which are changing the climate, perhaps disastrously so; protect biological diversity; reverse the destruction of forests everywhere; conserve soils; and reduce the human footprint to levels consonant with the carrying capacity

This article was originally published in 1988.

of Earth. They must learn how to use energy and materials with great efficiency. They must learn how to utilize solar energy in all of its forms. They must rebuild the economy in order to eliminate waste and pollution. They must learn how to manage renewable resources for the long run. They must begin the great work of repairing, as much as possible, the damage done to the Earth in the past 250 years of industrialization. And they must do all of this while reducing economic inequities.

For the most part, however, we are still educating the young as if there were no planetary emergency. Remove computers and a scattering of courses and programs throughout the catalog, and the present curriculum looks a lot like that of the 1950s. The crisis we face is first and foremost one of mind, perception, and values; hence, it is a challenge to those institutions presuming to shape minds, perceptions, and values, and that makes it an educational challenge. More of the same kind of education can only make things worse. This is not an argument against education but, rather, an argument for the kind of education that prepares people for lives and livelihoods suited to a planet with a biosphere that operates by the laws of ecology and thermodynamics.

The skills, aptitudes, and attitudes necessary to industrialize the Earth, however, are not necessarily the same as those that will be needed to heal the Earth or to build durable economies and good communities. Resolution of the great ecological challenges of the next century will require us to reconsider the substance, process, and purpose of education at all levels and to do so, in the words of Yale University historian Jaroslav Pelikan, "with an intensity and ingenuity matching that shown by previous generations in obeying the command to have dominion over the planet" (Pelikan 1992, 21). But Pelikan himself doubts whether the university "has the capacity to meet a crisis that is not only ecological and technological, but ultimately educational and moral" (Pelikan 1992, 21–22). Why should this be so? Why should those institutions charged with the task of preparing the young for the challenges of life be so slow to recognize and act on the major challenges of the coming century?

A clue can be found in a book by Derek Bok, a former president of Harvard University, who wrote:

> Our universities excel in pursuing the easier opportunities where established academic and social priorities coincide. On the other hand, when social needs are not clearly recognized and backed by adequate financial support, higher education has often failed to respond as effectively as it might, even to some of the most important challenges facing America.

> Armed with the security of tenure and the time to study the world with care, professors would appear to have a unique opportunity to act as society's scouts to signal impending problems. . . . Yet rarely have members of the academy succeeded in discovering emerging issues and bringing them vividly to the attention of the public. What Rachel Carson did for risks to the environment, Ralph Nader for consumer protection, Michael Harrington for problems of poverty, Betty Friedan for women's rights, they did as independent critics, not as members of a faculty. (Bok 1990, 108)

This observation, appearing on page 108 of Bok's book, is not mentioned thereafter. It should have been on page 1 and would have provided the grist for a better book. Had Bok gone further, he might have been led to ask whether the same charge of lethargy might be made against those presuming to lead American education. Bok might then have been led to rethink old and unquestioned assumptions about liberal education. For example, John Henry Newman, in his classic *The Idea of a University*, drew a distinction between practical and liberal learning that has influenced education from his time to our own. Liberal knowledge, according to Newman, "refuses to be informed by any end, or absorbed into any art" (Newman 1982, 81); knowledge is liberal if "nothing accrues of consequence beyond the using" (Newman 1982, 82). Furthermore, Newman stated that "liberal education and liberal pursuits are exercises of mind, of reason, of reflection" (Newman 1982, 80). All else he regarded as practical learning, which Newman believed has no place in the liberal arts. To this day, Newman's distinction between practical and liberal knowledge is seldom transgressed in liberal arts institutions. Is it any wonder that faculty, mindful of the penalties for transgressions, do not often deal boldly with the issues that Bok describes? I do not wish to take faculty off the hook, but I would like to note that educational institutions, more often than not, reward indoor thinking, careerism, and safe conformity to prevailing standards. Educational institutions are not widely known for encouraging boat rockers, and I seriously doubt that Bok's own institution would have awarded tenure to Rachel Carson, Ralph Nader, or Michael Harrington.

Harvard philosopher and mathematician Alfred North Whitehead had a different view of the liberal arts. "The mediocrity of the learned world," he wrote in 1929, could be traced to its "exclusive association of learning with book-learning" (Whitehead 1967, 51). Whitehead went on to say that real education requires "first-hand knowledge," by which he meant an intimate connection between the mind and "material creative activity." Others, such as John Dewey and J. Glenn Gray, reached similar conclusions. "Liberal education," Gray wrote, "is least dependent on formal instruction. It can be pursued in the kitchen, the workshop, on the ranch or farm . . . where we learn wholeness in response to others" (Gray 1984, 81). A genuinely liberal education, in other words, ought to be liberally conducted, aiming to develop the full range of human capacities. And institutions dedicated to the liberal arts ought to be more than simply agglomerations of specializations.

Had Bok proceeded further, he would have had to address the loss of moral vision throughout education at all levels. In ecologist Stan Rowe's words, the university has

> shaped itself to an industrial ideal—the knowledge factory. Now it is overloaded and top-heavy with expertness and information. It has become a know-how institution when it ought to be a know-why institution. Its goal should be deliverance from the crushing weight of unevaluated facts, from bare-bones cognition or ignorant knowledge: knowing in fragments, knowing without direction, knowing without commitment. (Rowe 1990, 129)

Many years ago William James saw this coming and feared that the university might one day develop into a "tyrannical Machine with unforeseen powers of exclusion and corruption" (James 1987, 113). We are moving along that road and should ask why this has come about and what can be done to reverse course.

One source of the corruption is the marriage between the universities and power and commerce. It was a marriage first proposed by Francis Bacon, but not consummated until the later years of the twentieth century. But marriage, implying affection and mutual consent, is perhaps not an accurate metaphor. This is instead a cash relationship, which began with a defense contract here and a research project there. At present more than a few university departments still work as adjuncts of the Pentagon and even more as adjuncts of industry in the hope of reaping billions of dollars in fields such as genetic engineering, nanotechnologies, agribusiness, and computer science. Even where this is not true, it is difficult to escape the conclusion that much of what passes for research, as historian Page Smith wrote, is "essentially worthless . . . busywork on a vast almost incomprehensible scale" (Smith 1990, 7).

Behind the glossy facade of the modern academy there is often a vacuum of purpose waiting to be filled by whomever and whatever. For example, the college of agriculture at a nearby land-grant university of note claims to be helping "position farmers for the future." But when asked what farming would be like in the twenty-first century, the dean of the college replied by saying, "I don't know." When asked, "How can you [then] position yourself for it?" the Dean replied, "We have to try as best we can to plan ahead" (Logsdon 1994, 74). This reminds me of the old joke in which the airline pilot reports to the passengers that he has good news and bad news. The good news is that the flight is ahead of schedule. The bad news is that we're lost. And in a time of eroding soils and declining rural communities, "turf grass management" is the hot new item at the college of agriculture.

Finally, had Bok so chosen, he would have been led to question how we define intelligence and what that might imply for our larger prospects. At the heart of our pedagogy and curriculum, one finds cleverness confused with intelligence. Cleverness, as I understand it, tends to fragment things and to focus on the short run. The epitome of cleverness is the specialist whose intellect and person have been shaped by the overdevelopment of one intellectual capacity, what Nietzsche once called an "inverted cripple." Ecological intelligence, on the other hand, requires a broader view of the world and a long-term perspective. Cleverness can be adequately computed by the Scholastic Aptitude Test and the Graduate Record Examination, but intelligence is not so easily measured. In time I think we will come to see that true intelligence tends to be integrative and often works slowly while one mulls things over.

The modern fetish with smartness is no accident. The highly specialized, narrowly focused intellect fits the demands of instrumental rationality built into the industrial economy, and for reasons described by Brooks Adams many years ago, "capital has preferred the specialized mind and that not of the highest quality, since it has found it profitable to set quantity before quality to the limit the market will endure. Capitalists have never insisted upon raising an educational standard, save in science and mechanics, and the relative overstimulation of the scientific mind has now become an actual menace to order." (Smith 1984, 116) The demands of building good communities within a sustainable society will require more than the specialized, one-dimensional mind and more than instrumental cleverness.

Looking ahead to the twenty-first century, education must be guided by more comprehensive and ecologically solvent standards for truth. The architects of the modern worldview assumed that those things that could be weighed, measured, and counted were more true than those that could not be quantified. If it could not be counted, in other words, it did not count. Cartesian philosophy was full of potential ecological mischief, a potential that has become reality. Descartes' philosophy separated man from nature, stripped all intrinsic value from nature, and then proceeded to divide mind and body. Descartes was, at heart, an engineer, and his legacy to the environment of our time is the cold passion to remake the world as if we were merely remodeling a machine. Feelings and intuition were tossed out, as were those fuzzy, qualitative parts of reality, such as esthetic appreciation, loyalty, friendship, sentiment, empathy, and charity. Descartes' assumptions were neither as simple nor as inconsequential as they might have appeared in his lifetime (1596–1650).

If sustainability is our aim, we will need a broader conception of science and a more inclusive rationality that joins empirical knowledge with the same emotions that make us love and sometimes fight. Philosopher Karl Polanyi (1958) described this as "personal knowledge," by which he meant knowledge that calls forth a wider range of human perceptions, feelings, and intellectual powers than those presumed to be narrowly "objective." Personal knowledge, according to Polanyi,

> is not made but discovered. . . . It commits us, passionately and far beyond our comprehension, to a vision of reality. Of this responsibility we cannot divest ourselves by setting up objective criteria of verifiability—or falsifiability, or testability. . . . For we live in it as in the garment of our own skin. Like love, to which it is akin, this commitment is a "shirt of flame", blazing with passion and, also like love, consumed by devotion to a universal demand. Such is the true sense of objectivity in science. (Polanyi 1958, 64)

Cartesian science rejects emotion but cannot escape it. Emotion and passion are embedded in all knowledge, including the most ascetic scientific knowledge driven by the passion for objectivity. Descartes had it wrong. There is no way to separate feeling from knowledge. There is no way to separate object from subject. There is no good way and no good reason to separate mind or body from its ecological and emotional context. And some persons, with good evidence, are coming to suspect that intelligence is not a human monopoly at all but woven throughout the animal world and perhaps beyond. Science without emotional valence can give us no reason to appreciate the sunset, nor can it give us any purely objective reason to value life. These must come from deeper sources.

As a result of unquestioned assumptions that human domination of nature is good; that the growth of economy is natural; that all knowledge, regardless of its consequences, is equally valuable; and that material progress is our right, we suffer a kind of cultural immune deficiency anemia that renders us unable to resist the seductions of technology, convenience, and short-term gain.

The modern curriculum teaches little about citizenship and responsibilities and a great deal about individualism and rights. The ecological emergency, however, can be resolved only if enough people come to hold a larger idea of what it means to be a citizen. But a pervasive cynicism about our higher potentials and collective possibilities works against us. Even my most idealistic students, for example, often confuse self-interest with selfishness, which makes it possible to equate Mother Teresa and Donald Trump, each merely doing "their thing." This is not just a social and political problem. The ecological emergency is about the failure to comprehend our citizenship in the biotic community. From the modern perspective we cannot see clearly how utterly dependent we are on the "services of nature" and on the wider community of life. Our political language gives little hint of this dependence. As it is now used, the word *patriotism*, for example, is devoid of ecological content. But logically it should include any and all threats to our land, forests, air, water, wildlife, and health, including those from within. To abuse natural "resources," to erode soils, to destroy natural diversity, to waste, to take more than one's fair share, to fail to replenish what has been used someday must be regarded as equivalent to an attack on the country from without. And "politics" once again must come to mean, in Vaclav Havel's words, "serving the community and serving those who will come after us" (Havel 1992, 6).

There is a widespread, and mostly unquestioned, assumption that our future is one of constantly evolving technology and that this is always and everywhere a good thing. Those who question this faith are dismissed as Luddites by people who, as far as I can tell, know little or nothing about the real history of Luddism. Faith in technology is built into nearly every part of the curriculum. When pressed, however, true believers describe progress to mean, not human, political, or cultural improvement, but a kind of technological juggernaut. Technological fundamentalism, like all fundamentalisms, deserves to be challenged. Is technological change taking us where we want to go? What effect does it have on our imagination and particularly on our social, political, and moral imagination? What effect does it have on our ecological prospects? George Orwell once warned that the "logical end" of technological progress "is to reduce the human being to something resembling a brain in a bottle" (Orwell 1958,

201). Decades later some propose to develop the necessary technology to "download" the contents of the brain into a machine/body (Moravic 1988). Orwell's nightmare is coming true and in no small part because of research conducted in our most prestigious universities. Such research stands in sharp contrast to our real needs. We need decent communities, good work to do, loving relationships, stable families, the knowledge necessary to restore what we have damaged, and ways to transcend our inherent self-centeredness. Our needs, in short, are those of the spirit; yet, our imagination and creativity are overwhelmingly aimed at things that as often as not degrade spirit, nature, and true economy.

Ecological education, in Leopold's words, is directed toward changing our "intellectual emphasis, loyalties, affections, and convictions" (Leopold 1966, 246). It requires breaking free of old pedagogical assumptions, of the straitjacket of discipline-centric curriculum, and even of confinement in classrooms and school buildings. Ecological education means changing: the substance and process of education contained in curriculum, how educational institutions actually work, the physical architecture of schools and colleges, and most important, the purposes of learning.

Chapter 24

What Is Education For?

(1990)

AUTHOR'S NOTE 2010: *Delivered as a commencement address at Arkansas College—now Lyon College—in May 1990. The numbers are dated but still roughly accurate, and the point of the essay is still valid.*

IF TODAY IS A TYPICAL DAY on planet Earth, we will lose 116 square miles of rain forest, or about an acre a second. We will lose another 72 square miles to encroaching deserts, the results of human mismanagement and overpopulation. We will lose 40 to 250 species, but no one knows the actual number. Today the human population will increase by 250,000. And today we will add 2700 tons of chlorofluorocarbons and 15 million tons of carbon dioxide to the atmosphere. Tonight the Earth will be a little hotter, its waters more acidic, and the fabric of life more threadbare. By year's end the numbers are staggering: The total loss of rain forest will equal an area the size of the state of Washington; expanding deserts will equal an area the size of the state of West Virginia; and the global population will have risen by more than 70 million. By the year 2000 a sizeable fraction of the life-forms extant on the planet in the year 1900 will be extinct or in jeopardy.

The truth is that many things on which our future health and prosperity depend are in dire jeopardy: climate stability, the resilience and productivity of natural systems, the beauty of the natural world, and biological diversity. It is worth noting that this is not the work of ignorant

This article was originally published in 1990.

people. Rather, it results from the work by people with BAs, BSs, LLBs, MBAs, and PhDs. Elie Wiesel once made the same point, noting that the designers and perpetrators of Auschwitz, Dachau, and Buchenwald—the Holocaust—were the heirs of Kant and Goethe, widely thought to be the best educated people on Earth. But their education did not serve as an adequate barrier to barbarity. What was wrong with their education? In Wiesel's (1990) words, "it emphasized theories instead of values, concepts rather than human beings, abstraction rather than consciousness, answers instead of questions, ideology and efficiency rather than conscience." I believe that the same could be said of our education. Toward the natural world it too emphasizes theories, not values; abstraction rather than consciousness; neat answers instead of questions; and technical efficiency over conscience. It is a matter of no small consequence that the only people who have lived sustainably on the planet for any length of time could not read or, like the Amish, do not make a fetish of reading. My point is simply that education is no guarantee of decency, prudence, or wisdom. More of the same kind of education will only compound our problems. This is not an argument for ignorance but rather a statement that the worth of education must now be measured against the standards of decency and human survival—the issues now looming so large before us in the twenty-first century. It is not education but education of a certain kind that will save us.

What went wrong with contemporary culture and education? We can find insight in literature, including Christopher Marlowe's portrayal of Faust who trades his soul for knowledge and power, Mary Shelley's Dr. Frankenstein who refuses to take responsibility for his creation, and Herman Melville's Captain Ahab who says, "All my means are sane, my motive and my object mad." In these characters we encounter the essence of the modern drive to dominate nature.

Historically, Francis Bacon's proposed union between knowledge and power foreshadowed the contemporary alliance between government, business, and knowledge that has wrought so much mischief. Galileo's separation of the intellect foreshadowed the dominance of the analytical mind over that part given to creativity, humor, and wholeness. And in Cartesian epistemology, one finds the roots of the radical separation of self and object. Together these three laid the foundations for modern education, foundations that now are enshrined in myths that we have come to accept without question. Let me suggest six.

First, there is the myth that ignorance is a solvable problem. Ignorance is not a solvable problem; it is rather an inescapable part of the human condition. We cannot comprehend the world in its entirety. The advance of knowledge always carried with it the advance of some form of ignorance. For example, in 1929 the knowledge of what a substance like chlorofluorocarbons (CFCs) would do to the stratospheric ozone and climate stability was a piece of trivial ignorance as the compound had not yet been invented. But in 1930 after Thomas Midgely Jr. discovered CFCs, what had been a piece of trivial ignorance became a critical life-threatening gap in human understanding of the biosphere. Not until the early 1970s did anyone think to ask "What does this substance do to what?" In 1986 we discovered that CFCs had created a hole in the ozone over the South Pole the size of the lower 48 U.S. states; by the early 1990s, CFCs had created a worldwide reduction of ozone. With the discovery of CFCs, knowledge increased, but like the circumference of an expanding circle, ignorance grew as well.

A second myth is that with enough knowledge and technology, we can, in the words of *Scientific American* (1989), "manage planet earth." Higher education has largely been shaped by the drive to extend human domination to its fullest. In this mission, human intelligence may have taken the wrong road. Nonetheless, managing the planet has a nice ring to it. It appeals to our fascination with digital readouts, computers, buttons, and dials. But the complexity of Earth and its life systems can never be safely managed. The ecology of the top inch of topsoil is still largely unknown, as is its relationship to the larger systems of the biosphere. What might be managed, however, is us: human desires, economies, politics, and consumption. But our attention is caught by those things that avoid the hard choices implied by politics, morality, ethics, and common sense. It makes far better sense to reshape ourselves to fit a finite planet than to attempt to reshape the planet to fit our infinite wants.

A third myth is that knowledge, and by implication human goodness, is increasing. An information explosion, by which I mean a rapid increase of data, words, and paper, is taking place. But this explosion should not be mistaken for an increase in knowledge and wisdom, which cannot be measured so easily. What can be said truthfully is that some knowledge is increasing while other kinds of knowledge are being lost. For example, David Ehrenfeld has pointed out that biology departments no longer hire faculty in such areas as systematics, taxonomy, or ornithology (Ehrenfeld personal communication). In other words, important knowledge is being lost because of the recent overemphasis on molecular biology and genetic engineering, which are more lucrative but not necessarily more important areas of inquiry. Despite all of our advances in some areas, we still do not have anything like the science of land health that Aldo Leopold called for a half century ago.

It is not just knowledge in certain areas that we are losing but also vernacular knowledge, by which I mean the knowledge that people have of their places. According to Barry Lopez,

> it is the chilling nature of modern society to find an ignorance of geography, local or national, as excusable as an ignorance of hand tools; and to find the commitment of people to their home places only momentarily entertaining, and finally naive. [I am] forced to the realization that something strange, if not dangerous, is afoot. Year by year the number of people with firsthand experience in the land dwindles. Rural populations continue to shift to the cities. . . . In the wake of this loss of personal and local knowledge, the knowledge from which a real geography is derived, the knowledge on which a country must ultimately stand, has come something hard to define but I think sinister and unsettling. (Lopez 1989a, 55)

The modern university does not consider this kind of knowledge worth knowing except to record it as an oddity, "folk culture." Instead, it conceives its mission as that of adding to what is called "the fund of human knowledge" through research. What can be said of research? Historian Page Smith gives one answer:

> The vast majority of so-called research turned out in the modern university is essentially worthless. It does not result in any measurable benefit to anything or anybody. It does not push back those omnipresent "frontiers of knowledge" so confidently evoked; it does not in the main result in greater health or happiness among the general populace or any particular segment of it. It is busywork on a vast, almost incomprehensible scale. It is dispiriting; it depresses the whole scholarly enterprise; and most important of all, it deprives the student of what he or she deserves—the thoughtful and considerate attention of a teacher deeply and unequivocally committed to teaching. (Smith 1990, 7)

In the confusion of data with knowledge is a deeper mistake that learning will make us better people. But learning, as Loren Eiseley once said, is endless and "in itself . . . will never make us ethical men" (Eiseley 1979,

284). Ultimately, it may be the knowledge of the good that is most threatened by all of our other advances. All things considered, it is possible that we are becoming more ignorant of the things we must know to live well and sustainably on the Earth.

In thinking about the kinds of knowledge and the kinds of research that we will need to build a sustainable society, we need to make a distinction between intelligence and cleverness. True intelligence is long-range and aims toward wholeness. Cleverness is mostly short-range and tends to break reality into bits and pieces. Cleverness is personified by the functionally rational technician armed with know-how and methods but without a clue about the higher ends technique should serve. The goal of education should be to connect intelligence with an emphasis on whole systems and the long range with cleverness, which involves being smart about details.

A fourth myth of higher education is that we can adequately restore that which we have dismantled. I am referring to the modern curriculum. We have fragmented the world into bits and pieces called disciplines and subdisciplines, hermetically sealed from other such disciplines. As a result, after 12 or 16 or 20 years of education, most students graduate without any broad, integrated sense of the unity of things. The consequences for their personhood and for the planet are large. For example, we routinely produce economists who lack the most rudimentary understanding of ecology or thermodynamics. This explains why our national accounting systems still do not subtract the costs of biotic impoverishment, soil erosion, poisons in our air and water, and resource depletion from gross national product. We add the price of the sale of a bushel of wheat to the gross national product while forgetting to subtract the three bushels of topsoil lost to grow it. As a result of incomplete education, we have fooled ourselves into thinking that we are much richer than we are. The same point could be made about other disciplines and subdisciplines that have become hermetically sealed from life itself.

Fifth, there is a myth that the purpose of education is to give students the means for upward mobility and success. Thomas Merton once identified this as the "mass production of people literally unfit for anything except to take part in an elaborate and completely artificial charade" (Merton 1985, 11). When asked to write about his own success, Merton responded by saying that "if it so happened that I had once written a best seller, this was a pure accident, due to inattention and naïveté, and I would take very good care never to do the same again" (Merton 1985, 11).

His advice to students was to "be anything you like, be madmen, drunks, and bastards of every shape and form, but at all costs avoid one thing: success."

The plain fact is that the planet does not need more successful people. But it does desperately need more peacemakers, healers, restorers, storytellers, and lovers of every kind. It needs people who live well in their places. It needs people of moral courage willing to join the fight to make the world habitable and humane. And these qualities have little to do with success as our culture defines it.

Finally, there is a myth that America represents the pinnacle of human achievement. This, of course, is arrogance of the worst sort and a gross misreading of history and anthropology. Recently, this view has taken the form that we won the Cold War. Communism failed because it produced too little at too high a cost. But capitalism has also failed because it produces too much, shares too little, also at too high a cost to our children and grandchildren. Communism failed as an ascetic morality. Capitalism has failed because it destroys morality altogether. This is not the happy world that any number of feckless advertisers and politicians describe. We have built a world of sybaritic wealth for a few and Calcuttan poverty for a growing underclass. At its worst, it is a world of crack on the streets, insensate violence, anomie, and the most desperate kind of poverty. The fact is that we live in a disintegrating culture. Ron Miller stated it this way:

> Our culture does not nourish that which is best or noblest in the human spirit. It does not cultivate vision, imagination, or aesthetic or spiritual sensitivity. It does not encourage gentleness, generosity, caring, or compassion. Increasingly in the late twentieth century, the economic-technocratic-statist worldview has become a monstrous destroyer of what is loving and life-affirming in the human soul. (Miller 1989, 2)

Rethinking Education

Measured against the agenda of human survival, how might we rethink education? Let me suggest six principles. First, all education is environmental education. By what is included or excluded, students are taught that they are part of or apart from the natural world. To teach economics, for example, without reference to the laws of thermodynamics or ecology is to teach a fundamentally important ecological lesson: that physics and ecology have nothing to do with the economy. It just happens to be dead wrong. The same is true throughout the curriculum.

A second principle comes from the Greek concept of Paideia. The goal of education is not mastery of subject matter but mastery of one's person. Subject matter is simply the tool. Much as one would use a hammer and a chisel to carve a block of marble, one uses ideas, knowledge, and experience to forge one's own personhood. For the most part we labor under a confusion of ends and means, thinking that the goal of education is to stuff all kinds of facts, techniques, methods, and information into the student's mind, regardless of how and with what effect it will be used. The Greeks knew better.

Third, I propose that knowledge carries with it the responsibility to see that it is well used in the world. The results of a great deal of contemporary research bear resemblance to those foreshadowed by Mary Shelley: monsters of technology and its by-products for which no one takes responsibility or is even expected to take responsibility. Whose responsibility is Love Canal? Chernobyl? Ozone depletion? The *Exxon Valdez* oil spill? Climate destabilization? Each of these tragedies was possible because of knowledge created for which no one was ultimately responsible. This may finally come to be seen for what I think it is: a problem of scale. Knowledge of how to do vast and risky things has far outrun our ability to use it responsibly. Some of this knowledge cannot be used responsibly, safely, and to consistently good purposes.

Fourth, we cannot say that we know something until we understand the effects of this knowledge on real people and their communities. I grew up near Youngstown, Ohio, which was largely destroyed by corporate decisions to "disinvest" in the economy of the region. In this case MBA graduates, educated in the tools of leveraged buyouts, tax breaks, and capital mobility, have done what no invading army could do: They destroyed an American city with total impunity and did so on behalf of an ideology called the "bottom line." But the bottom line for society includes other costs: those of unemployment, crime, higher divorce rates, alcoholism, child abuse, lost savings, and wrecked lives. In this instance what was taught in the business schools and economics departments did not include the value of good communities or the human costs of a narrow, destructive economic rationality that valued efficiency and economic abstractions above people and community.

My fifth principle follows and is drawn from William Blake. It has to do with the importance of "minute particulars" and the power of examples over words. Students hear about global responsibility while being educated in institutions that often spend their budgets and invest their endowments in the most irresponsible things. The lessons being taught are those of hypocrisy and ultimately despair. Students learn, without anyone ever telling them, that they are helpless to overcome the frightening gap between ideals and reality. What is desperately needed are (1) faculty and administrators who provide role models of integrity, care, and thoughtfulness and (2) institutions capable of embodying ideals wholly and completely in all of their operations.

Finally, I propose that the way in which learning occurs is as important as the content of particular courses. Process is important for learning. Courses taught as lecture courses tend to induce passivity. Indoor classes create the illusion that learning only occurs inside four walls, isolated from what students call, without apparent irony, the "real world." Dissecting frogs in biology classes teaches lessons about nature that no one in polite company would profess. Campus architecture is crystallized pedagogy that often reinforces passivity, monologue, domination, and artificiality. My point is simply that students are being taught in various and subtle ways beyond the overt content of courses.

Reconstruction

What can be done? Lots of things, beginning with the goal that no student should graduate from any educational institution without a basic comprehension of things like the following:

- the laws of thermodynamics
- the basic principles of ecology
- carrying capacity
- energetics
- least-cost, end-use analysis
- limits of technology
- appropriate scale
- sustainable agriculture and forestry
- steady state economics
- environmental ethics

I would add to this list of analytical and academic things, practical things necessary to the art of living well in a place: growing food; building shelter; using solar energy; and a knowledge of local soils, flora, fauna, and the local watershed. Collectively, these are the foundation for the capacity to distinguish between health and disease, development and growth, sufficient and efficient, optimum and maximum, and "should do" from "can do." In Aldo Leopold's words, does the graduate know that "he is only a cog in an ecological mechanism? That if he will work with that mechanism his mental wealth and his material wealth can expand indefinitely? But that if he refuses to work with it, it will ultimately grind him to dust"? And Leopold asked, "If education does not teach us these things, then what is education for?" (Leopold 1953, 64).

Chapter 25

Some Thoughts on Intelligence

(1992)

Ours is about the most ignorant age that can be imagined.
Erwin Chargaff

No other society, to my knowledge, has made such a fetish of intelligence as has modern America. Indeed we have what philosopher Mary Midgley calls a veritable "cult of intelligence" administered by tribes of experts whose function is to measure it, raise it, write books about it, and make those purportedly without it feel inferior. But exactly what is intelligence? More to the point, what is intelligence as measured against the standards of biological diversity and human longevity on Earth? And what might such answers say about how we go about the work of conservation education?

I have no credentials to raise such questions. I could not distinguish the Minnesota Multiphasic Personality Inventory from multiple phases of anything in Minnesota. Nonetheless, I believe that, in the main, the evidence now indicates that we do not know very much about intelligence and that from the perspective of the Earth, much of what we presume to know may be wrong, which is to say that it is not intelligent enough. I am also inclined to agree that the modern stereotype of an intelligent person is wrong. As Wendell Berry puts it, "the prototypical modern intelligence

This article was originally published in 1992.

seems to be that of the Whiz Kid—a human shape barely discernable in [a] fluff of facts" (Berry 1983, 77). What we call intelligence and what we test for and reward in schools and colleges is something else, more akin to cleverness. True intelligence, as I understand it, has to do with the long run and is mostly integrative, whereas cleverness is mostly preoccupied with the short run and tends to fragment things. This distinction has serious consequences for our willingness and ability to conserve biological diversity.

Although I do not think it is possible to give an adequate definition of intelligence, I believe it is possible to describe certain characteristics of it. First, people acting or thinking with intelligence are good at separating cause from effect. Geographer I. G. Simmons, for example, tells the story of an eighteenth-century protopsychiatrist who developed an infallible method of distinguishing the sane from the insane. Those to be diagnosed were locked in a room with water taps on one side and a supply of mops and buckets on the other. He then turned on the water and watched: Those he considered mad ran for the mops and buckets; the sane walked over and turned off the taps (Simmons 1989, 334). I keep a file of "mop and bucket" proposals by persons well paid and honored for their reputed intelligence, the contents of which range from the ridiculous through the absurd to the potentially criminal. The common characteristic of these is the recurrent inability to ask questions having to do with big causes and large consequences.

A second and related characteristic of intelligence is the ability to separate know-how from know-why. Atomic physicist Enrico Fermi is quoted as saying to a skeptic before the first atomic test, "Don't bother me with your conscientious scruples; after all, the thing's superb physics!" (Rowe 1990, 129). I also understand that some believed that the detonation of an atomic bomb might set off a chain reaction that would destroy the entire planet. It was done nonetheless. However superb the physics, the human and ecological consequences of the bomb have been disastrous, and we have not seen the last of it. The bomb is only one illustration of the kind of obsession that feeds on cleverness without internal or external restraint. We are capable of doing many more clever things than true intelligence would have us do. But obsession to do whatever is possible regardless of whether it is desirable is no longer unusual in the modern university, which has become, in Rowe's words, a "know how" rather than a "know why" kind of institution. Its stock in trade is "ignorant knowledge," by which Rowe means "knowing in fragments, knowing without direction, knowing without commitment" (Rowe 1990, 129). Its graduates "score high on means tests but low on ends tests." And the means are derived from paradigms that are "dead wrong . . . life-denying, fostering sickness instead of health" (Rowe 1990). Intelligent people would reverse the order, first asking, "Why?" and "For what reason?"

Real intelligence often works slowly and is close to or even synonymous with what we call wisdom. Intelligent people, according to Mary Midgley, excel in a different range of faculties: "They possess strong imaginative sensibility—the power to envisage possible goods that the world does not yet have and to see what is wrong with the world as it is. They are good at priorities, at comparing various goods, at asking what matters most. They have a sense of proportion, and a nose for the right directions" (Midgley 1990, 41). People possessed of such faculties Midgley calls "wise or sensible rather than clever or smart." They know when enough is enough.

Wendell Berry suggests a third characteristic of intelligence: the "good order or harmoniousness of his or her surroundings." By this standard "any statistical justification of ugliness or violence is a revelation of stupidity" (Berry 1983, 77). This means that the consequences of one's actions are a measure of intelligence and the plea of ignorance is no good defense. Because some consequences cannot be predicted, the exercise of intelligence requires forbearance and a sense of limits. In other words it does not presume to act beyond a certain scale on which effects can be determined and unpredictable consequences would not be catastrophic. In Berry's words, intelligent people (and civilizations) do not assume "that we can first set demons at large, and then, somehow become smart enough to control them" (Berry 1983, 65). If there is such a thing as a societal IQ, what we call "developed" societies would be judged retarded by Berry's standard. Overflowing landfills, climate destabilization, befouled skies, eroded soils, polluted rivers, acid rain, and radioactive wastes suggest ample attainments for early admission into some intergalactic school for learning-disabled species.

A fourth characteristic of intelligent action and thought is that it does not violate the bounds of morality. In other words, it does not, in the name of some alleged higher good, demand the violation of life, community, or decency. Intelligent behavior is consonant with moderation, loyalty, justice, compassion, and truthfulness, not for ethereal theological reasons but because these are fundamental to living well. Morality is long-term practicality that recognizes our limits, fallibility, and ignorance. On the other hand, the cultivation of vice leads, as E. F. Schumacher once wrote, to "a collapse of intelligence." Driven by vice a person "loses the power of seeing things as they really are . . . in their roundness and wholeness" (Schumacher 1973, 29). Intellect driven by vice cannot lead to intelligent action or thought. Said differently, real intelligence depends upon character as much as it does on mental horsepower. The corruption of character, as Emerson wrote in his essay "Nature" in 1836, leads in turn to "the corruption of language . . . new imagery ceases to be created, and old words are perverted to stand for things which are not. . . . In due time the fraud is manifest, and words lose all power to stimulate the understanding or the affections" (Emerson 1972, 32–33).

From these characteristics, I conclude that it is possible for a person to be clever without being very intelligent, or as Walker Percy put it, to "get all A's and still flunk life." Furthermore, whole civilizations can be simultaneously clever and stupid, by which I mean that they might be able to perform amazing technological feats while being unable to solve their most basic public problems. Perhaps these go together. As exhibit A, consider our phenomenal and growing information technology side by side with our decaying inner cities, insensate violence, various addictions, rising public debt, and the destruction of nature all around us. Can it be that we are in fact becoming both more clever and less intelligent? If so, why?

I am tempted to round up all of the usual members of the rogue's gallery, from Descartes ("I think therefore I am") through all of the peddlers of instrumental rationality, artificial intelligence, and unfettered curiosity, all of whom are eminently blameworthy. But they are only partial manifestations of a deeper cause having to do with the very origins of intelligence.

Could it be that what Aldo Leopold referred to as the "integrity, stability, and beauty" of nature is the wellspring of human intelligence? Could it be that the conquest of nature, however clever, is in fact a war against the source of mind? Could it be that the systematic homogenization of nature inherent in contemporary technology and economics is undermining human intelligence? If so, biological diversity is important to us, not only as a source of wonder drugs and miracle fruits, but as the source of what made us human. We have good reason to believe that human intelligence could not have evolved on the moon—a landscape devoid of biological diversity. We also have good reason to believe that the sense of awe toward the creation had a great deal to do with the origin of language and why early humans wanted to talk, sing, and write poetry in the first place. Elemental things like flowing water, wind, trees, clouds, rain, mist, mountains, landscape, animal behavior, changing seasons, the night sky, and the mysteries of the life cycle gave birth to thought and language. For this reason I think it not possible to unravel the Creation without undermining human intelligence as well. The issue is not so much about what nature can do for us as resources as it is about the survival of human intelligence cut off from its source.

Cleverness would have us advance a narrowly defined, short-term, and anemic self-interest at all costs and at all risks. But cleverness, pure intellect, is just not intelligent enough. Its final destination is madness. True intelligence would lead us, on the contrary, to stabilize the climate and protect the web of life, but for reasons that go beyond the calculation of self-interest. The surest sign of the maturity of intelligence is the evolution of biocentric wisdom, by which I mean the capacity to nurture and shelter life—a fitting standard for a species calling itself *Homo sapiens*.

What can educators do to foster real intelligence? One view is that we should not try, because the best we can do is to help students avoid being stupid (Postman 1988, 87). I think we should prevent stupidity where possible, but I also think we can do more. First, we can question the standard model of pre-ecological intelligence and encourage students to think the matter out for themselves, including the matter of collective intelligence. Second, we can reward intelligence in all sorts of ways without necessarily penalizing cleverness. Third, we can develop the kind of firsthand knowledge of nature from which real intelligence grows. This means breaking down walls made by clocks, bells, rules, academic requirements, and a tired, indoor pedagogy. I am proposing a jailbreak that would put young people outdoors more often. "No child left inside," as Richard Louv puts it. Fourth, we can liberalize the liberal arts to include ecological competence in areas of restoration ecology, agriculture, forestry, ecological engineering, landscape design, and solar technology. Fifth, we can suspend the implicit belief that a PhD is a sign of intelligence and draw those who have demonstrated a high degree of applied ecological intelligence, courage, and creativity (farmers, foresters, naturalists, ranchers, restoration ecologists, urban ecologists, landscape planners, citizen activists) into education as mentors and role models. Finally, we can attempt to teach the things that one might imagine the Earth would teach us: silence, humility, holiness, connectedness, courtesy, beauty, celebration, giving, restoration, obligation, and wildness.

∴ Chapter 26 ∴

Ecological Literacy

(1992)

LITERACY IS THE ABILITY to read. Numeracy is the ability to count. Ecological literacy, according to Garrett Hardin, is the ability to ask "What then?" Considerable attention is properly being given to our shortcomings in teaching the young to read, count, and compute, but not nearly enough is being given to ecological literacy. Reading, after all, is an ancient skill. And for most of the twentieth century we have been busy adding, subtracting, multiplying, dividing, and now computing. But "What then?" questions have not come easy for us despite all of our formidable advances in other areas. Napoleon did not ask the question until he had reached Moscow, by which time no one could give any good answer except "Let's get outta here." If Custer asked the question, we have no record of it. His last known words at Little Big Horn were "Hurrah, boys, now we have them." And economists, who are certainly both numerate and numerous, have not asked the question often enough. Asking "What then?" on the west side of the Niemen River, or at Fort Laramie, would have saved a lot of trouble. For the same reason, "What then?" is also an appropriate question to ask before the last rain forests disappear, before the growth economy consumes itself into oblivion, and before we have warmed the planet too much.

The failure to develop ecological literacy is a sin of omission and of commission. Not only are we failing to teach the basics about the Earth, and how it works, but we are in fact teaching a large amount of stuff that is simply wrong. By failing to include ecological perspectives in any number of subjects, we are teaching students that ecology is unimportant

---

This article was originally published in 1992.

to history, politics, economics, society, and so forth. From television they learn that the Earth is theirs for the taking. The result is a generation of ecological yahoos without a clue about why the color of the water in their rivers is related to their food supply, or why storms are becoming more severe as the climate is unbalanced. The same persons, as adults, will create businesses, vote, have families, and above all, consume. If they come to reflect on the discrepancy between the splendor of their private lives and the realities of life in a hotter, more toxic and violent world, as ecological illiterates they will have roughly the same success as one trying to balance a checkbook without knowing arithmetic.

To become ecologically literate, one must certainly be able to read and, I think, even like to read. Ecological literacy also presumes an ability to use numbers and the ability to know what is countable and what is not, which is to say the limits of numbers. But these are indoor skills. Ecological literacy also requires the more demanding capacity to observe nature with insight, a merger of landscape and mindscape. "The interior landscape," in Barry Lopez's words, "responds to the character and subtlety of an exterior landscape; the shape of the individual mind is affected by land as it is by genes" (Lopez 1989b, 65). The quality of thought is related to the ability to relate to "where on this Earth one goes, what one touches, the patterns one observes in nature—the intricate history of one's life in the land, even a life in the city, where wind, the chirp of birds, the line of a falling leaf, are known" (Lopez 1989b, 65). The fact that this kind of intimate knowledge of our landscapes is rapidly disappearing can only impoverish our mental landscapes as well. People who do not know the ground on which they stand have no way to understand the difference between health and disease in the nature around them and its relation to their own health.

If literacy is driven by the search for knowledge, ecological literacy is driven by the sense of wonder, the sheer delight in being alive in a beautiful, mysterious, bountiful world. The darkness and disorder that we have brought to that world give ecological literacy an urgency it lacked a century ago. We can now look over the abyss and see the end of it all. Ecological literacy begins in childhood. "To keep alive his inborn sense of wonder," a child, in Rachel Carson's words, "needs the companionship of at least one adult who can share it, rediscovering with him the joy, excitement and mystery of the world we live in" (Carson 1984, 45). The sense of wonder is rooted in the emotions or what E. O. Wilson has called "biophilia," which is simply the affinity for the living world. The nourishment of that affinity is the beginning point for the sense of kinship with life, without which literacy of any sort will not help much. This is to say that even a thorough knowledge of the facts of life and of the threats to it will not save us in the absence of the feeling of kinship with life of the sort that cannot entirely be put into words.

There are, I think, several reasons why ecological literacy has been so difficult for Western culture. First, it implies the ability to think broadly, to know something of what is hitched to what. This ability is being lost in an age of specialization. Scientists of the quality of Rachel Carson or Aldo Leopold are rarities who must buck the pressures toward narrowness and also endure a great deal of professional rejection and hostility. By inquiring into the relationship between chlorinated hydrocarbon pesticides and bird populations, Rachel Carson was asking an ecolate question. Many others failed to ask, not because they did not like birds, but because they had not, for whatever reasons, thought beyond the conventional categories. To do so would have required that they relate their food system to the decline in the number of birds in their neighborhood. This means that they would have had some direct knowledge of farms and farming practices and were also paying attention to birds in the neighborhood. To think in ecolate fashion presumes a breadth of experience with healthy natural systems, both of which are increasingly rare. It also presumes that the persons be willing and able to "think at right angles" to their particular specializations, as Leopold put it.

Ecological literacy is difficult, second, because we have come to believe that education is solely an indoor activity. A good part of it, of necessity, must be, but there is a price. William Morton Wheeler once compared the naturalist with the professional biologist in these words: "[The naturalist] is primarily an observer and fond of outdoor life, a collector, a classifier, a describer, deeply impressed by the overwhelming intricacy of natural phenomena and reveling in their very complexity." The biologist, on the other hand, "is oriented toward and dominated by ideas, and rather terrified or oppressed by the intricate hurly-burly of concrete, sensuous reality . . . he is a denizen of the laboratory. His besetting sin is oversimplification and the tendency to undue isolation of the organisms he studies from their natural environment" (Wheeler 1962). Since Wheeler wrote, ecology has become increasingly specialized and, one suspects, remote from its subject matter. Ecology, like most learning worthy of the effort, is an applied subject. Its goal is not just a comprehension of how the world works but, in the light of that knowledge, a life lived accordingly. The same is true of theology, sociology, political science, and most other subjects that grace the conventional curriculum.

The decline in the capacity for esthetic appreciation is a third factor working against ecological literacy. We have become comfortable with all kinds of ugliness and seem incapable of effective protest against its purveyors: urban developers, businessmen, government officials, television executives, timber and mining companies, utilities, and advertisers. But ugliness is not just an esthetic problem; it signals a more fundamental disharmony between people and between people and the land. Ugliness is, I think, the surest sign of disease, or what is now being called "unsustainability." Show me the hamburger stands, neon ticky-tacky strips leading toward every city in America, and the shopping malls, and I'll show you devastated rain forests, a decaying countryside, a politically dependent population, and toxic waste dumps. It is all of a fabric. And this is the heart of the matter. To see things in their wholeness is politically threatening. To understand that our manner of living, so comfortable for some, is linked to cancer rates in migrant laborers in California, the disappearance of tropical rain forests, 50,000 toxic dumps across the U.S.A., and the depletion of the ozone layer is to see the need for a change in our way of life. To see things whole is to see both the wounds we have inflicted on the natural world in the name of mastery and those we have inflicted on ourselves and on our children for no good reason, whatever our stated intentions. Real ecological literacy is radicalizing in that it forces us to reckon with the roots of our ailments, not just with their symptoms. For this reason, it can revitalize and broaden the concept of citizenship to include membership in a planet-wide community of humans and living things.

And how does this striving for community come into being? There is no one answer, but there are certain common elements. First, in the lives of most, if not all, people who define themselves as environmentalists, there is experience in the natural world at an early age. Leopold came to know birds and wildlife in the marshes and fields around his home in Burlington, Iowa, before his teens. David Brower, as a young boy on long walks over the Berkeley hills, learned to describe the flora to his nearly blind mother. Second, and not surprisingly, there is often an older teacher or mentor as a role model: a grandfather, a neighbor, an older brother, a parent, or a teacher. Third, there are seminal books that explain, heighten, and say what we have felt deeply but not said so well. In my own life, Rene Dubos and Loren Eiseley served this function of helping to bring feelings to articulate consciousness. Ecological literacy is becoming more difficult, I believe, not because there are fewer books about nature, but because there is less opportunity for the direct experience of it. Fewer people grow up on farms or in rural areas where access is easy and where it is easy to learn a degree of competence and self-confidence toward the natural world. Where the ratio of the human-created environment to the purely natural world exceeds some point, the sense of place can only be a sense of habitat. One finds the habitat familiar and/or likeable but without any real sense of belonging in the natural world. A sense of place requires more direct contact with the natural aspects of a place, with soils, landscape, and wildlife. This sense is lost as we move down the continuum toward the totalized urban environment where nature exists in tiny, isolated fragments by permission only. Said differently, this is an argument for more urban parks, summer camps, greenbelts, wilderness areas, public seashores. If we must live in an increasingly urban world, let's make it one of well-designed compact cities that include trees, river parks, meandering greenbelts, and urban farms where people can see, touch, and experience nature in a variety of ways. In fact, no other cities will be sustainable in a greenhouse world.

The goal of ecological literacy as I have described it has striking implications for that part of education that must occur in classrooms, libraries, and laboratories. To the extent that most educators have noticed the environment, they have regarded it as a set of problems which are (1) solvable (unlike dilemmas, which are not) by (2) the analytic tools and methods of reductionist science which (3) create value-neutral, technological remedies that often create even worse side effects. Solutions, therefore, originate at the top of society, from governments and corporations, and are passed down to a passive citizenry in the form of laws, policies, and technologies. The results, it is assumed, will be socially, ethically, politically, and humanly desirable, and the will to live and to sustain a humane culture can be preserved in a technocratic society. In other words, business can go on as usual. Assuming no need for an ecologically literate and ecologically competent public, people most often regard environmental education as an extra in the curriculum, not as a core requirement pervading the entire educational process.

Clearly, some parts of the crisis can be accurately described as problems. Some of these can be solved by technology, particularly those that require increased resource efficiency. It is a mistake, however, to think that all we need is better technology, not an ecologically literate and competent public that understands the relation between its well-being and the health of the natural systems.

For this to occur, we must rethink both the substance and the process of education at all levels. What does it mean to educate people to live sustainably, going, in Aldo Leopold's words, from "conqueror of the land-community to plain member and citizen of it"? However it is applied in practice, the answer will rest on six foundations.

The first is the recognition that all education is environmental education. By what is included or excluded, emphasized or ignored, students learn that they are a part of or apart from the natural world. Through all education we inculcate the ideas of careful stewardship or carelessness. Conventional education, by and large, has been a celebration of all that is human to the exclusion of our dependence on nature. As a result, students frequently resemble what Wendell Berry has called "itinerant professional vandals," persons devoid of any sense of place or stewardship, or inkling of why these are important.

Second, environmental issues are complex and cannot be understood through a single discipline or department. Despite a decade or more of discussion and experimentation, interdisciplinary education remains an unfulfilled promise. The failure occurred, I submit, because it was tried within discipline-centric institutions. A more promising approach is to reshape institutions as trans-disciplinary laboratories that include components such as agriculture, solar technologies, forestry, land management, wildlife, waste cycling, architectural design, and economics. Part of the task, then, of Earth-centered education is the study of interactions across the boundaries of conventional knowledge and experience.

Third, for inhabitants, education occurs in part as a dialogue with a place and has the characteristics of good conversation. Formal education happens mostly as a monologue of human interest, desires, and accomplishments that drowns out all other sounds. It is the logical outcome of the belief that we are alone in a dead world of inanimate matter, energy flows, and biogeochemical cycles. But true conversation can occur only if we acknowledge the existence and interests of the other. In conversation, we define ourselves, but in relation to another. The quality of conversation does not rest on the brilliance of one or the other person. It is more like a dance in which the artistry is mutual.

In good conversation, words represent reality faithfully. And words have power. They can enliven or deaden, elevate or degrade, but they are never neutral, because they affect our perception and ultimately our behavior. The use of words such as *resources*, *manage*, *channelize*, *engineer*, *produce*, and *geoengineer* makes our relation to nature a monologue rather than a conversation. The language of nature includes the sounds of animals, whales, birds, insects, wind, and water—a language more ancient and basic than human speech. Its books are the etchings of life on the face of the land. To hear this language requires patient, disciplined study of the natural world. But it is a language for which we have an affinity.

Good conversation is unhurried. It has its own rhythm and pace. Dialogue with nature cannot be rushed. It will be governed by cycles of day and night, the seasons, the pace of procreation, and by the larger rhythm of evolutionary and geologic time. Human sense of time is increasingly frenetic, driven by clocks, computers, and revolutions in transportation and communication. Good conversation has form, structure, and purpose. Conversation with nature has the purpose of establishing, in Wendell Berry's words; "What is here? What will nature permit here? What will nature help us do here?" (Berry 1987, 146). The form and structure of any conversation with the natural world is that of the discipline of ecology as a restorative process and healing art.

Fourth, it follows that the way education occurs is as important as its content. Students taught environmental awareness in a setting that does not alter their relationship to basic life-support systems learn that it is sufficient to intellectualize, emote, or posture about such things without having to live differently. Environmental education ought to change the way people live, not just how they talk. The best learning occurs in response to real needs and the life situation of the learner. The radical distinctions typically drawn between teacher and student, between the school and the community, and between areas of knowledge are dissolved. Real learning is participatory and experiential, not just didactic. The flow can be two ways—between teachers, who function as facilitators, and students, who are expected to be active agents in defining what is learned and how.

Fifth, experience in the natural world is both an essential part of understanding the environment and an important source of intellectual clarity. Experience, properly conceived, trains the intellect to observe the land carefully and to distinguish between health and its opposite. Direct experience is an antidote to abstract, indoor learning, demanding a disciplined and observant intellect. But nature, in Emerson's words, is also "the vehicle of thought" as a source of language, metaphor, and symbol. Natural diversity may well be the source of much of human creativity and intelligence. If so, the simplification and homogenization of ecosystems can only result in a lowering of human intelligence.

Sixth, education relevant to the challenge of building a sustainable society will enhance the learner's competence with natural systems. For reasons once explained by Whitehead and Dewey, practical competence is an indispensable source of good thinking. Good thinking proceeds from the friction between a thoughtful and well-prepared mind and real problems. Aside from its effects on thinking, practical competence will be essential if sustainability requires, as I think it does, that people must take an active part in rebuilding their homes, businesses, neighborhoods, communities, and towns. Shortening supply lines for food, energy, water, and materials—while recycling waste locally—implies a high degree of competence not necessary in a society dependent on central vendors and experts.

If these can be taken as the foundations of Earth-centered education, what can be said of its larger purpose? In a phrase, it is that quality of mind that seeks out connections. It is the opposite of the specialization and narrowness characteristic of most education. The ecologically literate person has the knowledge necessary to comprehend interrelatedness, and an attitude of care or stewardship. Such a person would also have the practical competence required to act on the basis of knowledge and feeling. Competence can only be derived from the experience of doing and the mastery of what philosopher Alasdair MacIntyre describes as a "practice."

Ecological literacy, further, implies a broad understanding of how people and societies relate to each other and to natural systems and how they might do so sustainably. It presumes both an awareness of the interrelatedness of life and knowledge of how the world works as a physical system. To ask, let alone answer, "What then?" questions presumes an understanding of concepts such as carrying capacity, overshoot, Liebig's law of the minimum, thermodynamics, trophic levels, energetics, and succession. Ecological literacy presumes that we understand our place in the story of evolution. It is to know that our health, well-being, and ultimately our survival depend on working with, not against, natural forces. The basis for ecological literacy, then, is the comprehension of the interrelatedness of life grounded in the study of natural history, ecology, and thermodynamics. It is to understand that "there ain't no such thing as a free lunch"; "you can never throw anything away"; and "the first law of intelligent tinkering is to keep all of the pieces." It is also to understand, with Leopold, that we live in a world of wounds senselessly inflicted on nature and on ourselves.

A second stage in ecological literacy is to know something of the speed of the crisis that is upon us. It is to know magnitudes, rates, and trends of population growth, species extinction, soil loss, deforestation, desertification, climate change, ozone depletion, resource exhaustion, air and water pollution, toxic and radioactive contamination, resource and energy use—in short, the vital signs of the planet and its ecosystems. Becoming ecologically literate is to understand the human enterprise for what it is: a sudden and brief eruption of a single species in the vastness of evolutionary time.

Ecological literacy requires a comprehension of the dynamics of the modern world. The best starting place is to read the original rationale for the domination of nature found in the writings of Bacon, Descartes, and Galileo. Here one finds the justification for the union of science with power and the case for separating ourselves from nature in order to control it more fully. To comprehend the idea of controlling nature, one must fathom the sources of the urge to power and the paradox of rational means harnessed to insane ends portrayed in Marlowe's *Doctor Faustus*, Mary Shelley's *Frankenstein*, Melville's *Moby Dick*, and Dostoevsky's *Legend of the Grand Inquisitor*.

Ecological literacy, then, requires a thorough understanding of the ways in which people and whole societies have become destructive. The ecologically literate person will understand how the causes of our predicament can be traced to economic and social structures, religion, science, politics, technology, patriarchy, culture, agriculture, and garden variety orneriness.

The diagnosis of the causes of our plight is only half of the issue. But before we can address solutions, there are several issues that demand clarification. "Nature," for example, is variously portrayed as "red in tooth and claw" or, like the Disney film *Bambi*, full of sweet little critters. Economists see nature as natural resources to be used; the backpacker, as a wellspring of transcendent values. We are no longer clear about our own nature, whether we are made in the image of God, or are merely a machine or computer, or animal. These are not trivial, academic issues. Unless we can make reasonable distinctions between what is natural and what is not, and what difference that difference makes, we are liable to be at the mercy of the engineers who want to remake all of nature, including our own.

Environmental literacy also requires a broad familiarity with the development of ecological consciousness. It is not yet clear whether the science of ecology will be "the last of the old sciences, or the first of the new." As the former, ecology is the science of efficient resource management. As the first of the new sciences, ecology is the basis for a broader search for pattern and meaning. As such, it cannot avoid issues of values, and the ethical questions most succinctly stated in Leopold's "The Land Ethic."

The study of environmental problems is an exercise in futility unless it is regarded as only a preface to the study, design, and implementation of solutions. The concept of sustainability implies a radical change in the institutions and patterns that we have come to accept as normal. It begins with ecology as the basis for the redesign of technology, cities, farms, and educational institutions and with a change in metaphors from mechanical to organic, industrial to biological. As part of the change we will need alternative measures of well-being such as those proposed by Amory Lovins (least-cost end-use analysis), H. T. Odum (energy accounting), and John Cobb (index of sustainable welfare). Sustainability also implies a different approach to technology, one that gives greater priority to those that are smaller in scale, are less environmentally destructive, and rely on the free services of natural systems. Not infrequently, technologies with these characteristics are also highly cost-effective, especially when the economic playing field is level.

If sustainability represents a minority tradition, it is nonetheless a long one dating back at least to Jefferson. Students should not be considered ecologically literate until they have read Thoreau, Kropotkin, Muir, Albert Howard, Alfred North Whitehead, Gandhi, Schweitzer, Aldo Leopold, Lewis Mumford, Rachel Carson, E. F. Schumacher, and Wendell Berry. There are alternatives to the present patterns that have remained dormant or isolated, not because they did not work, were poorly thought out, or were impractical, but because they were not tried. In contrast to the directions of modern society, this tradition emphasizes democratic participation, the extension of ethical obligations to the land community, careful ecological design, simplicity, competence with natural systems, the sense of place, holism, decentralization of whatever can best be decentralized, and human-scaled technologies and communities. It is a tradition dedicated to the search for patterns, unity, and connections between people of all ages, races, nationalities, and generations and between people and the natural world. This is a tradition grounded in the belief that life is sacred and not to be carelessly expended on the ephemeral. It is a tradition that challenges militarism, injustice, ecological destruction, and authoritarianism, while supporting all of those actions that lead to real peace, fairness, sustainability, and people's right to participate in those decisions that affect their lives. Ultimately, it is a tradition built on a view of ourselves as finite and fallible creatures living in a world limited by natural laws. The contrasting Promethean view, given force by the success of technology, holds that we should remove all limits, whether imposed by nature, human nature, or morality. Its slogan is found emblazoned on the advertisements of the age: "you can have it all" (Michelob beer) or "your world should know no limits" (Merrill Lynch). The ecologically literate citizen will recognize these immediately for what they are: the stuff of epitaphs. Ecological literacy leads in other, and more durable, directions toward prudence, stewardship, and the celebration of the Creation.

⁓ Chapter 27 ⁓

Place and Pedagogy

(1992)

THOREAU WENT TO LIVE by an ordinary pond on the outskirts of an unremarkable New England village, "to drive life into a corner, and reduce it to its lowest terms." Thoreau did not "research" Walden Pond; rather, he went to live, as he put it, "deliberately." Nor did he seek the far-off and the exotic, but the ordinary, "the essential facts of life." He produced no particularly usable data, but he did live his subject carefully, observing Walden, its environs, and himself. In the process he revealed something of the potential lying untapped in the commonplace, in our own places, in ourselves, and the relation between all three.

In contemporary jargon, Thoreau's excursion was "interdisciplinary." *Walden* is a mosaic of philosophy, natural history, geology, folklore, archeology, economics, politics, education, and more. He did not restrict himself to one academic pigeonhole. His "discipline" was as broad as his imagination and as specific as the $28.12 he spent for his house. Thoreau lived his subject. *Walden* is more than a diary of what he thought; it is a record of what he did and what he experienced. If, as Whitehead put it, "the learned world . . . is tame because it has never been scared by the facts," one finds little that is tame in *Walden*. For Thoreau, the facts, including both Walden Pond and himself, goaded, tempered, and maybe even scared his intellect. Nor is this the timid objective observer whose personhood and intellect remain strangers to each other. For Thoreau, philosophy was important enough "to live according to its dictates . . . to solve some of the problems of life, not only theoretically, but practically."

This article was originally published in 1992.

Ultimately, Thoreau's subject matter was Thoreau: his goal, wholeness; his tool, Walden Pond; and his methodology, simplification.

Aside from its merits as literature or philosophy, *Walden* is an antidote to the idea that education is a passive, indoor activity occurring between the ages of 6 and 21. In contrast to the tendencies to segregate disciplines, and to segregate intellect from its surroundings, *Walden* is a model of the possible unity between personhood, pedagogy, and place. For Thoreau, Walden was more than his location. It was a laboratory for observation and experimentation; a library of data about geology, history, flora, and fauna; a source of inspiration and renewal; and a testing ground for the man. *Walden* is a dialogue between a man and a place, not a monologue. In a sense, Walden Pond wrote Thoreau. His genius, I think, was to allow himself to be shaped by his place, to allow it to speak with his voice.

Other than as a collection of buildings where learning is supposed to occur, place has no particular standing in contemporary education. The typical college or university is organized around bodies of knowledge coalesced into disciplines. Sorting through a college catalogue, you are not likely to find courses dealing with the ecology, hydrology, geology, history, economics, politics, energy use, food policy, waste disposal, and architecture of the campus or its region. Nor are you likely to find many courses offering enlightenment to modern scholars in the art of living well in a place. The typical curriculum is reminiscent of Kierkegaard's comment after reading the vast, weighty corpus of Hegel's philosophy, that Hegel had "taken care of everything except perhaps for the question of how one was to live one's life." Similarly, a great deal of what passes for knowledge is little more than abstraction piled on top of abstraction, disconnected from tangible experience, real problems, and the places where we live and work. It is utopian, which means "nowhere."

The importance of place in education has been overlooked for many reasons. One is the ease with which we miss the immediate and mundane. Those things nearest at hand are often the most difficult to see. Second, for purists, place is a nebulous concept. Even so, Thoreau spent little time trying to define the precise boundaries of his place, nor was it necessary to do so. *Walden* is a study of an area small enough to be easily walked over in a day and still observed carefully. Place is defined by its human scale: a household, neighborhood, community, 40 acres, 1000 acres.

Place is nebulous to educators because to a great extent we are a deplaced people for whom our immediate places are no longer sources of food, water, livelihood, energy, materials, friends, recreation, or sacred inspiration. We are, as Raymond Dasmann noted long ago, "biosphere people," supplied with all these and more from places around the world that are largely unknown to us, as are those to which we consign our toxic and radioactive wastes, garbage, sewage, and industrial trash. We consume a great deal of time and energy going somewhere else. The average American moves 10 times in a lifetime and spends countless hours at airports and on highways going to places that look a great deal like those just left behind. Our lives are lived amidst the architectural expressions of de-placement: the shopping mall, apartment, neon strip, freeway, glass office tower, and homogenized development—none of which encourage much sense of rootedness, responsibility, and belonging.

Third, place by definition is specific, yet our mode of thought is increasingly abstract. The danger of abstraction lies partly in what Whitehead described as the "fallacy of misplaced concreteness," the confusion of our symbols with reality. Words and theories take on a life of their own, independent of the reality they purport to mirror, often with tragic results. At its worst, as Lewis Mumford describes it,

> the abstract intelligence, operating with its own conceptual apparatus, in its own self-restricted field is actually a coercive instrument: an arrogant fragment of the full human personality, determined to make the world over in its own oversimplified terms, willfully rejecting interests and values incompatible with its own assumptions, and thereby depriving itself of any of the cooperative and generative functions of life—feeling, emotion, playfulness, exuberance, free fantasy—in short, the liberating sources of unpredictable and uncontrollable creativity. (Mumford 1966, 10)

By capturing only a fragment of reality, unrelieved abstraction inevitably distorts perception. By denying genuine emotion, it distorts and diminishes human potentials. For the fully abstracted mind, all places become "real estate" or mere natural resources, their larger economic, ecological, social, political, and spiritual possibilities lost to the purely and narrowly utilitarian.

The idea that place could be a significant educational tool was proposed by John Dewey in an essay written in 1897. Dewey suggested that "we make each of our schools an embryonic community . . . with types of occupations that reflect the life of the larger society." He intended to broaden the focus of education, which he regarded as too "highly specialized, one-sided, and narrow." He proposed to make the school, its relations with the larger community, and all of its internal functions into curriculum.

The regional survey, which reflected a broader conception of the role of place in education, was developed by Lewis Mumford in the 1940s. In Mumford's words, the regional survey was

> not something to be added to an already crowded curriculum. It is rather (potentially) the backbone of a drastically revised method of study, in which every aspect of the sciences and the arts is ecologically related from the bottom up, in which they connect directly and constantly in the student's experience of his region and his community. Regional survey must begin with the infant's first exploration of his dooryard and his neighborhood; it must continue to expand and deepen, at every successive stage of growth until the student is capable of seeing and experiencing above all, of relating and integrating and directing the separate parts of his environment, hitherto unnoticed or dispersed. (Mumford 1946, 151–52)

The regional survey (Mumford cites Walden as a classic example) involved the intensive study of the local environment by specialists and every member of the community, including schoolchildren. As the focal point for education, the regional survey was intended to create habits of thinking across disciplines, promote cooperation, and dissolve distinctions between facts and values, the past and the future, and nature and human society. Beyond education, Mumford regarded the regional survey as the basis for rational coordination and planning and as a vehicle for widespread public participation.

The integration of place into education is important, first, because it requires the combination of intellect with experience. The typical classroom is an arena for lecture and discussion, both of which are important to intellectual growth. The study of place, however, involves complementary dimensions of intellect: direct observation, investigation, experimentation, and skill in the application of knowledge. The latter is regarded merely as "vocational education." But for Mumford and Dewey, practical and manual skills were an essential aspect of experience and good thinking and essential to the development of the whole person. Both regarded the acquisition of manual skills as vitally important in sharpening the intellect. John Dewey again:

> We cannot overlook the importance for educational purposes of the close and intimate acquaintance got with nature at first hand, with real things

> and materials, with the actual processes of their manipulation, and the knowledge of their special necessities and uses. In all this there [is] continual training of observation, of ingenuity, constructive imagination, of logical thought, and of the sense of reality acquired through firsthand contact with actualities. The educative forces of the domestic spinning and weaving, of the sawmill, the gristmill, the cooper ship, and the blacksmith forge were continuously operative. (Dewey 1981, 457)

Whitehead similarly writes:

> There is a coordination of senses and thought, and also a reciprocal influence between brain activity and material creative activity. In this reaction the hands are peculiarly important. It is a moot point whether the human hand created the human brain, or the brain created the hand. Certainly, the connection is intimate and reciprocal. (Whitehead 1967, 50)

In the reciprocity between thinking and doing and in application to specific places and problems, knowledge loses much of its abstractness and becomes tangible and direct.

Second, integration of place into education is important because the study of place is relevant to the problems of overspecialization. Mumford's remedy for the narrow, under-dimensioned mind is the requirement to balance analysis with synthesis. This cannot be accomplished by adding courses to an already overextended curriculum or by fine-tuning a system designed to produce specialists. It can be done only by reconceptualizing the purposes of education in order to promote diversity of thought and a wider understanding of interrelatedness. Places are laboratories of diversity and complexity, mixing social functions and natural processes. A place has a human history and a geologic past. It is a part of an ecosystem with a variety of microsystems; it is a landscape with a particular flora and fauna. Its inhabitants are part of a social, economic, and political order: they import or export energy materials, water, and wastes; they are linked by innumerable bonds to other places. A place cannot be understood from the vantage point of a single discipline or specialization. It can be understood only on its terms as a complex mosaic evolving as part of still larger systems. The classroom and indoor laboratory are ideal environments in which to narrow reality in order to focus on bits and pieces. The study of place, by contrast, enables us to widen the focus to examine the interrelationships between disciplines and to lengthen our perception of time.

It is important that learning not stop at the point of mere intellectual comprehension. Students should be encouraged to act on the basis of information from the survey to identify a series of projects to promote greater self-reliance, interdisciplinary learning, and physical competence, such as policies for food, energy, architecture, and waste. These provide opportunities for intellectual and experiential learning involving many different disciplines working on tangible problems. If the place also includes natural areas, forests, streams, and agricultural lands, the opportunities for environmental learning multiply accordingly.

Finally, for Mumford and Dewey, much of the pathology of contemporary civilization was related to the disintegration of the small community. Dewey (1981) wrote in 1927, "The invasion and partial destruction of the life of the [local community] by outside uncontrolled agencies is the immediate source of the instability, disintegration and restlessness which characterize the present epoch." The study of place, then, has a third significance in reeducating people in the art of living well where they are. But there is a distinction to be made between residing and dwelling in a place. A resident is a temporary occupant, putting down few roots and investing little, knowing little, and perhaps caring little for the immediate locale beyond its ability to gratify. As both a cause and an effect of displacement, the resident lives in an indoor world of office building, shopping mall, automobile, apartment, and suburban house and watches television an average of 4 hours each day. The inhabitant, in contrast, "dwells," as Ivan Illich puts it, in an intimate, organic, and mutually nurturing relationship with a place. Good inhabitance is an art requiring detailed knowledge of a place, the capacity for observation, and a sense of care and rootedness. Residence requires cash and a map. A resident can reside almost anywhere that provides an income. Inhabitants bear the marks of their places, whether rural or urban, in patterns of speech, through dress and behavior. Uprooted, they get homesick. Historically, inhabitants are less likely to vandalize their or others' places. They also tend to make good neighbors and honest citizens. They are, in short, the bedrock of the stable community and neighborhood that Mumford, Dewey, and Jefferson regarded as the essential ingredient of democracy.

Paul Shepard explains the stability of inhabitants as a consequence of the interplay between the psyche and a particular landform. "Terrain structure," he argues, "is the model for the patterns of cognition." The physical and biological patterns of a place are imprinted on the mind so the "cognition, personality, creativity, and maturity—all are in some way tied to particular gestalts of space" (Shepard 1977, 22–32). Accordingly, the child must have an opportunity to "soak in a place, and the adolescent and adult must be able to return to that place to ponder the visible substrate of his own personality." Hence, knowledge of a place—where you are and where you come from—is intertwined with knowledge of who you are. Landscape, in other words, influences mindscape. Since it diminishes the potential for maturation and inhabitance, the desecration of places is psychologically destructive as well. If Shepard is right, and I believe that he is, we are paying a high price for the massive rearrangement of the North American landscape over the past century.

For de-placed people, education in the arts of inhabitation is partly remedial learning: the unlearning of old habits of waste and dependency. It requires the ability to perceive and utilize the potentials of a place. One of the major accomplishments of the past several decades has been the rediscovery of how much ordinary people can do for themselves in small places. The significance of this fact coincides with the growing recognition of the ecological, political, and economic costs and the vulnerability of large-scale centralized systems, whether publicly or privately controlled. Smaller-scale technologies are often cheaper and more resilient, and they do not undermine democratic institutions by requiring the centralization of capital, expertise, and political authority. Taken together, they vastly expand the potential of ecologically designed, intensively developed places to meet human needs on a sustained basis.

Education for reinhabitation must also instill an applied ethical sense toward habitat. Again, Leopold's standard—"a thing is right when it tends to preserve the integrity, stability, and beauty of the biotic community. It is wrong when it tends otherwise"—is, on balance, a clear enough standard for most decisions about the use we make of our places. From the standpoint of education, the stumbling block to development of an ethic of place is not the complexity of the subject; it is the fact, as Leopold put it, "that our educational system is headed away from . . . an intense consciousness of land."

Critics might argue that the study of place would be inherently parochial and narrowing. If place were the entire focus of education, it certainly could be. But the study of place would be only a part of a larger curriculum which would include the study of relationships between places as well. For Mumford, place was simply the most immediate of a series of layers leading to the entire region as a system of small places. But parochialism is not the result of what is studied as much as how it is studied. Lewis Thomas, after all, was able to observe the planet in the workings of a single cell.

At issue is our relationship to our own places. What is the proper balance between mobility and rootedness? Indeed, are rootedness and immobility synonymous? How long does it take for one to learn enough about a place to become an inhabitant and not merely a resident? However one chooses to answer these questions, the lack of a sense of place, our "cult of homelessness," is endemic, and its price is the destruction of the small community and the resulting social and ecological degeneracy. We are not the first footloose wanderers of our species, but we wander on a larger and more destructive scale.

We cannot solve such deep problems quickly, but we can begin learning how to reinhabit our places, as Wendell Berry says, "lovingly, knowingly, skillfully, reverently," restoring context to our lives in the process. For a world growing short of many things, the next sensible frontiers to explore are those of the places where we live and work.

∽ Chapter 28 ∾

The Liberal Arts, the Campus, and the Biosphere

(1990)

AUTHOR'S NOTE 2010: *This essay in the* Harvard Educational Review *was an early statement of the rationale for what has grown into the green campus movement. It was inspired by our experience at Meadowcreek Project in the 1980s, a study we did of the food system at Hendrix College which was based on a report from the Rocky Mountain Institute by Bill Browning and Hunter Lovins ("A Trail of Two Hamburgers") and April Smith's master's thesis on the environmental impacts of the University of California, Los Angeles ("In Our Backyard").*

DEBATES ABOUT THE CONTENT and purposes of education are mostly conducted among committees of the learned conditioned to such fare. Allan Bloom changed all of that in 1987 by writing a best seller on the subject (Bloom 1987). Professor Bloom, as far as I can tell, believes that questions about the content of education (i.e., curriculum) were settled some time ago—perhaps once and for all with Plato, but certainly no later than Nietzsche. Subsequent elaborations, revisions, and refinements have worked great mischief with the high

This article was originally published in 1990.

culture he purports to defend. Bloom's discontent focuses on American youth. He finds them empty, intellectually slack, and morally ignorant. The "soil" of their souls is "unfriendly" to the higher learning. And he thinks no more highly of their music and sexual relationships.

In Professor Bloom's ideal academy, students of a higher sort would spend a great deal of time reading the Great Books, a list no longer universally admired. Bloom's avowed aim is to "reconstitute the idea of an educated human being and establish a liberal education again." But after 344 pages of verbal pyrotechnics—some illuminating the landscape, others merely the psyche of Professor Bloom—he leaves us only with some variation on the Great Books approach to education. The classics, he argues, "provide the royal road to the students' hearts . . . their gratitude [for being so exposed] is . . . boundless." Exclusion of the classics, he thinks, has culminated in an "intellectual crisis of the greatest magnitude which constitutes the crisis of our civilization" (Bloom 1987). Lesser minds might have related the crisis to more pedestrian causes, such as violence, nuclear weapons, technology, overpopulation, or injustice. No matter. All of this was revealed to Professor Bloom while on the faculty at Cornell during the student uprising in 1969. One may reasonably infer that Professor Bloom and his Great Books were not treated kindly. One may also infer that Professor Bloom has neither forgiven nor forgotten.

Bloom has been widely attacked as a snob and as having totally misunderstood what America is all about. In his defense, there is no reasonable case to be made against the inclusion of ancient wisdom in any good, liberal education. Nor can there be any good argument against the "idea of an educated human being." But questions about which ancient wisdom we might profitably consult, and about the intellectual and moral qualities of the educated person, have not been settled once and for all with Professor Bloom's book. At the end we know a great deal of what Professor Bloom is against, some of which is justified, but little of what he is for.

His vagueness about ends suggests that Professor Bloom, without saying so, regards education as an end in itself. In a time of global turmoil, what transcendent purposes will Bloom's academy serve? In a time of great wrongs, what injustices does he wish to right? In an age of senseless violence, what civil disorders and dangers does he intend to resolve? In a time of anomie and purposelessness, what higher qualities of mind and character does he propose to cultivate? A careful reading of *The Closing of the American Mind* (Bloom 1987) offers little insight about such matters.

Rather, it is indicative of the closure of the purely academic mind to ecological issues.

For all of his conspicuous erudition, Professor Bloom seems to regard the liberal arts as an abstraction. For example, rather than merely "reconstitute the idea" of educated human beings, why not actually educate a large number of them? Likewise, his reverence for the classics is not accompanied by any suggestion of how they might illuminate the major issues of our day. The effect is, ironically, to render them both sacred and unusable, except for purposes of conspicuous pedantry. It also distorts our understanding of the origins of some of humanity's best thinking. Many of what are now described as classics were produced by the friction of extraordinary minds wrestling with the problems of their day, which is to say that they were relevant in their time. Plato wrote *The Republic* in part as a response to the breakdown of civic order in fourth-century Athens. Locke wrote his *Two Treatises* partly to justify the English civil war. Only in hindsight does their work appear to have the immaculate qualities that they certainly lacked at birth. The progress of human thought has been hard fought, uneven, and erratic. If our descendants five centuries hence regard any books of our era as classics, they will be those that grappled with and illuminated the major issues of our time, in a manner that illuminates theirs. Beyond complaints about education, Professor Bloom does not offer an opinion about what these issues may be. He sounds rather like a fussy museum curator, irate over gum wrappers on the floor.

Amidst growing poverty, environmental deterioration, and violence in a nuclear-armed world, Professor Bloom is silent about how his version of the liberal arts would promote global justice, heal the breach with the natural world, promote peace, and restore meaning in a technocratic world. On the contrary, he arrogantly dismisses those concerned about such issues. Yet, ironically, if our era adds any "classics" to the library of human thought, they will, more likely than not, be written about these subjects.

It is now widely acknowledged that the classics of the Western tradition are deficient in certain respects. First, having been mostly composed by white males, they exclude the vast majority of human experience. Moreover, there are problems that this tradition has not successfully resolved, either because they are of recent origin or because they were regarded as unimportant. In the latter category is the issue of the human role in the natural world. Search as one may through Plato, Aristotle, and the rest of the authors of the Great Books, there is not much said about it. With a few exceptions such as Hesiod, Cicero, Spinoza, and St. Francis (who wrote no "Great Book"), what wisdom we have from Western sources begins with the likes of Thoreau and George Perkins Marsh in the middle years of the last century. Whatever timeless qualities human nature may or may not have, Western culture has not offered much enlightenment on the appropriate relationship between humanity and its habitat. Nor does Professor Bloom.

Professor Bloom, I believe, has also missed something basic about education. Whitehead put it this way: "First-hand knowledge is the ultimate basis of intellectual life. . . . The second-handedness of the learned world is the secret of its mediocrity. It is tame because it has never been scared by the facts" (Whitehead 1967, 51). An immersion in the classics, however valuable for some parts of intellectual development, risks no confrontation with the facts of life. The aim of education is not the ability to score well on tests, do well in games like Trivial Pursuit, or even to quote the right classic on the appropriate pedagogical occasion. The aim of education is life lived to its fullest. A study of the classics is one tool among many to this end.

The purpose of a liberal education has to do with the development of the whole person. J. Glenn Gray describes this person as "one who has fully grasped the simple fact that his self is fully implicated in those beings around him, human and nonhuman, and who has learned to care deeply about them" (Gray 1984, 34). Accordingly, its function is the development of the capacity for clear thought and compassion in the recognition of the interrelatedness of life.

And what do these mean in an age of violence, injustice, ecological deterioration, and nuclear weapons? What does wholeness mean in an age of specialization? It is perhaps easier to begin with what they do not mean. We do not lack for bad models: the careerist, the "itinerant professional vandal" devoid of any sense of place, the yuppie, the narrow specialist, the intellectual snob. In different ways, these all-too-common role models lack the capacity to relate their autobiography to the unfolding history of their time in a meaningful, positive way. They simply cannot speak to the urgent needs of the age, which is to say that they have been educated to be irrelevant. They have not grasped their implicatedness in the larger world, nor have they learned to care deeply about anything beyond themselves. To the extent that this has become the typical product of our educational institutions, it is an indictment of enormous gravity. Professor Bloom's emphasis on the classics and preservation of high culture does not remedy this dereliction in any obvious way.

It might be possible to dismiss Professor Bloom as a harmless crank were it not for the wide impact of his book, and because he has become a spokesman for the powerful. The problem is not with Professor Bloom's ideas, which are toothless enough. The danger lies in the combination of vagueness, surliness, and the large number of things that he does not say. The result is that *Closing* can be cited by any number of ill-informed proponents of bad causes wanting to exit the twentieth century backwards. Bloom has not provided any coherent vision of the liberal arts relevant to our time. What he does offer is a sometimes insightful cultural critique in combination with a mummified curriculum with the distinct aroma of formaldehyde.

Reconstruction: The Task of the Liberal Arts

The mission of the liberal arts in our time is not merely to inculcate a learned appreciation for the classics, as Bloom would have it, or to transmit "marketable skills," as many others propose, but to develop balanced, whole persons. Wholeness, first, requires the integration of the personhood of the student: the analytic mind with feelings, the intellect with manual competence. Failure to connect mind and feelings, in Gray's words, "divorces us from our own dispositions at the level where intellect and emotions fuse" (Gray 1984, 84–85). A genuinely liberal education will also connect the head and the hands. Technical education and liberal arts have been consigned to different institutions. This division creates the danger that students in each, in Gray's words, "miss a whole area of relation to the world" (Gray 1984, 81). For liberal arts students, it also undermines an ancient source of good thought: the friction between an alert mind and practical experience. Abstract thought, "mere book learning," in Whitehead's words, divorced from practical reality and the facts of life, promotes pedantry and mediocrity. It also produces half-formed or deformed persons: thinkers who cannot do, and doers who cannot think. Students typically leave 16 years of formal education without ever having mastered a particular skill or without any specific manual competence, as if the act of making anything other than term papers is without pedagogic or developmental value.

Second, an education in the liberal arts must overcome what Whitehead termed "the fatal disconnection of subjects." The contemporary curriculum continues to divide reality into a cacophony of subjects that are seldom integrated into any coherent pattern. Whitehead's point bears repeating: there is only one subject for education: "life in all its manifestations." Yet we routinely unleash specialists on the world, armed with expert knowledge but untempered by any inkling of the essential relatedness of things. Worse, specialization undermines the ability to communicate "plainly, in the common tongue." The academy, with its disciplines, divisions, and multiplying professional jargons, has come to resemble not so much a university as a cacophony of different jargons. I do not believe that Whitehead overstated the case. Disconnectedness in the form of excessive specialization is fatal to comprehension because it removes knowledge from its larger context. Collection of data supersedes understanding of connecting patterns, which is, I believe, the beginning of wisdom. It is no accident that connectedness is central to the meaning of the Greek root words for both *ecology* (*oikos*) and *religion* (*religio*).

A third task of the liberal arts is to provide a sober view of the world, but without inducing despair. Many college freshmen are shocked by the knowledge that this is not the happy world described by the advertising and entertainment industries and by any number of feckless politicians. This is a time of danger, terrorism, anomie, suffering, crack on the streets, changing climate, war, hunger, homelessness, toxic pollution, desertification, poverty, and the permanent threat of Armageddon. Ours is the age of paradox. The modern obsession to control nature through science and technology is resulting in a less predictable and less bountiful natural world. Material progress was supposed to have created a more peaceful world. Instead, the twentieth century was a time of unprecedented bloodshed, in which 200 million died, and the years ahead, perhaps, will be an age of terrorism. Our economic growth has multiplied wants, not satisfactions. Amidst a staggering quantity of artifacts—what economists call abundance—there is growing poverty of the most desperate sort. How many student counseling services convey this sense of peril? Or obligation? The often-cited indifference and apathy of students is, I think, a reflection of the prior failure of educators and educational institutions to stand for anything beyond larger and larger endowments and an orderly campus. The result is a growing gap between the real world and the academy, and between the attitudes and aptitudes of its graduates and the needs of their time.

Finally, a genuine liberal arts education will equip a person to live well in a place. To a great extent, formal education now prepares its graduates to reside, not to dwell. The difference is important. The resident is a temporary and rootless occupant who mostly needs to know where the banks and stores are in order to plug in. The inhabitant and a particular habitat cannot be separated without doing violence to both. The sum total of violence wrought by people who do not know who they are because they do not know where they are is the global environmental crisis. To reside is to live as a transient and as a stranger to one's place, and inevitably to some part of the self. The inhabitant and place mutually shape each other. Residents, shaped by outside forces, become merely "consumers" supplied by invisible networks that damage their places and those of others. The inhabitant and the local community are parts of a system that meets real needs for food, materials, economic support, and sociability. The resident's world, on the contrary, is a complicated system that defies order, logic, and control. The inhabitant is part of a complex order that strives for harmony between human demands and ecological processes. The resident lives in a constant blizzard of possibilities engineered by other residents. The life of the inhabitant is governed by the boundaries of sufficiency, by organic harmony, and by the discipline of paying attention to minute particulars. For the resident, order begins from the top and proceeds downward as law and policy. For the inhabitant, order begins with the self and proceeds outward. Knowledge for the resident is theoretical and abstract, akin to training. For inhabitants, knowledge in the art of living aims toward wholeness. Those who dwell can only be skeptical of those who talk about being global citizens before they have attended to the minute particulars of living well in their place.

Liberal Arts and the Campus

This brings me to the place where learning occurs, the campus. Do students in liberal arts colleges learn connectedness there or separation? Do they learn "implicatedness" or noninvolvement? And do they learn that they are "only cogs in an ecological mechanism," as Aldo Leopold put it, or that they are exempt from the duties of any larger citizenship in the community of life? A genuine liberal arts education will foster a sense of connectedness, implicatedness, and ecological citizenship and will provide the competence to act on such knowledge. In that kind of place, can such an education occur? The typical campus is the place where knowledge of other things is conveyed. Curriculum is mostly imported from other locations, times, and domains of abstraction. The campus as land, buildings, and relationships is thought to have no pedagogic value, and for those intending to be residents it need have none. It is supposed to be attractive and convenient without also being useful and instructive. A "nice" campus is one whose lawns and landscape are well manicured and whose buildings are kept clean and in good repair by a poorly paid maintenance crew. From distant and unknown places the campus is automatically supplied with food, water, electricity, toilet paper, and whatever else. Its waste and garbage are transported to other equally unknown places.

And what learning occurs on a "nice" campus? First, without anyone saying as much, students learn the lesson of indifference to the ecology of their immediate place. Four years in a place called a campus culminates in no great understanding of the place, or in the art of living responsibly in that or any other place. I think it significant that students frequently refer to the outside world as the "real world" and do so without any feeling that this is not as it should be. The artificiality of the campus is not unrelated to the mediocrity of the learned world of which Whitehead complained. Students also learn indifference to the human ecology of the place and to certain kinds of people: those who clean the urinals, sweep the floors, haul out the garbage, and collect beer cans on Monday morning. Indifference to a place is a matter of attention. The campus and its region are seldom brought into focus as a matter of practical study. To do so raises questions of the most basic sort. How does it function as an ecosystem? From where do its food, energy, water, and materials come and at what human and ecological cost? Where do its waste and garbage go? At what costs? What relation does the campus have to the surrounding region? What is the ecological history of the place? What ecological potentials does it have? What are the dominant soil types? Flora and fauna? And what of its geology and hydrology?

The study of place cultivates the habit of careful, close observation, and with it the ability to connect cause and effect. Aldo Leopold described the capacity in these terms:

> Here is an abandoned field in which the ragweed is sparse and short. Does this tell us anything about why the mortgage was foreclosed? About how long ago? Would this field be a good place to look for quail? Does short ragweed have any connection with the human story behind yonder graveyard? If all the ragweed in this watershed were short would that tell us anything about the future of floods in the stream? About the future prospects for bass or trout? (Leopold 1966, 210)

Second, students learn that it is sufficient only to learn about injustice and ecological deterioration without having to do much about them, which is to say, the lesson of hypocrisy. They hear that the vital signs of the planet are in decline without learning to question the de facto energy, food, materials, and waste policies of the very institution that presumes to induct them into responsible adulthood. Four years of consciousness raising proceeds without connection to those remedies close at hand. Hypocrisy undermines the capacity for constructive action and so contributes to demoralization and despair.

Third, students learn that practical incompetence is de rigueur, since they seldom are required to solve problems that have consequences except for their grade point average. They are not provided opportunities to implement their stated values in practical ways or to acquire the skills that would let them do so at a later time. Nor are they asked to make anything, it being presumed that material and mental creativity are unrelated. *Homo faber* and *Homo sapiens* are two distinct species, the former being an inferior sort that subsisted between the Neanderthal era and the founding of Harvard. The losses are not trivial—the satisfaction of good work and craftsmanship, the lessons of diligence and discipline, and the discovery of personal competence. After 4 years of the higher learning, students have learned that it is all right to be incompetent and that practical competence is decidedly inferior to the kind that helps to engineer leveraged buyouts and create tax breaks for people who do not need them. This is a loss of incalculable proportions both to the personhood of the student and to the larger society. It is a loss to their intellectual powers and moral development that can mature only by interaction with real problems. It is a loss to the society burdened with a growing percentage of incompetent people, ignorant of why such competence is important.

The conventional campus has become a place where indoor learning occurs as a preparation for indoor careers. The young of our advanced society are increasingly shaped by the shopping mall, the freeway, the television, and the computer. They regard nature, if they see it at all, as through a rearview mirror receding in the haze. We should not be astonished, then, to discover rates of ecological literacy in decline, at the very time that that literacy is most needed.

The Upshot

Every educational institution processes not only ideas and students but resources, taking in food, energy, water, materials and discarding organic and solid wastes. The sources (mines, wells, forests, farms, feedlots) and sinks (landfills, toxic dumps, sewage outfalls) are the least-discussed places in the contemporary curriculum. For the most part, these flows occur out of sight and mind of both students and faculty. Yet they are the most tangible connections between the campus and the world beyond. They also provide an extraordinary educational opportunity. The study of resource flows transcends disciplinary boundaries; it connects the foreground of experience with the background of larger issues and more distant places; and it joins empirical research on existing behavior and its consequences with the study of other and more desirable possibilities.

The study of institutional resource flows is aimed to determine how much of what comes from where, and with what human and ecological consequences. How much electricity from what power plants burning how much fuel extracted from where? What are the sources of food in the campus dining hall? Is it produced "sustainably" or not? Are farmers or laborers fairly paid or not? What forests are cut down to supply the college with paper? Are they replanted? Where does toxic waste from labs go? Or solid wastes? Why is there waste at all?

The study of actual resource flows must be coupled with the study of alternatives that may be more humane, ethically solvent, ecologically sustainable, cheaper, and better for the regional economy. Are there other and better sources of food, energy, materials, water? The study of potentials must also address issues of conservation. How much does the institution waste? How much energy, water, paper, and material can be conserved? What is the potential for recycling paper, glass, aluminum, and other materials? Can organic wastes be composted on-site or recycled through solar aquatic systems? At what cost? Can the institution shift its buying power from national marketing systems to support local economies? How? In what areas? How quickly? Can the landscape be designed for educational rather than decorative purposes? To what extent can good landscaping minimize energy spent for cooling and heating?

To address these and related questions, the Meadowcreek Project (a nonprofit organization I'd cofounded in 1979) conducted studies of the food systems of Hendrix College in Conway, Arkansas, and Oberlin College in Ohio. Both institutions are served by nationwide food-brokering networks that are not sustainable and that tend to undermine regional economies. In the Hendrix study, for example, students discovered that the college was buying only 9 percent of its food within the state. Beef came from Amarillo, Texas; rice from Mississippi. Yet the college is located in a cattle and rice-farming region. Both studies uncovered ample opportunities for the institutions to expand purchases of locally grown products. Not infrequently, these are fresher and less likely to be contaminated with chemicals, and not surprisingly, they are cheaper because transportation costs are lower. In conducting the research, which involved travel to the farms and feedlots throughout the United States that supply the campus, students confronted basic issues in agriculture, social ethics, environmental quality, economics, and politics. They were also involved in the analysis of existing buying patterns while having to develop feasible alternatives in cooperation with college officials. The results were action-oriented, interdisciplinary, and aimed to create practical results. Both colleges responded cooperatively in the implementation of plans to increase local buying. In the Hendrix case, in-state purchases doubled in the year following the study. Through video documentaries and articles in the campus newspaper, the studies became part of a wider campus dialogue. Finally, the willingness of both colleges to support local economies helped to bridge the gap between the institutions and their locality in a way no public relations campaign could have done.

Conclusion

The study of institutional resource flows can lead to three results. The first is a set of policies governing food, energy, water, materials, architectural design, landscaping, and waste flows that meet standards for sustainability. A campus energy policy, for example, would set standards for conservation, while directing a shift toward the maximum use of both passive and active solar systems for hot water, space conditioning, and electricity. A campus food policy would give high priority to local and regional organic sources. A materials policy would aim to minimize solid waste and recycling. An architectural policy governing all new construction and renovation would give priority to solar design and the use of nontoxic and locally available building materials. A landscape policy would stress the use of trees for cooling and windbreaks and as a means to offset campus CO_2 emissions. Decorative landscaping would be replaced by "edible landscaping." A campus waste policy, aimed to close waste loops, would lead to the development of on-site composting and the exploration of biological alternatives for handling waste water.

The study of campus resource flows and the development of campus policies would lead to a second and more important result: the reinvigoration of a curriculum around the issues of human survival—a plausible foundation for the liberal arts. This emphasis would become a permanent part of the curriculum through research projects, courses, seminars, and the establishment of interdisciplinary programs in resource management or environmental studies. By engaging the entire campus community in the study of resource flows, debate about the possible meanings of sustainability, the design of campus resource policies, and curriculum innovation, the process would carry with it the potential to enliven the educational process. I can think of few disciplines throughout the humanities, social sciences, and sciences without an important contribution to this debate.

Third, the study and its implementation as policy and curriculum would be an act of real leadership. Nearly every college and university claims to offer "excellence" in one way or another. Mostly the word is invoked by unimaginative academic officials who want their institution to be like some other. But prestige, like barnacles on the hull of a ship, often limits institutional velocity and mobility. Real excellence in an age of cataclysmic potentials consists neither in imitation nor timidity. College and university officials with courage and vision have the power to lead in the transition to a sustainable future. Within their communities, their institutions have visibility, respect, and buying power. What they do matters to a large number of people. How they spend their institutional budget counts for a great deal in the regional economy. Through alumni, they reach present leaders. Through students they reach those of the future. All of which is to say that colleges and universities are leverage institutions. They can help create a humane and livable future, rather than remaining passively on the sidelines, poised to study the outcome.

Those who presume to defend the liberal arts in the fashion of Allan Bloom ironically have undersold them. A genuinely liberal education will produce whole persons with intellectual breadth, able to think at right angles to their major field; practical persons able to act competently; and persons of deep commitment, willing to roll up their sleeves and join the struggle to build a humane and sustainable world. They will not be merely well-read. Rather, they will be ecologically literate citizens able to distinguish health from its opposite and to live accordingly. Above all, they will make themselves relevant to the crisis of our age, which in its various manifestations is about the care, nurturing, and enhancement of life. And life is the only defensible foundation for a liberal education.

PART 5

On Energy and Climate

∻ AUTHOR'S NOTE 2010: ∻

"Man's conquest of nature," C. S. Lewis once wrote, was an illusion. "All of nature's apparent reverses have been but tactical withdrawals. We thought we were beating her back when she was luring us on" (Lewis 1947). What Jacob Bronowski once called "the ascent of man" has been powered by carbon in soils and forests, and more recently by ancient sunlight in the form of coal and oil. Every advance along the way seemed to be permanent. But nature, as Lewis has it, was luring us on, and now the trap is nearly sprung in the form of spiraling climatic destabilization, ocean acidification, and loss of species. We have precious little time to stabilize the Earth's vital signs and move civilization to safer ground. Climate destabilization is the largest challenge to global civilization ever, but with luck it may prove to be an opportunity to build the foundation for a more durable and decent world order.

∽ Chapter 29 ∼

Pascal's Wager and Economics in a Hotter Time

(1992)

IN WEIGHING THE QUESTION concerning the existence of God, seventeenth-century philosopher and mathematician Blaise Pascal (1941) proceeded in a manner perhaps instructive for other and more mundane questions. "Reason," he declared, "can decide nothing here." Nonetheless, "you must wager. It is not optional." You have, he believed, "two things to lose, the true and the good; and two things to stake, your reason and your will, your knowledge and your happiness; and your nature has two things to shun, error and misery." What would you lose by believing that God exists and living a life accordingly? Pascal's answer was, "If you gain, you gain all; if you lose, you lose nothing." By doing so you would become "faithful, honest, humble, grateful, generous, a sincere friend, truthful." The opposite decision, that God did not exist, and a life lived in pursuit of "poisonous pleasures, glory and luxury," whatever its short-term gains, would bring misery. In other words, if you chose not to believe and it turned out that God did exist, you would have hell to pay. On the other hand, if God did not exist and you had lived a life of faith, you would have sacrificed only a few fleeting pleasures but

This article was originally published in 1992.

gained much more. Pascal's argument for faith, then, rested on the sturdy foundation of prudential self-interest aimed to minimize risk.

The world now faces a somewhat analogous choice. On one side a large number of scientists believe that the planet is warming rapidly. If we continue to spew out heat-trapping gases, such as carbon dioxide, methane, chlorofluorocarbons, and nitrous oxide, these scientists say, we will warm the planet intolerably within the next century. The consequences of dereliction and procrastination may include killer heat waves, drought, sea level rise, superstorms, vast changes in forests and biota, considerable economic dislocation, and increases in disease: a passable description of hell. But like Pascal's wager, no one can say with absolute certainty what will happen until the consequences of our choice, whatever they may be, are upon us. Nonetheless, "we must wager. It is not optional" (Pascal 1941).

Others, however, claim to have looked over the brink and have decided that hell may not be so bad after all, or at least that we should research the matter further. Yale University economist William Nordhaus (1990b), for example, believes that a hotter climate will mostly affect "those sectors [of the economy] that interact with unmanaged ecosystems" such as agriculture, forestry, and coastal activities. The rest of the economy, including that which operates in what Nordhaus (1990b) called "a carefully controlled environment," which includes shopping malls and presumably the activities of economists, will barely notice that things are considerably hotter. "The main factor to recognize," Nordhaus asserted, "is that the climate has little economic impact upon advanced industrial societies" (Nordhaus 1990a, 193).

Nordhaus concluded that "approximately 3 percent of U.S. national output originates in climate-sensitive sectors and another 10 percent in sectors modestly sensitive to climatic change." There may even be, he noted, beneficial side effects of global warming: "The forest products industry may also benefit from CO_2 fertilization." (It is, I think, no mistake that he did not say "forest" but rather "forest products industry.") Construction, he thinks, will be "favorably affected" as will "investments in water skiing." In sum, Nordhaus's "best guess" is that the impact of a doubling of carbon dioxide "is likely to be around one-fourth of one percent of national income." He admits the estimate has a "large margin of error" (Nordhaus 1990a, 195).

Nordhaus, however, wishes not to be thought to favor climate change. Rather, the point he tried to make is that "those who paint a bleak picture of desert Earth devoid of fruitful economic activity may be exaggerating the injuries and neglecting the benefits of climate change" (Nordhaus 1990b, 196). Whether a hotter Earth, but one not "devoid of fruitful economic activity," might, however, be devoid of poetry, laughter, sidewalk cafés, forests, or even economists he does not say. But he did note that there are a number of technological responses to our plight, including "climate engineering . . . shooting particulate matter [books on economics?] into the stratosphere to cool the earth or changing cultivation patterns in agriculture." Nordhaus, an economist, gave no estimate of the costs, benefits, or even feasibility of these "options." He did, however, estimate the cost of reducing carbon dioxide emissions by 50 percent as $180 billion per year. Faced with such costs, Nordhaus expressed the view that "societies may choose to adapt," which in his words means "population migration, capital relocation, land reclamation, and technological change" (Nordhaus 1990b, 201), solutions for which he again has given no cost estimate. What about those who cannot adapt, migrate, buy expensive remedies, or relocate their capital? Nordhaus does not say, and one suspects that he does not say because he has not thought much about it.

The complications Nordhaus has noticed have to do with "how to discount future costs and how to allow for uncertainty." A discount rate of, say, 8 percent or higher would lead us to do nothing about warming for a few decades while the problem grows gradually or perhaps rapidly worse. A rate of 4 percent or less "would give considerable weight today to climate changes in the late twenty-first century." What is Nordhaus's solution? "The efficient policy," he argued, "would be to invest heavily in high-return capital now and then use the fruits of those investments to slow climate change in the future" (Nordhaus 1990b, 205). He described this as a "sensible compromise" between what he asserts is a "*need* for economic growth" and "the *desire* for environmental protection" [emphasis added], that is, one more binge, virtue later.

To his credit, Nordhaus has acknowledged that "most climatologists think that the chance of unpleasant surprises rises as the magnitude and pace of climatic change increases" (Nordhaus 1990b, 206). He has also noted that the discovery of the ozone hole came as a "complete surprise," suggesting the possibility of more surprises ahead. But in the end he has come down firmly in favor of what he calls "modest steps" that "avoid any precipitous and ill-designed actions that [we] may later regret," actions that he does not specify, making it impossible to know whether they would be in fact precipitous, ill-designed, or regrettable. Nordhaus has stated the belief that "reducing the risks of climatic change is a worthwhile objective" but one, in his opinion, not more important than "factories and equipment, training and education, health and hospitals, transportation and communications, research and development, housing and environmental protection" (Nordhaus 1990b, 209) and so forth. He seems not to have noticed the close relationship between heat, drought, and climate instability, on one hand, and the economy, public health, human behavior under stress, and even what he has called "environmental protection," on the other.

One might dismiss Nordhaus's analysis as an aberration were it not characteristic of the recklessness masquerading as caution that prevails in the highest levels of government and business here and elsewhere and were he not as influential at these levels as he certainly is. Nordhaus's views on global warming are neither an aberration within his profession nor without consequence where portentous choices are made. Nordhaus's opinions about global warming, for example, weighed heavily in the 1991 report issued by the National Academy of Sciences' Adaptation Panel (National Academy of Sciences 1991). The panel, which included Nordhaus, approached global warming as an investment problem requiring the proper discount rate. However, for those whose interests were discounted, such as the poor and future generations, the problem appears differently, as one of power and intergenerational responsibilities. The panel, moreover, assumed a great deal about the adaptability of complex, mass, technological societies under what may be extreme conditions. In citing "the proven adaptability of farmers," for example, are they referring to the 4 million failed farms in the past 50 years? Or to those 1.5 million farms presently at or close to the margin? Or are they referring to the overdependence of agriculture and food distribution systems on the very fossil energy sources that are now heating the Earth? Or perhaps to present rates of soil loss and groundwater depletion due to current farm practices? Can farmers adapt if warming is sudden? Since people live "in both Riyadh and Barrow," the panel drew the implication that humans are almost infinitely adaptable, while admitting that some cities will have to be abandoned and people in poorer countries may be substantially harmed. The panel smartly hedged its bets by admitting that the warming could be sudden and catastrophic but quickly dismissed these possibilities. They did not ask what could happen beyond their 50-year horizon, nor did they ask about the effects on American society of making such portentous decisions in the same way that investment decisions are made about building bridges or shopping malls. It is therefore a matter of concern that such analysis gives considerable aid and comfort to those with all too much to gain by ignoring the risks involved in climate change or the benefits of a farsighted energy policy. Accordingly, we should attempt to understand how such thought comes to pass, whose ends it serves, and what consequences it risks.

By comparison, it is instructive to note that atmospheric physicists, climate experts, and biologists agree almost without exception that the theory of global warming is beyond dispute. It is widely agreed that heat-trapping gases in the atmosphere do in fact trap heat. If we put enough of these in the atmosphere, we will trap a great deal of heat. There is further agreement that if the warming turns out to be rapid, the consequences will in all probability be widely catastrophic, even though we cannot predict these with absolute certainty. Disagreement focuses on matters having to do with rates, thresholds, and the effects of feedbacks that might enhance or retard rates of warming. However these are decided, there is no doubt at all that by increasing heat-trapping gases to levels higher than any in the past 600,000 years and at rates far more rapid than characteristic of past climate shifts, we are conducting an unprecedented experiment with the Earth and its biota. This experiment need not, and should not, be carried out. But like Pascal's wager, certainty about the consequences will come only after all bets are called in.

Given what is at stake, errors of fact and logic committed by Nordhaus and the Adaptation Panel deserve close attention. For example, the belief that decline in agriculture and forestry would be of little consequence because they are only 3 percent of the U.S. economy is equivalent to believing that since the heart is only a few percent of bodyweight, it can be removed or damaged without consequences for one's health. Both Nordhaus and the Adaptation Panel regard the economy as linear and additive without straws that break the back of the camel, surprises, thresholds of catastrophe, or even places where angels would fear to go. The biological facts underlying the research are also suspect. There are many reasons to believe that "CO_2 fertilization" will not enhance farm and forest productivity as Nordhaus and the Adaptation Panel believe. Changes in rainfall, temperature, and biological conditions would more likely reduce growth. Higher temperatures mean higher rates of respiration, hence the release of still more carbon and methane. The rate of climate change may well be many times faster than that to which plants and animals can adapt. This will mean at some unknown date a dieback of forests and the release of even more carbon through fire and rapid decay. It will also mean a sharp reduction in biological diversity.

Economic estimates used by Nordhaus and the Adaptation Panel are also questionable. Both ignore a large and growing body of evidence that the actions necessary to minimize global warming would be good for the economy, human health, and the land. Studies by the U.S. Environmental Protection Agency, the Electric Power Research Institute, and independent researchers [Author's note 2010: and more recently McKinsey & Company] all point to the same conclusion: energy efficiency, which reduces the emission of carbon dioxide, is not only inexpensive, it is in fact a prerequisite of economic vitality. The U.S. economy is roughly one half as energy efficient as that of the Japanese. This fact translates into a 5 percent cost disadvantage for comparable U.S. goods and services (Lovins 1990). Instead of an annual cost that Nordhaus estimates at $180 billion, more reliable studies have shown a net savings of approximately $200 billion from improvements in energy efficiency. This, in Lovins's words, is not a free lunch but a lunch we are paid to eat. However, estimates by Nordhaus or the Adaptation Panel do not include the costs of relocating millions of people, the costs of failing to do so, the costs entailed in diking coasts, the costs of international conflicts over water, the costs of importing food when the plains states become drier, or the costs of changes in diseases due to climate change. Nor does Nordhaus or the Adaptation Panel say what the cost might be if global warming turns out to be rapid and full of even worse surprises.

The practice of discounting the future creates other costs that cannot be quantified but that will be assessed. If they had included the preferences of, say, the third generation hence in the equation, their conclusions would have been quite different. Nordhaus and the Adaptation Panel chose not to do so, however, by assuming that investments in more of the same kinds of activities that created the problem in the first place were "worthy goals." On closer examination, most of these will intensify the problem of global warming and dig us in still deeper while ignoring opportunities to invest in energy efficiency and renewables that would reduce the emission of heat-trapping gases in an economically sound manner.

The economic estimates of Nordhaus and the Adaptation Panel are not to be trusted, because their economy is an abstraction independent of biophysical realities, comparable, say, to an airline pilot who regarded the law of gravity as merely an interesting but untested theory. Their economics are not to be trusted because they fail to acknowledge the vast and unknowable complexity of planetary systems, which cannot be "fixed" by any technology without courting other risks. Their economics cannot be trusted because they are not very good economics. They have ignored the relationship between economic prosperity and energy efficiency, as well as that between energy efficiency and the emission of greenhouse gases. Their economics are not to be trusted because the problem of global warming is not first and foremost one of economics, as they believe, but rather one of judgment, wisdom, and love for the Creation. Their economics cannot be trusted because they do not include flesh-and-blood people who, under conditions of a rapidly changing climate, will not act with the rationality presumed in abstract models concocted in air-conditioned rooms. Real people stressed by heat, drought, economic decline, and perhaps worse will curse and kill more often and celebrate and love less often. And they will mourn the loss of places disfigured by heat, drought, and death that were once familiar, restoring, and consoling.

Finally, the economics of Nordhaus and the Adaptation Panel cannot be trusted because they would have us risk this and more for another decade or two of business as usual, which as we now know does not mean sustainable prosperity or basic fairness. This is a foolish risk for reasons Pascal described well. If it turns out that global warming would have been severe and we forestalled it by becoming more energy efficient and making a successful transition to renewable energy, we will have avoided disaster. If, however, it turns out that factors as yet unknown minimized the severity and impact of warming while we became more energy efficient in the belief that it might be otherwise, we will not have saved the planet, but we will have reduced acid rain, improved air quality, decreased oil spills, reduced the amount of strip-mining, reduced our dependence on imported oil and thereby improved our balance of payments, become more technologically adept, and improved our economic competitiveness. In either case we will have set an instructive and farsighted precedent for our descendants and for the future of the Earth. If we gain, we gain all; if we "lose," we still gain a great deal. On the other hand, if we do as Nordhaus and the members of the Adaptation Panel would have us do, and the warming proves to be rapid, there will be hell to pay.

Chapter 30

The Carbon Connection

(2007)

Having seen pictures of the devastation did not prepare me for the reality of New Orleans. Mile after mile of wrecked houses, demolished cars, piles of debris, twisted and downed trees, and dried mud everywhere. We stopped every so often to look into abandoned houses in the ninth ward and along the shore of Lake Pontchartrain to see things close up: mud lines on the walls, overturned furniture, moldy clothes still hanging in closets, broken toys, a lens from a pair of glasses . . . once cherished and useful objects rendered into junk. Each house with a red circle painted on the front to indicate results of the search for bodies. Some houses showed the signs of desperation: holes punched through ceilings as people tried to escape rising water. The smell of musty decay was everywhere, overlaid with an oily stench. Despair hung like Spanish moss in the dank, hot July air.

Ninety miles to the south, the Louisiana delta is rapidly sinking below the rising waters of the Gulf. This is no "natural" process, but rather the result of decades of mismanagement of the lower Mississippi that became federal policy after the great flood of 1927. Sediment that built the richest and most fecund wetlands in the world is now deposited off the continental shelf—part of an ill-conceived effort to tame the river. The result is that the remaining wetlands, starved for sediment, are both eroding and compacting, sinking below the water and perilously close to no return. Oil extraction has done most of the rest by cutting channels that crisscross the marshlands, allowing the intrusion of salt water and storm surges. Wakes from boats have widened the original channels considerably fur-

---

This article was originally published in 2007.

ther, unraveling the ecology of the region. The richest fishery in North America and a unique culture that once thrived in the delta are disappearing and with it the buffer zone that protects New Orleans from hurricanes. "Every 2.7 miles of marsh grass," in Mike Tidwell's words, "absorbs one foot of a hurricane's storm surge" (Tidwell 2003, 57).

And the big hurricanes will come. Kerry Immanuel, an MIT scientist and once greenhouse skeptic, researched the connection between rising levels of greenhouse gases in the atmosphere, warmer sea temperatures, and the severity of storms. He's a skeptic no longer for reasons he described in *Nature* (Immanuel 2005). The hard evidence on this and other parts of climate science has moved beyond the point of legitimate dispute. Carbon dioxide, the prime greenhouse gas, is at the highest level in at least the last 650,000 years. CO_2 continues to accumulate by more than 2.5 parts per million per year, edging closer and closer to what some scientists believe is the threshold of runaway climate change. British scientist James Lovelock compares our situation to being on a boat upstream from Niagara Falls with the engines about to fail (Midgley 2006, vii).

If this were not enough, the evidence now shows a strong likelihood that sea levels will rise more rapidly than previously thought. The third report of the Intergovernmental Panel on Climate Change (2001) predicted about a 1-meter rise in the twenty-first century, but more recent evidence puts this figure at 6 to 7 meters—the result of accelerated melting of the Greenland ice sheet and polar ice, along with the thermal expansion of water.

Nine hundred miles to the northeast as a sober crow would fly it, Massey Energy, Arch Coal, and other companies are busy leveling the mountains of Appalachia to get at the upper seams of coal in what was one of the most diverse and relatively undisturbed forests in the U.S. and one of the most diverse ecosystems anywhere. Throughout the coalfields of West Virginia and Kentucky they have already leveled 456 mountains across 1.5 million acres and intend to damage a good bit more. Coal is washed on-site, leaving billions of gallons of a dilute asphalt-like gruel laced with toxic flocculants and heavy metals. An estimated 225 such containment ponds are located over abandoned mines in West Virginia, held back from the communities below only by earthen dams prone to failure either by collapse or by draining down through old mine tunnels that honeycomb the region. One did fail in October 11, 2000, in Martin County, Kentucky, when the slurry broke through a thin layer of shale and into mines and out into hundreds of miles of streams and rivers. The result was the permanent destruction of waterways and property values of people living in the wake of an ongoing and mostly ignored disaster. This is typical of the coalfields. They are a third world colony within the United States; a national sacrifice zone in which fairness, decency, and the rights of old and young alike are discarded as so much overburden on behalf of the national obsession with "cheap" electricity. For his role in trying to enforce even the flimsy laws that might have held Massey Energy slightly accountable for its flagrant and frequent malfeasances, the Bush administration tried unsuccessfully to fire Jack Spadaro from his position as a mine safety inspector in the Interior Department but eventually forced him to retire.

Jack is in the first plane to take off from Yeager Field in Charleston, along with the chief attorney for the largest corporation in the world. Hume Davenport, founder of SouthWings, Inc., is the pilot of the four-seat Cessna. The ground recedes below us as we pass over Charleston and the Kanawha River lined with barges hauling coal to power plants along the Ohio River and points more distant. Quickly on the horizon to the west is the John Amos plant owned by American Electric Power that, by one estimate, releases more mercury to the environment than any other facility in the U.S. as well as hundreds of tons of sulfur oxides, hydrogen sulfide, and CO_2. For a few minutes we can see the deep green of wrinkled Appalachian hills below, but very soon the first of the mountaintop removal sites appears. It is followed by another and then another. The pattern of ruin spreads out below us for many miles stretching to the far horizon on all points of the compass. From a mile above, trucks with 12-foot-diameter tires and draglines that could pick up two Greyhound buses at a single bite look like Tonka toys in a sandbox. What is left of Kayford Mountain comes into sight. It is surrounded by leveled mountains, and a few still being leveled. "Overburden," the mining industry term for dismantled mountains, is dumped into valleys covering hundreds of miles of streams—an estimated 1500 miles in the past 25 years. Many more miles will be buried if the coal companies have their way. Coal slurry ponds loom above houses, towns, and even elementary schools. When the earthen dams break on some dark rainy night, those below will have little if any warning before the deluge hits.

Jack Spadaro is our guide to the devastation. He is a heavyset, rumpled, and bearded man with the knack for describing outrageous things calmly and with clinical precision. A mining engineer by profession, he spent several frustrating decades trying to enforce the laws, such as they are, against an industry with friends in high places in Charleston, Congress, and the White House. In a flat, unemotional monotone he describes what we are seeing below. Aside from the destruction of the Appalachian forest, the math is all wrong. The slopes are too steep, the impoundments too large. The angles of slope, dam, weight, and proximity of houses and towns are the geometry of tragedies to come. He points out the Marsh Fork elementary school situated close to a coal-loading operation and below a huge impoundment back up the hollow. In the event of a dam failure, the evacuation plan calls for the principal to use a bullhorn to initiate the evacuation of the children ahead of the 50-foot wall of slurry that will be moving at maybe 60 miles an hour. If all works according to the official evacuation plan, they will have 2 minutes to get to safety, but there is no safe place for them to go. And so it is in the coalfields—ruin at a scale for which there are no adequate words; ecological devastation to the far horizon of topography and time. We say that we are fighting for democracy elsewhere, but no one in Washington or Charleston seems aware that we long ago deprived some of our own of the rights to life, liberty, and property.

On the circle back to Yeager Field in Charleston, Tom Hyde, a corporate attorney, calls this a "tragedy." We all nod, knowing the word does not quite describe the enormity of the things we've just seen or the cold-blooded nature of it. In our 1-hour flight we saw maybe 1 percent of the destruction now metastasizing through four states. Until recently it was all but ignored by the national media. But we have known of the costs of mining at least since Harry Caudill published *Night Comes to the Cumberlands* in 1963, but we have yet to summon the moral energy to resolve the problem or pay the full costs of the allegedly cheap electricity that we use.

Under the hot afternoon sun we board a 15-person van to drive out to the edge of the coalfields to see what it looks like on the ground. On the way to Kayford Mountain, we take the interstate south from Charleston and exit at a place called Sharon onto winding roads that lead to mining country. Trailer parks, small evangelical churches, truck repair shops, and small often lovingly tended houses line the road intermixed with those abandoned long ago when underground mining jobs disappeared. The two-lane paved road turns to gravel and climbs toward the top of the hollow and Kayford Mountain. Within a mile or two the first valley fill appears. It is a green V-shaped insertion between wooded hills. Reading the signs made by water coursing down its face, Jack Spadaro notes that this one will soon fail. Valley fills are mountains turned upside down: rocky mining debris, trees illegally buried, along with what many locals believe to be more sinister things brought in by unmarked trucks in the dead of night. He adds that some valley fills may contain as much as 500 million tons of blasted mountains and run for as long as 6 miles. We ascend the slope toward Kayford, passing by the no-trespassing signs that appear around the gate that leads to the mining operations.

Larry Gibson, a diminutive bulldog of a man fighting for his land, meets us at the summit, really a small peak on what was once a long ridge. The family has been on Kayford since the eighteenth century, operating a small coal mine. Larry is the proverbial David fighting Goliath, but he has no slingshot unless it is that of moral authority spoken with a fierce, inborn eloquence. Those traits and the raw courage he shows every day have made Larry a poster child for the movement, with his picture in *Vanity Fair*, *National Geographic*, and other newsstand magazines. Larry's land has been saved so far because he made 40 acres of it into a park and has fought tooth and nail to save it from the lords of Massey Energy. They have leveled nearly everything around him and have punched holes underneath Kayford because the mineral rights below and the ownership of the surface were long ago separated in a shameless scam perpetrated on illiterate and trusting mountain people.

Larry describes what has happened, using a model of the area that comes apart more or less like the mountains around him have been dismantled. As he talks, he illustrates what has happened by taking the model apart piece by piece, leaving the top of Kayford rather like a knob sticking up amidst the encircling devastation. So warned, we walk down the country lane to witness the advancing ruin. Fifteen of us stand for maybe half an hour on the edge of the abyss, watching giant bulldozers and trucks at work below us. Plumes of dust from the operations rise up several thousand feet. The next set of explosive charges is ready to go on an area about the size of a football field. Every day some 3 million pounds of explosives are used in the 11 counties south of Charleston. This is a war zone. The mountains are the enemy, profits from coal the prize, and the local residents and all those who might have otherwise lived here or would have been re-created here are the collateral damage.

We try to wrap our minds around what we are seeing, but words do no justice to the enormity of it. The oldest mountains on Earth are being turned into gravel for a pittance, their ecologies radically simplified, forever. Perhaps as a defense mechanism from feeling too much or being overwhelmed by what we've seen, we talk about lesser things. In the late afternoon drive back to Charleston, we pass by the coal-loading facilities along the Kanawha River. Mile after mile of barges lined up to haul coal to hungry Ohio River power plants, the umbilical cord between mines, mountains, and us—the consumers of cheap electricity.

Over dinner that night we hear from two residents of Mingo County who describe what it is like to live in the coalfields. Without forests to absorb rainwater, flash floods are a normal occurrence. A 3-inch rain can become a 10-foot wall of water cascading off the flattened mountains and down the hollows. The mining industry calls these "acts of God," and the thoroughly bought public officials agree, leaving the victims with no recourse. Groundwater is contaminated by coal slurry and the chemicals used to make coal suitable for utilities. Well water is so acidic that it dissolves pipes and plumbing fixtures. Cancer rates are off the charts, but few in Charleston or Washington care enough to notice. Coal companies are major buyers of politicians, and the head of Massey Energy, Donald Blankenship, has been known to spend lots of money to buy precisely the kind of representatives he likes—the sort that can accommodate themselves to exploitation of land and people and the profits to be made from it. His campaign to ravage the rest of West Virginia is titled "For the Sake of the Kids."

Pauline and Carol from the town of Sylvester, both in their seventies, are known as the "dust busters" because they go around the town wiping down flat surfaces with white cloths that are then covered with coal dust from a nearby loading facility. These are presented as evidence of foul air at open hearings to the irritated and unmovable servants of the people. Black lung and silicosis disease is now common among young and old alike exposed to the dust from surface operations but who have never set foot in a mine. They have little or no voice in government; they are considered to be expendable. Pauline, a fiercely eloquent woman, whose husband was wounded and captured by the Germans in the Battle of the Bulge in 1944, rhetorically asks, "Is this what he fought for?" The clock reads 9:30 PM; we quit for the day.

To permanently destroy millions of acres of Appalachia in order to extract maybe 20 years of coal is not just stupid; it is a derangement at a scale for which we as yet do not have adequate words, let alone the good sense and the laws to stop it. Unlike deep mining, mountaintop removal employs few workers. It is destroying the wonders of the mixed mesophytic forest of northern Appalachia once and for all, including habitat for dozens of endangered species. It contaminates groundwater with toxics and heavy metals and renders the land permanently uninhabitable and unusable. Glib talk of the economic potential of flatter places for commerce of one kind or another is just that: glib talk. Coal companies' efforts to plant grass and a few trees here and there are like putting lipstick on a corpse. The fact of the matter is that one of the most diverse and beautiful ecosystems in the world is being destroyed and rendered uninhabitable forever, along with the lives and culture of the people who have stayed behind in places like Sylvester and Kayford. We justify this on the grounds of necessity and cost. But virtually every competent independent study of energy use done in the past 30 years has concluded that we could cost-effectively eliminate half or more of our energy use and strengthen our economy, lower costs of asthma and lung disease, raise our standard of living, and improve environmental quality. More complete accounting of the costs of coal would also include the rising tide of damage and insurance claims attributable to climate change. Some say that if we don't burn coal, the economy will collapse and we will all have to go back to the caves. But with wind and solar power growing by 25 percent plus per year and the technology of energy efficiency advancing rapidly, we have good options that make burning coal unnecessary. And before long we will wish that we had not destroyed so much of the capacity of the Appalachian forests and soils to absorb the carbon that makes for bigger storms and more severe heat waves and droughts.

No one in a position of authority in West Virginia politics, excepting that noble patriarch of good sense, Ken Hechler, asks the obvious questions. How far does the plume of heavy metals coming from coal-washing operations go down the Kanawha, Ohio, and Mississippi and into the drinking water of communities elsewhere? What other economy, based on the sustainable use of forests, nontimber products, ecotourism, and human craft skills, might flourish in these hills? What is the true cost of "cheap" coal? Why do the profits from coal mining leave the state? Why is so much of the land owned by absentee corporations like the Pocahontas Land Company? Once you subtract the permanent ecological ruin and crimes against humanity, there really isn't much to add, as a country song once put it. Those touting "clean coal" ought to spend some time in the coalfields and talk to the residents in order to understand what those words really mean at the point of extraction. And for those who assume that the carbon from burning coal can be safely and permanently sequestered underground at an affordable cost, I have oceanfront property to sell you in Arizona.

Nearly 1000 miles separate the coalfields of West Virginia from the city of New Orleans and Gulf coast, yet they are a lot closer than that. The connection is carbon. Coal is mostly carbon, and for every ton burned, 3.6 tons of CO_2 eventually enters the atmosphere, raising global temperatures, warming oceans and thereby creating bigger storms, melting ice, and raising sea levels. For every ton of coal extracted from the mountains, perhaps 100 tons of what is tellingly called "overburden" is dumped, burying streams and filling the valleys and hollows of West Virginia, Kentucky, and Tennessee. And between the hills of Appalachia and the sinking land of the Louisiana coast, tens of thousands of people living downwind from coal-fired power plants die prematurely each year from inhalation of small particles of smoke laced with heavy metals that penetrate deeply into lungs.

Like all life-forms, we search out great pools of carbon to perpetuate ourselves. It is our mismanagement of carbon that threatens the human future, and this is an old story. Humans have long fought for the control of carbon found in rich soils and deep forests and later in fossil fuels. The root of all evil does not begin with money, but with the carbon in its various forms that money can buy. The exploitation of carbon is the original sin, leading quite possibly to the heat death of a great portion of life on Earth, including us. This is what James Lovelock calls "the revenge of Gaia" (Lovelock 2006).

Chapter 31

2020: A Proposal

(2000)

AUTHOR'S NOTE 2010: *This essay first appeared as my column in* Conservation Biology *and subsequently in* The Chronicle of Higher Education.

We all live by robbing Asiatic coolies, and those of us who are "enlightened" all maintain that those coolies ought to be set free; but our standard of living, and hence our "enlightenment" demands that the robbery shall continue.

GEORGE ORWELL

BY A LARGE MARGIN 1998 was the warmest year ever recorded. The previous year was the second warmest. A growing volume of scientific evidence indicates that, given present trends, the combustion of fossil fuels, deforestation, and poor land-use practices will cause a major, and perhaps self-reinforcing, shift in global climate (Houghton 1997). With climatic change will come severe weather extremes, super storms, droughts, killer heat waves, rising sea levels, spreading disease, accelerating rates of species loss, and collateral political, economic, and social effects that we cannot imagine. We are conducting, as Roger Revelle once noted, a one-time experiment on the Earth that cannot be reversed and should not be run.

The debate about climatic change has, to date, been mostly about scientific facts and economics, which is to say a quarrel about unknowns and numbers. On one side are those (greatly appreciated by the fossil fuel industry) who argue that we do not yet know enough to act and

---

This article was originally published in 2000.

that acting prematurely would be prohibitively expensive. On the other side are those who argue that we do know enough to act and that further procrastination will make subsequent action both more difficult and less efficacious. In an election year in the United States, which happens to be the largest emitter of greenhouse gases, the issue is not likely to be discussed in any constructive manner. And the U.S. Congress, caught in a miasma of ideology and partisanship, is in deep denial, unable to ratify the Kyoto Accord that called for a 7 percent reduction of 1990 CO_2 levels by 2012. Even that level of reduction, however, would not be enough to stabilize climate.

To see our situation more clearly, we need a perspective that transcends the minutiae of science, economics, and current politics. Since the effects, whatever they may be, will fall most heavily on future generations, understanding their likely perspective on our present decisions would be useful to us now. And how are future generations likely to regard various positions in the debate about climatic change? Will they applaud the precision of our economic calculations that discounted their prospects to the vanishing point? Will they think us prudent for delaying action until the last-minute scientific doubts were quenched? Will they admire our heroic devotion to inefficient cars and sport utility vehicles, urban sprawl, and consumption? Hardly. They are more likely, I think, to judge us much as we now judge the parties in the debate on slavery prior to the Civil War.

Stripped to its essentials, defenders of the idea that humans can hold other humans in bondage developed four lines of argument. First, citing Greek and Roman civilization, some justified slavery by arguing that the advance of human culture and freedom had always depended on slavery. "It was an inevitable law of society," according to John C. Calhoun, "that one portion of the community depended upon the labor of another portion over which it must unavoidably exercise control" (Miller, W. L., 1998, 132). And "freedom," the editor of the *Richmond Inquirer* once declared, "is not possible without slavery" (Oakes 1998, 141). This line of thought, discordant when appraised against other self-evident doctrines that "all men are created equal," is a tribute to the capacity of the human mind to simultaneously accommodate antithetical principles. Nonetheless, it was used by some of the most ardent defenders of "freedom" right up to the Civil War.

A second line of argument was that slaves were really better off living here in servitude than they would have been in Africa. Slaves, according to Calhoun, "had never existed in so comfortable, so respectable, or so civilized a condition as that which [they] enjoyed in the Southern States" (Miller, W. L., 1998, 132). The "happy slave" argument fared badly with the brute facts of slavery that became vivid for the American public only when dramatized by Harriet Beecher Stowe in *Uncle Tom's Cabin* published in 1852.

A third argument for slavery was cast in cost-benefit terms. The South, it was said, could not afford to free its slaves without causing widespread economic and financial ruin. This argument put none too fine a point on the issue; slavery was simply a matter of economic survival for the ruling race.

A fourth argument, developed most forcefully by Calhoun, held that slavery, whatever its liabilities, was up to the various states, and the federal government had no right to interfere with it because the Constitution was a compact between independent political units. Beneath all such arguments, of course, lay bedrock contempt for human equality, dignity, and freedom. Most of us, in a more enlightened age, find such views repugnant.

While the parallels are not exact between arguments for slavery and those used to justify inaction in the face of prospective climatic change, they are, perhaps, sufficiently close to be instructive. First, those saying that we do not know enough yet to limit our emission of greenhouse gases argue that human civilization, by which they mean mostly economic growth for the already wealthy, depends on the consumption of fossil fuels. We, in other words, must take substantial risks with our children's future for a purportedly higher cause: the material progress of civilization now dependent on the combustion of fossil fuels. Doing so, it is argued, will add to the stock of human wealth that will enable subsequent generations to better cope with the messes that we will leave behind.

Second, proponents of procrastination now frequently admit the possibility of climatic change but argue that it will lead to a better world. Carbon enrichment of the atmosphere will speed plant growth, enabling agriculture to flourish, increasing yields, lowering food prices, and so forth. Further, while some parts of the world may suffer, a warmer world will, on balance, be a nicer and more productive place for succeeding generations.

Third, some, arguing from a cost-benefit perspective, assert that energy conservation and solar energy are simply too expensive now. We must wait for technological breakthroughs to reduce the cost of energy efficiency and a solar-powered world. Meanwhile we continue to expand our dependence on fossil fuels, thereby making any subsequent transition still more expensive and difficult.

Finally, arguments for procrastination are grounded in a modern-day version of states' rights and extreme libertarianism, which makes squandering fossil fuels a matter of individual rights, devil take the hindmost.

The fit between slavery and our present use of fossil fuels is by no means perfect, but it is close enough to be suggestive. Of course we do not intend to enslave subsequent generations, but we will leave them in bondage to degraded climatic and ecological conditions that we will have created. Further, they will know that we failed to act on their behalf with alacrity even after it became clear that our failure to use energy efficiently and develop alternative sources of energy would severely damage their prospects. In fact, I am inclined to think that our dereliction will be judged as a more egregious moral lapse than that which we now attribute to slave owners. For reasons that one day will be regarded as no more substantial than those supporting slavery, we knowingly bequeathed the risks and results of climatic change to all subsequent generations, everywhere. If not checked soon, that legacy will include severe droughts, heat waves, famine, changing disease patterns, rising sea levels, and political and economic instability. It will also mean degraded political, economic, and social institutions burdened by bitter conflicts over declining supplies of fossil fuels, water, and food. It is not far-fetched to think that human institutions, including democratic governments, will break under such conditions.

Other similarities exist. Both the use of humans as slaves and the use of fossil fuels allow those in control to command more work than would otherwise be possible. Both inflate wealth of some by robbing others. Both systems work only so long as something is underpriced: the devalued lives and labor of a bondsman or fossil fuels priced below their replacement costs. Both require that some costs be ignored: those to human beings stripped of choice, dignity, and freedom or the cost of environmental "externalities," which cast a long shadow on the prospects of our descendants. In the case of slavery, the effects were egregious, brutal, and immediate. But massive use of fossil fuels simply defers the costs, different but no less burdensome, onto our descendants, who will suffer the consequences with no prospect of manumission. Slavery warped the politics and cultural evolution of the South. But our dependence on fossil fuels has also warped and corrupted our politics and culture in ways too numerous to count. Slaves could be manumitted, but the growing numbers of victims of global warming have no reprieve. We leave behind steadily worsening conditions that cannot be altered in any time span meaningful to humans.

Both slavery and fossil fuel–powered industrial societies require a mass denial of responsibility. Slave owners were caught in a moral quandary. Their predicament, in James Oakes's words, was "the product of a deeply rooted psychological ambivalence that impels the individual to behave in ways that violate fundamental norms even as they fulfill basic desires" (Oakes 1998, 120). Regarding slavery, George Washington confessed, "I shall frankly declare to you that I do not like even to think, much less talk, of it." As one Louisiana slave owner put it, "a gloomy cloud is hanging over our whole land." Many wished for some way out of a profoundly troubling reality. Instead of finding a decent way out, however, the South created a culture of denial around the institutions of bondage. They were enslaved by their own system until it came crashing down around them in the Civil War.

We, too, find ourselves in a quandary. A poll conducted for the American Geophysical Union revealed that most Americans believe that global warming is real and that its consequences will be tragic and irreversible. But the response of Congress and much of the business community has been to deny that the problem exists and continue with business as usual. Proposals for higher gasoline taxes, increasing fuel efficiency, or limits on use of automobiles, for example, are regarded as politically impossible as the abolition of slavery in the 1830s. Unless we take appropriate steps soon, our system, too, will end badly.

We now know that heated arguments made for the enslavement of human beings were both morally wrong and self-defeating. The more alert knew this early on. Benjamin Franklin noted that slaves "pejorate the families that use them; the white children become proud, disgusted with labor, and being educated in idleness, are rendered unfit to get a living by industry" (Finley 1980, 100). Thomas Jefferson knew all too well that slavery degraded slaves and slave owners alike, while providing no sustainable basis for prosperity in an emerging capitalist economy. In a rough parallel, it is possible that the extravagant use of fossil fuels has become a substitute for intelligence, exertion, design skill, and foresight. On the other hand, we have every reason to believe that vastly improved energy efficiency and an expeditious transition to a solar-powered society would be to our advantage, morally and economically. Energy efficiency could lower our energy bill in the U.S. alone by as much as $200 billion per year (Hawken et al. 1999). It would reduce environmental impacts associated with mining, processing, transportation, and combustion of fossil fuels and promote better technology. Elimination of subsidies for fossil fuels, nuclear power, and automobiles would save tens of billions each year (Myers 1998). In other words, the "no regrets" steps necessary to avert the possibility of severe climatic change, taken for sound ethical reasons, are the same steps we ought to take for reasons of economic self-interest. History rarely offers such a clear convergence of ethics and self-interest.

If we are to take this opportunity, however, we must be clear that the issue of climatic change is not, first and foremost, a matter of economics, technology, or science but, rather, a matter of principle that is best seen from the vantage point of our descendants. The same historical period that gave us slavery also gave us the principles necessary to abolish it. What Thomas Jefferson called "remote tyranny" was not merely tyranny remote in space but in time as well—what Bill McDonough has termed "intergenerational remote tyranny." In a letter to James Madison written in 1789, Jefferson argued that no generation had the right to impose debt on its descendants, for were it to do so, the future would be ruled by the dead, not the living.

A similar principle applies in this instance. Drawing from Jefferson, Aldo Leopold, and others, such a principle might be stated thusly:

> No person, institution, or nation has the right to participate in activities that contribute to large-scale, irreversible changes of the Earth's biogeochemical cycles or undermine the integrity, stability, and beauty of the Earth's ecologies—the consequences of which would fall on succeeding generations as an irrevocable form of remote tyranny.

That principle will likely fall on uncomprehending ears in Congress and in most corporate boardrooms. Who, then, will act on it? Who ought to act? Who can lead? What institutions represent the interests of our children and succeeding generations on whom the cost of present inaction will fall? At the top of my list are those that purport to educate and thereby to equip the young for useful and decent lives. Education is done in many ways, the most powerful of which is by example. The example the present generation needs most from those who propose to prepare them for responsible adulthood is a clear signal that their teachers and mentors are themselves responsible and will not, for any reason, encumber their future with risk or debt—ecological or economic. And they need to know that our commitment is more than just talk. This principle can be stated in these words:

> The institutions that purport to induct the young into responsible adulthood ought themselves to operate responsibly, which is to say that they should not act in ways that might plausibly undermine the world their students will inherit.

Accordingly, I propose that every school, college, and university stand up and be counted on the issue of climatic change by beginning now to develop plans to reduce and eventually eliminate or offset the emission of heat-trapping gases by the year 2020. Opposition to such a proposal will, predictably, follow along three lines. The first line of objection will arise from those who argue that we do not yet know enough to act. In other words, until the threat of climatic change is clear beyond any possible doubt (and also less easily reversed), we cannot act. Presumably, these same people do not wait until they smell smoke in the house at 2 AM to purchase fire insurance. A "no regrets" strategy relative to the far-from-remote possibility of climatic change is, by the same logic, a way to insure our descendants against the possibility of disaster otherwise caused by our carelessness.

A second line of objection will come from those who will argue that even so, educational institutions on their own cannot afford to act. To be certain, there will be initial expenses, but there are also quick savings from reducing energy use. In fact, done smartly, implementation of energy efficiency and solar technology can save money. Moreover, it is now possible to use energy service companies that will finance the work and pay themselves from the stream of savings, making the transition budget neutral. The real problem here has less to do with costs than with moral energy and the failure to imagine possibilities in places where imagination and creativity are reportedly much valued.

A third kind of objection will come from those who agree with the overall goal of stabilizing climate but will argue that our business is education, not social change. This position is premised on the quaint belief that what occurs in educational institutions must be uncontaminated by contact with the affairs of the world and that we have no business objecting to how that world does its business. It is further assumed that education occurs only in classrooms and must be remote from anything having practical consequences. Were the effort to eliminate the use of fossil fuels, however, done as a 20-year effort in which students worked with faculty, staff, administration, energy engineers, and technical experts, the educational and institutional benefits would be substantial.

How might the abolition of fossil fuels occur? In outline, the basic steps are straightforward, requiring

- thorough audit of current institutional energy use;
- preparation of detailed engineering plans to upgrade energy efficiency and eliminate waste;
- development of plans to harness renewable energy sources sufficient to meet campus energy needs by 2020;
- competent implementation.

These steps ought to engage students, faculty, administration, staff, and representatives of the surrounding community. They ought to be taken publicly as a way to educate a broad constituency about the consequences of our present course and the possibilities and opportunities for change.

The longer-term goal of this effort is to begin, from the grassroots, the long-delayed transition to energy efficiency and solar power. Perhaps our leaders will follow one day when they are wise enough to distinguish the public interest from narrow short-run private interests. Someday, too, all of us will come to understand that true prosperity neither permits nor requires bondage of any human being, in any form, for any reason, now or ever.

Chapter 32

Baggage: The Case for Climate Mitigation

(2009)

Adapt to species loss, ice sheet disintegration, increased intensity of floods, storms, droughts and fires? Such talk is disingenuous and futile. For the sake of justice and equity, for our children, grandchildren and nature we have no choice but to focus on mitigation.

JAMES HANSEN

ON JUNE 24 OF 1812 Napoleon invaded Russia but with no very clear idea of what he intended to do. His motives, we can assume, included the usual testosterone-driven potpourri of territorial expansion, plunder, power, adventure, and glory. The opponent was said to be the czar of Russia, Alexander I, a mercurial sort much given to religious zealotry and the conviction that he was but a humble instrument of God or vice versa.

From the beginning, the campaign was difficult. Water, forage for horses, and food were scarce. Storms turned roads into quagmires one day, and on the next, soldiers baked in the extreme summer heat. To make matters worse, a confused Alexander did not give adequate opportunity for manly combat. Instead the Russian armies led by the capable General Mikhail Barclay de Tolly avoided battle by retreating eastward toward

This article was originally published in 2009.

Moscow, drawing Napoleon's Grande Armée deeper into the endless Russian plains. The result was to lengthen French supply lines, rendering them vulnerable to attack and the normal breakdowns of horse-drawn transport. Only on September 7 did Barclay's replacement, the elderly General Mikhail Kutuzov, deign to give combat on the outskirts of Moscow at the village of Borodino. The battle distinguished neither general, but Napoleon prevailed in a manner of speaking, and the way to Moscow lay open.

On arrival, however, Napoleon and his Grande Armée discovered two-thirds of the city had been burned by the retreating Russians, and the exotic glories, pleasures, and practical usefulness of Moscow were thereby considerably diminished. Nonetheless they set about with considerable alacrity to loot what remained and settled in to a more or less uneventful 5-week occupation. With no enemy willing to give battle, and longing for the delights of warmer, safer, and more civilized places, Napoleon decided to go home. A considerably less grand Grande Armée departed Moscow October 19, weighted down with everything of value that its soldiers could haul—jewelry, women's finery, household furnishings, artwork, musical instruments—booty of every sort and description. One participant saw "soldiers wheeling barrows loaded with everything they had been able to pile on them . . . their senseless greed had closed their eyes to the fact that two thousand miles and many battles lay between them and their destination" (de Ségur 1980, 136).

The long journey home was not Napoleon's finest hour. First, by its brutality and arrogance the Grande Armée had managed to ignite the hostility of the Russian peasantry, a fairly difficult thing to do. The result was unending guerrilla attacks on Napoleon's flanks and rear. Second, winter set in with a vengeance. Temperatures plummeted below zero and stayed there. Under assault by peasant guerrillas, the Russian army, and bitter cold, the once formidable soldiers began to shed their plunder. Roads westward were littered with candelabras, women's finery, furniture, artwork, and assorted treasures for hundreds of miles. As the situation became desperate, Napoleon's soldiers became more like a mob and threw away everything that was not absolutely essential to life and limb and the westward stampede. For most, however, it was too late. Of the nearly 600,000 men who invaded Russia, fewer than 100,000 got out alive.

The story is perhaps useful to illustrate what harsh reality can do to clarify priorities. Sometimes you can't take it all with you. Wishful thinking and denial do not change the weather. Sometimes you get out of a jam by the narrowest of margins, if you get out at all. But it is always smarter to avoid them in the first place.

~

The awareness that humans could alter the climate of Earth has dawned slowly on our consciousness. In 1896 Svante Arrhenius deflected his anguish over a failed marriage into remarkably tedious and, as it turned out, accurate calculations about the effect of CO_2 emissions on climate. It was an oddly therapeutic thing to do, but it had no more effect on public attention than the smallest cloud on a distant horizon. Another 69 years would pass before scientists warned a U.S. president of the potential for serious climate disruption, and still another 30 years would pass before the first report from the Intergovernmental Panel on Climate Change.

Facing climate destabilization, our choices are said to be adaptation, mitigation, and suffering. The suffering from climate change–driven weather events and rising seas has already begun and will likely grow more extreme in decades ahead but is beyond the scope of this article. Accordingly I will consider only adaptation and mitigation. The advocates of each appear to come from different scientific backgrounds. Adaptationists, I think, come mostly from backgrounds in wildlife conservation, agriculture, urban planning, and landscape architecture, while mitigationists represent the various branches of atmospheric and climate science. The differences are telling.

The argument for adaptation to the effects of climate change rests on a chain of logic that goes something like this:

1. Climate change is real but will be
2. slow and moderate enough to permit orderly adaptation to changes
3. that we can foresee and comprehend and which
4. will, in a few decades, plateau around a new, manageable stable state,
5. leaving the gains of the modern world mostly intact, albeit powered by advanced technology, wind, solar, and as yet undreamed technology.

In other words, the developed world can adapt to climatic changes without sacrificing much. The targets for adaptation include developing heat and drought-tolerant crops for agriculture, changing architectural standards to withstand greater heat and larger storms, and modifying infrastructure to accommodate larger storm events as well as prolonged heat and drought (Morello and Goodman 2009). These are imminently sensible and obvious measures that we must take. But beyond some point there are limits to what can be done and the places in which such measures can be effective. With predicted changes in temperature, rainfall, and sea level rise, it is not likely that we can "promote ecosystem resiliency" or adapt to such changes with "no regrets" as some suggest. To the contrary, ecological resilience and biological diversity will almost surely decline as climatic changes now under way accelerate, and going forward we will surely have a great many regrets—but of the "why did we not do more to stop it earlier" sort. Accordingly, more extreme adaptive measures called "geoengineering" are being discussed. These include proposals to fertilize oceans with iron to increase carbon uptake or injecting SO_2 into the upper atmosphere to increase the reflective albedo and thereby provide temporary cooling. But since the effects of geoengineering are largely unstudied and its risks largely unknown, it is a "true option of last resort," in the words of one analysis. The authors conclude that "the best and safest strategy for reversing climate change is to halt the buildup of greenhouse gases" (Victor et al. 2009, 76).

Proponents of mitigation, then, give priority to limiting the emission of heat-trapping gases as quickly as possible to reduce the eventual severity of climatic disruption. The essence of the case for mitigation is that

1. growing scientific evidence indicates that the effects of climate change will be greater and will occur faster than previously thought;
2. the duration of climate effects will last for thousands of years, not decades;
3. we are in a very tight race to avoid causing irreversible changes that would seriously damage or destroy civilization;
4. the effects of climate destabilization can be contained perhaps only by emergency action to stabilize and then reduce CO_2 levels.

Practically, climate mitigation means reversing the addition of carbon to the atmosphere by making a rapid transition to energy efficiency and renewable energy. Arguments for mitigation, in other words, are rather like those for turning the water off in an overflowing tub before mopping. Those advocating mitigation believe that we are in a race to reduce the forcing effects of heat-trapping gases before we cross various thresholds—some known, some unknown—tipping us into irretrievable disaster beyond the ameliorative effects of any conceivable adaptation.

Of course, neither adaptation nor mitigation alone will be sufficient, and sometimes they may overlap. But in a world of limited resources, money, and time, we will be forced often to choose between the two. In such choices, the major issues in dispute have to do with estimates of the pace, scale, and duration of climatic disruption. And here the scientific evidence tilts the balance strongly toward mitigation for five reasons.

First, the record shows that climate change (1) is occurring much faster than previously thought, (2) will affect virtually every aspect of life in every corner of Earth, and (3) will last far longer than we'd once believed (Archer 2009; Solomon et al. 2009). The small cloud that Arrhenius saw on the distant horizon in 1896 is growing into a massive storm dead ahead. The effects of climatic destabilization, in other words, will be global, pervasive, permanent, and steadily—or rapidly—worsening. Given the roughly 30-year lag between what comes out of our tailpipes and smokestacks and the climate change we see, today's climate change–driven weather effects are being driven by emissions that occurred in the late 1970s. What is in store 30 years ahead, when the forcing effects of our present 392 ppm CO_2 will be manifest? Or further out when, say, the warming and acidifying effects of 450 ppm CO_2 or higher on the oceans have significantly diminished their capacities to absorb carbon? No one knows for certain, but trends in predictive climate science suggest that they will be much worse than once thought.

The implications for climate response strategies are striking. For example, it is now obvious that impacts will change with higher levels of climate forcing, which is to say that they are targets that will often move faster than we can anticipate and will become manifest in surprising ways. To what climatic conditions do we adapt? What happens when previous adaptive measures become obsolete, as they will? Similarly, at every level of climate forcing, the changes will be difficult to anticipate, which raises questions of where and when to intervene effectively in complex, nonlinear ecological and social systems. Are there places in which no amount of adaptation will work for long? Given what is now known about the pace of sea level rise, for example, what adaptive strategies can possibly work in New Orleans or South Florida, or much of the U.S. East Coast or in those regions that will likely become progressively much hotter and dryer and perhaps one day mostly inhabitable under drastically worsened conditions?

Second, the implications of the choice between adaptation and mitigation fall, not just on those able perhaps to adapt, for a time, to climatic destabilization, but on those who lack the resources to adapt and on future generations who will have to live with the effects of whatever atmospheric chemistry we leave behind. The choice between mitigation and adaptation, in other words, is one about ethics and justice in the starkest form. A few wealthy communities in the developed world may be able to avoid the worst for a time, but unless the emission of heat-trapping gases is soon reduced everywhere, worsening conditions will hit hardest those least able to adapt. The same can be said far more emphatically about future generations.

There is, third, a "stitch in time saves nine" kind of economic argument for giving priority to mitigation. Stabilizing climate now will be expensive and fraught with difficulties for certain, but it will be much cheaper and easier to do it sooner than it will be later under much more economically difficult and ecologically harrowing conditions. Nicholas Stern (2007), for one, estimates "that if we don't act [soon], the overall costs and risks of climate change will be equivalent to losing at least 5% of global GDP each year, now and forever" (Stern 2007, xv).

Fourth, efforts to adapt to climate change will run against institutional barriers, established regulations, building codes, and a human tendency to react to, rather than anticipate, events. There are, in economist Robert Repetto's words, "many reasons to doubt whether adaptive measures will be timely and efficient, even in the U.S. where the capabilities exist" (Repetto 2008, 5). In the best of all possible worlds, effective adaptation to the changes to which we are already committed would be complicated and difficult. In the real world of procrastination, denial, politics, and paradox, however, anything like thorough adaptation is not likely. It will, rather, be piecemeal, partial, sometimes counterproductive, wasteful, temporary, and ultimately mostly ineffective. In contrast, measures pressing energy efficiency and renewable energy—as complicated as they are—are much more straightforward and measurable hence achievable. And they have the advantage of resolving the causes of the problem, which has to do with anthropogenic changes in the carbon cycle.

Finally, beyond some fairly obvious and prudent measures, federal, state, and foundation support for climate adaptation gives the appearance that we are doing something serious about the climatic catastrophe looming ahead. The political and media reality, however, is that efforts toward climatic adaptation will be used by those who wish to do as little as possible, to block doing what is necessary to avert catastrophe.

Climate scientist James Hansen believes that "our global climate is nearing tipping points. Changes are beginning to appear, and there is a potential for explosive changes with effects that would be irreversible—if we do not rapidly slow fossil fuel emissions over the next few decades" (Hansen pers. comm.). The conclusion, in economist Nicholas Stern's words, is that "adaptation will be necessary on a major scale, but the stronger and the more timely the mitigation, the less will be the challenge of adaptation" (Stern 2009, 71). In other words, adaptation must be a second priority to effective and rapid mitigation that contains the scale and scope of climatic destabilization. When they compete for funding and attention, the priority must be given to efforts toward a rapid transition to energy efficiency and deployment of renewable energy. Until we get our priorities right, the emission of greenhouse gases will continue to rise beyond the point at which humans could ever adapt. "The only true adaptation," in George Woodwell's words, "is mitigation" (Woodwell pers. comm.).

⁓

Napoleon made a series of bad decisions, beginning with that to invade Russia. But having done so and having gotten as far as Moscow in the fall of 1812, he made two decisions that proved fatal to his army and to the French empire. One was to tarry in Moscow for 5 weeks with the Russian winter approaching. The second was to permit his soldiers to load up with plunder that encumbered their escape, weighed down their knapsacks and wagons, undermined discipline, and diverted their attention from the serious business of escaping disaster.

Of course all metaphors and historical analogies have their limits. But rather like Napoleon's Grand Armée, we, too, are in a race. For our part, we were first warned of climate change over a century ago and have lingered in increasingly dangerous territory in the belief that we can return to safer ground on our terms with all of the booty seized at the apogee of the fossil-fueled industrial era. It's not likely that we can do so and return to safer ground. According to James Hansen et al. (2008), that means a rapid return to CO_2 levels of about 350–300 ppm. If we wait too long to prevent climate change, we will, perhaps sooner than later, create conditions beyond reach of any conceivable adaptive measures. With sea level rise now said to be on the order of 1 to 2 meters by 2100, for example, we cannot save many low-lying places and many species we would otherwise prefer to save. And sea levels and temperatures will not stabilize until long after the year 2100.

There will be unavoidable and tragic losses in the decades ahead, but far fewer if we act to contain the scope and scale of climate change now. That is to say that there is some baggage accumulated in the fossil fuel era of our recent history that we cannot take with us. No matter what we do to adapt, we cannot save some coastal cities, we will lose many species, and ecosystems will be dramatically altered by changes in temperature and rainfall. Our best course is to reduce the scale and scope of the problem with a sense of wartime urgency. And we better move quickly and smartly while the moving's good.

Chapter 33

Long Tails and Ethics: Thinking about the Unthinkable

(2010)

It is a mistaken belief that one can philosophize without having been compelled to philosophize by problems outside philosophy.
KARL POPPER

WE HAVE LONG LIVED in the faith that "nature does not set booby traps for unwary species," as Robert Sinsheimer (1978) once noted. Whether nature does or not, we humans do, and we have nearly trapped ourselves by exploiting large pools of carbon found in soils, forests, coal, oil, and gas. The result is a rapid change in the chemistry of the atmosphere, leading to rising temperatures, destabilization of virtually every part of the biosphere, and the looming prospect of global catastrophe. The effect of climatic disruption now gathering momentum is a tsunami of change that will roll across every corner of the Earth, affect every sector of every society, and worsen problems of insecurity, hunger, poverty, and societal instability. We live now in the defining moment of our species that will determine whether

This article was originally published in 2010.

we are smart enough, competent enough, and wise enough to escape from a global trap entirely of our own making.

The first scientific evidence that human activity could alter atmospheric chemistry came from the laborious calculations of Svante Arrhenius in 1896. Compared with the later findings of the Intergovernmental Panel on Climate Change, his numbers are surprisingly accurate. His overall conclusion, however, was less accurate. Arrhenius, a Swede, thought a warmer Earth to be a good thing on the whole, a conclusion that has not stood the test of time. But it would be another 69 years before the President's Science Advisory Committee in 1965 delivered the first official warning of the possible scale and scope of global warming (Weart 2003, 44).

Nearly a half century later, we know that global warming, in the words of John Holdren, President Obama's science advisor, "is already well beyond dangerous and is careening toward completely unmanageable" (Holdren 2008, 20). Further, the destabilization of climate is now believed to be more or less permanent in human timescales. Geophysicist David Archer puts it this way:

> The climate impacts of releasing fossil fuel CO_2 to the atmosphere will last longer than Stonehenge. Longer than time capsules, longer than nuclear waste, far longer than the age of human civilization so far. The CO_2 coming from a quarter of that ton will still be affecting the climate one thousand years from now, at the start of the next millennium. (Archer 2009, 1)

In other words, even if we were to stop emitting carbon immediately, sea levels would continue to rise for at least another thousand years and temperatures would continue to rise with collateral effects one can scarcely imagine (Solomon et al. 2009). In short, because of our past actions the Earth likely will become a hotter, more barren, and more capricious place for time spans we typically associate with the half-life of nuclear waste. The climatic destabilization we have incurred is not a solvable problem but a steadily worsening condition with which humans will have to contend for a long time to come. Early and effective action to end our use of coal, oil, and natural gas and switch to renewable energy can only contain the eventual scale, scope, and duration of climatic destabilization but will not remedy the situation in any way that could reasonably be called a solution. That's the science. But the gap between science and the public discourse about climate destabilization seems as wide and seemingly as unbridgeable as the Grand Canyon itself. We are, to say the least, quite unaccustomed to thinking about matters so total and so permanent.

We rely on analogies and metaphors to understand things otherwise inexplicable. But what analogies, metaphors, or manner of thinking clarifies the issues posed by climatic destabilization? We will first turn to the familiar beginning with the standard metaphor of our age rooted in the image of the machine—devices of our own making that are accordingly understandable, purposeful, and repairable. Machine thinking leads some to regard climate destabilization as a solvable problem and, of course, as an opportunity to build a better world. In one recent view, "solving climatic change" is described as a new pathway to prosperity. "We can have it all," the author opines, "growth in the economy, a thriving business environment, and a solution to the climate crisis." Would that it were so. Machine thinking is rooted in the Enlightenment era's faith in progress, so machines beget better machines that beget still better ones. And better machines and more cleverness, it is assumed, will restore climate stability without disrupting our manner of living. But the Earth and its enveloping atmosphere are not simply machines and accordingly are not repairable. Nor is their "malfunction" a solvable problem as we understand those words.

Reliance on the discipline of economics rooted in the metaphor of "invisible hands" doesn't clarify our plight much either. Humans are not the rational calculators assumed in economic models. And the common use of discounting marginalizes the prospect of future disasters, so a new shopping mall is privileged over investments that reduce the scale of catastrophe, say, 50 years hence. Neither are the "pre-analytic assumptions" about human mastery of nature, infinite substitutability of technology for scarce natural resources, and the beneficence of economic growth useful for adapting economic activity to the limits of the Earth.

What about Biblical narratives? There is, for one, a similarity of sorts between the story of Adam and Eve's eviction from paradise and that which we are now writing about our own self-eviction from the 10,000-year paradise that geologists call the Holocene into a hotter world that some call the Anthropocene. Perhaps a better story is to be found in narratives about End Times. Theologian Jack Miles (2001), for instance, wonders what we will do once we discover that achieving sustainability is beyond our capacities and that we are living in the End Times, although not as told by rabid End-Timers like Pastor Tim LaHaye, coauthor of the

"Left Behind" books. Would our demise turn out to be our finest hour or simply a nasty and brutish final scene?

Perhaps climate destabilization bears a resemblance to the issue of abortion writ large. Where the public debate about abortion has been focused on an individual fetus, climate destabilization carries with it the possibility of aborting many species forever and many generations of humans that would otherwise have lived. But in Jonathan Schell's words,

> how are we to comprehend the life or death of the infinite number of possible people who do not yet exist at all? . . . To kill a human being is murder, but what crime is it to cancel the numberless multitude of unconceived people? In what court is such a crime to be judged? Against whom is it committed? . . . What standing should they have among us? (Schell 2000, 116)

This is a case of what Hannah Arendt once called "radical evil," which Schell interprets as evil that "goes beyond destroying individual victims and, in addition, destroys the world that can in some way respond to—and thus in some measure redeem—the deaths suffered" (Schell 2000, 145). Climate destabilization, like nuclear war, has the potential to destroy all human life on Earth and in effect "murder the future" (Schell 2000, 168). But never having lived, those not born will not suffer, will know no deprivation, and can make no claims against those who aborted the opportunity they might otherwise have had to live. Willfully caused extinction is a crime as yet with no name. There would be no judge, no jury, no sentence—simply a void and a great silence that would once again descend on Earth.

There are other metaphors and analogies that we could summon to help us begin to comprehend the full gravity of our situation, but all will be found wanting in one way or another. We are now in the era that biologist E. O. Wilson has called "the bottleneck," for which we have no precedent and no very useful example. I have faith that humankind will emerge someday, chastened but improved. But deliverance will require more than astute science and a great deal more than smarter technology—both necessary but insufficient. Science can describe our situation down to parts per trillion and help to create better technologies, but it can give us no clear reason why we should want to survive, why we deserve to be sustained on Earth, or why we should worry about the lives or well-being of generations whose existence now hangs in the balance. That is, rather, the function of deeper senses that we catalog with words like *morality*, *ethics*, and *spirituality*. But what kind of morality or ethics is remotely adequate when measured against the time spans necessary to restabilize Earth systems? I do not know. But with each turn of the screw, it will be tempting to avoid asking such questions and give in to trade-offs that privilege the living and damn those who reside only in the abstraction we call the future. And, for sure, there is no easy or, perhaps, good case to be made for current destitution except a bit more of it for the wealthy.

I do not presume to know what the content of that morality might be. Whatever it is, I doubt that it will be born in "deep thinking" characteristic of the academy or from philosophers debating esoteric points of obscure doctrines. I think the birth will be harder than that: messy and painful, which is to say a philosophy born of necessity and of stories of real people caught in the acts of struggle, generosity, and failure. Perhaps it won't be philosophy at all but rather a kind of practical worldview that emerges from the recognition of realities we've created and with which humankind must now contend for centuries to come. Let me suggest three illustrations of such a process.

The first is taken from a friend who recently spent several months as a patient in a cancer ward. During hours of treatment, he witnessed the growth of community among his fellow cancer patients. Once reticent to say much about themselves, under the new reality of a life-threatening disease they gradually became more talkative and open to thinking about their lives and listening to the experiences of other patients. Living in the shadow of death, they were more open to ideas and people, including some that they formerly regarded as threatening or incomprehensible. They were less prone to arrogance and more sympathetic to the suffering of others. They were less sure of once strongly held convictions and more open to contrary opinions. No longer masters of their lives, their schedules, or even their bodies, many achieved a higher level of mastery by letting go of illusions of invulnerability, and, in the letting go, they reached a more solid ground for hope and the kind of humble but stubborn resilience necessary for beating the odds or at least for living their final days with grace.

Another possible narrative can be drawn from the experience of people overcoming addiction. Alcoholics Anonymous, for example, offers a 12-step process to overcome addiction that begins with self-awareness and leads to a public confession of the problem, a reshaping of intention, the stabilizing influence of a support group, and a reclaiming of self-mastery to higher ends. The power of this narrative line is in the similarity between substance addiction and its collateral damages and our societal addictions to consumption, entertainment, and energy and their destructive effects on our places, selves, and children.

A third narrative comes from the haunting story of the Native American Crow Chief Plenty Coups, told by philosopher Jonathan Lear. Under the onslaught of white civilization, the world of the Plains tribes collapsed and their accomplishments disappeared, along with their culture, sense of purpose, and meaning. At the end of his life, Plenty Coups told his story to a trapper, Frank Linderman, saying, "But when the buffalo went away the hearts of my people fell to the ground, and they could not lift them up again. After this nothing happened" (Lear 2006, 2). Of course many things happened, but without the traditional bearings by which they understood reality or themselves, nothing happened that the Crow people could interpret in a familiar framework. Lear describes Chief Plenty Coups' courageous efforts to respond to the collapse of his civilization with "radical hope" but without the illusion that they could ever recreate the world they had once known. There were others, like Sitting Bull, who pined for vengeance and a return to a past before the juggernaut of American civilization swept across the Plains. Likewise, Ghost Dancers hoped fervently to restore what had been, but Plenty Coups knew that the Crow culture, organized around the hunt and warfare, would have to become something inconceivably different. The courage necessary to fight had to be transformed into the courage to face and respond creatively and steadfastly to a new reality with "a traditional way of going forward" (Lear 2006, 154). What makes his hope radical, Lear says, "is that it is directed toward a future goodness that transcends the current ability to understand what it is. Radical hope anticipates a good for which those who have the hope as yet lack the appropriate concepts with which to understand it" (Lear 2006, 104).

It is clear by now that we have quite underestimated the magnitude and speed of the human destruction of nature, but the rapid destabilization of climate and the destruction of the web of life are just symptoms of larger issues, the understanding of which runs hard against our national psyche and the Western worldview generally. It is easier, I think, to understand the reality of dilemmas in places that have historic ruins and are overlaid with memories of tragedies and misfortunes that testify to human fallibility, ignorance, arrogance, pride, overreach, and sometimes evil. Amidst shopping malls, bustling freeways, and all of the accoutrements, paraphernalia, enticements, and gadgetry of a booming fantasy industry, it is harder to believe that sometimes things don't work out because they simply cannot or that limits to desire and ambition might really exist. When we hit roadblocks, we have a national tendency to blame the victim or bad luck but seldom the nature of the situation or our beliefs about it. What Spanish philosopher Miguel de Unamuno called "the tragic sense of life" has little traction just yet in the U.S. because it runs against the national character, and we don't read much philosophy anyway (de Unamuno 1977).

A tragic view of life is decidedly not long faced and resigned, but neither is it giddy about our possibilities. It is merely a sober view of things, freed from the delusion that humans should be about "the effecting of all things possible" or that science should put nature on the rack and torture secrets out of her, as we learned from Francis Bacon. It is a philosophy that does not assume that the world or people are merely machines or that minds and bodies are separate things, as we learned from Descartes. It is not rooted in the assumption that what can't be counted does not count, as Galileo believed. The tragic sense of life does not assume that we are separate atoms, bundles of individual desires, unrelated hence without obligation to others or what went before or those yet to be born. Neither does it assume that the purpose of life is to become as rich as possible for doing as little as possible, or that being happy is synonymous with having fun. The tragic view of life, on the contrary, recognizes connections, honors mystery, acknowledges our ignorance, has a clear-eyed view of the depths and heights of human nature, knows that life is riddled with irony and paradox, and takes our plight seriously enough to laugh at it.

Whether aware of it or not, all of us are imprinted with the stamp of Bacon and the others who shaped the modern worldview. The problem, however, is not that they were wrong but rather that we believed them too much for too long. Taken too far and applied beyond their legitimate domain, their ideas are beginning to crumble under the weight of history and the burden of a reality far more complex and wonder-filled than they knew and could have known. Anthropogenic climate destabilization is a symptom of something more akin to a cultural pathology. So, dig deep enough and the "problem" of climate is not reducible to the standard categories of technology and economics. It is not merely a problem awaiting solution by one technological fix or another. It is, rather, embedded in a larger matrix; a symptom of something deeper. Were we to "solve" the "problem" of climate change, our manner of thinking and being in the world would bring down other curses and nightmares now waiting in the wings. Perhaps it would be a nuclear holocaust, or terrorism, or a super plague, or as Sun Microsystems founder Bill Joy warns, an invasion of self-replicating devices like nanotechnologies, genetically engineered organisms, or machines grown smarter than us that will find us exceedingly inconvenient. There is no shortage of such plausible nightmares, and each is yet another symptom of a fault line so deep that we hesitate to call it by its right name.

The tragic sense of life accepts our mortality, acknowledges that we cannot have it all, and is neither surprised nor dismayed by human evil. The Greeks who first developed the dramatic art of tragedy knew that we are ennobled, not by our triumphs or successes, but by rising above failure and tragedy. Sophocles, for example, portrays Oedipus Rex as a master of the world—powerful, honored, and quite full of himself but also honest enough to search out the truth relentlessly. In his searching, Oedipus falls from the heights, and that is both his undoing and his making. Humbled, blind, old, and outcast, Oedipus is a far nobler man than he had been at the height of his kingly power. Tragedy, the Greeks thought, was necessary to temper our pride, to rein in the tug of hubris, and to open our eyes to hidden connections, obligations, and possibilities.

We are now engaged in a global debate about what it means to become "sustainable." But no one knows how we might secure our increasingly tenuous presence on the Earth or what that will require of us. We have good reason to suspect, however, that the word *sustainable* must imply something deeper than merely the application of more technology and smarter economics. It is possible and perhaps even likely that more of the same "solutions" would only compound our tribulations. The effort to secure a decent human future, I think, must be built on the awareness of the connections that bind us to each other, to all life, and to all life to come. And, in time, that awareness will transform our politics, laws, economy, lifestyles, and philosophies.

⁓ Chapter 34 ⁓

Hope (in a Hotter Time)

(2007)

Fraudulent hope is one of the greatest malefactors, even enervators, of the human race, concretely genuine hope its most dedicated benefactor.
Ernst Bloch

We like optimistic people. They are fun, often funny, and very often capable of doing amazing things otherwise thought to be impossible. Were I stranded on a life raft in the middle of the ocean and had a choice between an optimist and pessimist as a companion, I'd want an optimist, providing he did not have a liking for human flesh. Optimism, however, is often rather like a Yankee fan believing that the team can win the game when it's the bottom of the ninth, they're up by a run, with two outs, a two-strike count against a .200 hitter, and Mariano Rivera in his prime is on the mound. He or she is optimistic for good reason. The Red Sox fans, on the other hand, believe in salvation by small percentages and hope for a hit to get the runner home from second base and tie the game. Optimism is the recognition that the odds are in your favor; hope is the faith that things will work out whatever the odds. *Hope* is a verb with its sleeves rolled up. Hopeful people are actively engaged in defying the odds or changing the odds. Optimism leans back, puts its feet up, and wears a confident look, knowing that the deck is stacked.

I know of no good reason for anyone to be optimistic about the human future, but I know many reasons to be hopeful. How can one be optimistic,

This article was originally published in 2007.

for example, about global warming? First, it isn't a "warming," but rather a total destabilization of the planet brought on by the behavior of one species: us. Whoever called this "warming" must have worked for the advertising industry or the northern Siberian bureau of economic development. The Intergovernmental Panel on Climate Change—the thousand-plus scientists who study climate and whose livelihoods depend on authenticity, replicability, data, facts, and logic—put it differently. A hotter world means rising odds of

- more heat waves and droughts;
- more and larger storms;
- bigger hurricanes;
- forest dieback;
- changing ecosystems;
- more tropical diseases in formerly temperate areas;
- rising ocean levels, faster than once thought;
- losing many things nature once did for us;
- losing things like Vermont maple syrup;
- more and nastier bugs;
- food shortages due to drought, heat, and more and nastier bugs;
- more death from climate-driven weather events;
- refugees fleeing floods, rising seas, drought, and expanding deserts;
- international conflicts over energy, food, and water.

And, if we do not act quickly and wisely, runaway climate change to some new stable state most likely without humans.

Some of these changes are inevitable, given the volume of heat-trapping gases we've already put into the atmosphere. There is a lag of several decades between the emission of carbon dioxide and other heat-trapping gases, and the weather headlines, and there is still another lag until we experience their full economic and political effects. The sum total of the opinions of climate experts goes like this:

- We've already warmed the planet by 0.8°C.
- We are committed to another approximately 0.6°C warming.
- It's too late to avoid trauma.
- But it's probably not too late to avoid global catastrophe, which includes the possibility of runaway climate change.
- There are no easy answers or magic bullet solutions.
- It is truly a global emergency.

The fourth item above is anyone's guess, since the level of heat-trapping gases is higher than it has been in the past 650,000 years and quite likely for a great deal longer. We are playing a global version of Russian roulette, and no one knows for certain what the safe thresholds of various heat-trapping gases might be. Scientific certainty about the pace of climate change over the past three decades has a brief shelf life, but the pattern is clear. As scientists learn more, it's mostly worse than they previously thought. Ocean acidification went from being a problem a century or two hence to being a crisis in a matter of decades. Melting of the Greenland and the West Antarctic ice sheets went from being a distant likelihood to a nearer-term possibility of a century or two. The threshold of perceived safety went down from perhaps 560 ppm CO_2 to perhaps 450 ppm CO_2. And so forth.

Optimism in these circumstances is like whistling as one walks past the graveyard at midnight. There is no good case to be made for it, but the sound of whistling sure beats the sound of the rustling in the bushes beside the fence. But whistling doesn't change the probabilities one iota, nor does it much influence any goblins lurking about. Nonetheless, we like optimism and optimistic people. They soothe, reassure, and sometimes they motivate us to accomplish a great deal more than we otherwise might. But sometimes optimism misleads, and on occasion badly so. This is where hope enters.

Hope, however, requires us to check our optimism at the door and enter the future without illusions. It requires a level of honesty, self-awareness, and sobriety that is difficult to summon and sustain. I know a great many smart people and many very good people, but I know far fewer people who can handle hard truth gracefully without despairing. In such circumstances it is tempting to seize on anything that distracts us from unpleasant things. The situation is rather like that portrayed in the movie *A Few Good Men* in which Jack Nicholson playing a beleaguered Marine Corps officer tells the prosecuting attorney (Tom Cruise), "You can't handle the truth!" T. S. Eliot, less dramatically, noted the same tendency: "Human kind cannot bear very much reality" (*Four Quartets*, "Burnt Norton").

Authentic hope, in other words, is made of sterner stuff than optimism. It must be rooted in the truth as best we can see it, knowing that our vision is always partial. Hope requires the courage to reach farther, dig deeper, confront our limits and those of nature, work harder, and dream dreams. Optimism doesn't require much effort, since you're likely to win anyway, but hope has to hustle, scheme, make deals, and strategize. But how do we find authentic hope in the face of climate change, the biological holocaust now under way, the spread of global poverty, seemingly unsolvable human conflicts, terrorism, and the void of world leadership adequate to the issues?

I've been thinking about the difference between optimism and hope since being admonished recently to give a "positive" talk at a gathering of ranchers, natural resource professionals, and students. Presumably the audience was incapable of coping with the bad news it was assumed that I would otherwise deliver. I gave the talk that I intended to give and the audience survived, but the experience caused me to think more about what we say and what we can say to good effect about the kind of news that we reckon with daily.

The view that the public can only handle happy news, nonetheless, rests on a chain of reasoning that goes like this:

- We face problems which are solvable, not dilemmas which can be avoided with foresight but are not solvable, and certainly not losses which are permanent;
- people, and particularly students, can't handle much truth;
- so resolution of different values and significant improvement of human behavior otherwise necessary are impossible;
- greed and self-interest are in the driver's seat and always will be;
- so the consumer economy is here to stay;
- but consumers sometimes want greener gadgets;
- and capitalism can supply these at a goodly profit and itself be greened a bit, but not improved otherwise;
- and so . . . matters of distribution, poverty, and political power are nonstarters;
- therefore, the focus should be on problems solvable at a profit by technology and policy changes;
- significant improvement of politics, policy, and governance are unlikely and probably irrelevant because better design and market adjustments can substitute for governmental regulation and thereby eliminate most of the sources of political controversy—rather like Karl Marx's prediction of the withering away of the state.

Disguised as optimism, this approach is, in fact, pessimistic about our capacity to understand the truth and act well. So we do not talk about limits to growth, unsolvable problems, moral failings, unequal distribution of wealth within and between generations, emerging dangers, impossibilities, technology gone awry, or necessary sacrifices. "Realism" requires us to portray climate change as an opportunity to make a great deal of money, which it may be for some, but without saying that it might not be for most or mentioning its connections to other issues, problems, and dilemmas or the possibility that the four horsemen are gaining on us. We are not supposed to talk about coming changes in our "lifestyles," a telling and empty word implying fashion, not necessity or conviction.

Ultimately, this approach is condescending to those who are presumably incapable of facing the truth and acting creatively, courageously, and even nobly in dire circumstances. Solving climate change, for example, is reduced to a series of wedges representing various possibilities that would potentially eliminate so many gigatons of carbon without any serious changes in how we live. There is, accordingly, no wedge called "suck it up," because that is considered to be too much to ask of people who have been consuming way too much, too carelessly, for too long. The "American way of life" is thought to be sacrosanct. In the face of a global emergency, brought on in no small way by the profligate American way of life, few are willing to say otherwise. So we are told to buy hybrid cars but not asked to walk, travel by bikes, or go less often, even at the end of the era of cheap oil. We are asked to buy compact fluorescent light bulbs but not to turn off our electronic stuff or not buy it in the first place. We are admonished to buy green but seldom asked to buy less or repair what we already have or just make do. We are encouraged to build LEED-rated buildings that are used for maybe 10 hours a day for 5 days a week, but we are not told that we cannot build our way out of the mess we've made or to repair existing buildings. We are not told that the consumer way of life will have to be rethought and redesigned to exist within the limits of natural systems and better fitted to our human limitations. And so we continue to walk north on a southbound train, as Peter Montague once put it.

And maybe, told that its hindquarters are caught in a ringer, the public would panic or, on the other hand, become so despairing that it would stop doing what it otherwise would do that could save us from the worst outcomes possible. This is an old view of human nature epitomized in the work of Edward Bernays, a nephew of Sigmund Freud and the founder of modern advertising. Public order, he thought, had to be engineered by manipulating people to be dependent and dependable consumers. People who think too much or know too much were, in his view, a hazard to social stability.

Maybe this is true and maybe gradualism is the right strategy. Perhaps the crisis of climate and those of equity, security, and economic sustainability will yield to the cumulative effects of many small changes without any sacrifice at all. Maybe changes now under way are enough to save us. Maybe, small changes will increase the willingness to make larger changes in the future. And state-level initiatives in California, Florida, and northeastern states are changing the politics of climate. Wind and solar are growing at 40+ percent per year, taking us toward a different energy regime. A cap and trade bill will soon pass in Congress, and maybe that will be enough. Maybe we can win the game of climate roulette at a profit and never have to confront the nastier realities of global capitalism and inequity or confront the ecological and human violence that we've unleashed in the world. But I wouldn't bet the Earth on it.

For one, the remorseless working out of the big numbers gives us little margin for safety and none for delay in reducing CO_2 levels before we risk triggering runaway change. "Climate," as Wallace Broecker once put it, "is an angry beast and we are poking it with sticks," and we've been doing that for a while. So call it prudence, precaution, insurance, common sense, or what you will, but this ought to be regarded as an emergency like no other. Having spent any margin of error we might have had 30 years ago, we now have to respond fast and effectively or else. That's what the drab language of the fourth report from the Intergovernmental Panel on Climate Change is saying. What is being proposed, I think, is still too little, too late—necessary but not nearly sufficient. And it is being sold as "realism" by people who have convinced themselves that they have to understate the problem in order to be credible.

Second, climate roulette is part of a larger equation of exploitation of people and nature, violence, inequity, imperialism, and intergenerational exploitation, the parts of which are interlocked. In other words, heat-trapping gases in the atmosphere are a symptom of something a lot bigger. To deal with the causes of climate change, we need a more thorough and deeper awareness of how we got to the brink of destroying the human prospect and much of the planet. It did not happen accidentally but is, rather, the logical working out of a set of assumptions, philosophy, worldview, and unfair power relations that have been evident for a long time. The wars, gulags, ethnic cleansings, militarism, and the destruction of forests, wildlife, and oceans throughout the twentieth century were earlier symptoms of the problem. We've been playing fast and loose with life for a while now, and it's time to discuss the changes we must make in order to conduct the public business fairly and decently over the long haul.

The upshot is that the forces that have brought us to the brink of climate disaster and biological holocaust and are responsible for the spread of global poverty—the crisis of sustainability—remain mostly invisible and in charge of climate policy. The fact is that climate stability, sustainability, and security are impossible in a world with too much violence, too many weapons, too much unaccountable power, too much stuff for some, too little for others, and a political system that is bought and paid for behind closed doors. Looming climate catastrophe, in other words, is a symptom of a larger disease.

What do I propose? Simply this: that those of us concerned about climate change, environmental quality, and equity treat the public as intelligent adults who are capable of understanding the truth and acting creatively and courageously in the face of necessity—much as a doctor talking to a patient with a potentially terminal disease. There are many good precedents for telling the truth. Abraham Lincoln, for one, did not pander, condescend, evade, or reduce moral and political issues to economics, jobs, and happy talk. Rather he described slavery as a moral disaster for slaves and slave owners alike. Similarly, Winston Churchill in the dark days of the London blitzkrieg in 1940 did not talk about defeating Nazism at a profit and the joys of urban renewal. Instead he offered the British people only "blood, toil, tears, and sweat." And they responded with heart, courage, stamina, and sacrifice. At the individual level, faced with a life-threatening illness, people more often than not respond heroically. Every day, soldiers, parents, citizens, and strangers do heroic and improbable things in the full knowledge of the price they will pay.

Telling the truth means that the people must be summoned to a level of extraordinary greatness appropriate to an extraordinarily dangerous time. People, otherwise highly knowledgeable of the latest foibles of celebrities, must be asked to be citizens again, to know more, think more, take responsibility, participate publicly, and, yeah, suck it up. They will have to see the connections between what they drive and the wars we fight; the stuff they buy and crazy weather; the politicians they elect and the spread of poverty and violence. They must be taught to see connections between climate, environmental quality, security, energy use, equity, and prosperity. They must be asked to think and to see. As quaint and naive as that may sound, people have done it before and it's worked.

Telling the truth means that we will have to speak clearly about the causes of our failures that have led us to the brink of disaster. If we fail to deal with causes, there are no Band-Aids that will save us for long. The problems can in one way or another be traced to the irresponsible exercise of power that has excluded the rights of the poor, the disenfranchised, and every generation after our own. That this has happened is in no small way a direct result everywhere of money in politics, which has aided and abetted the theft of the public commons, including the airwaves where spreading misinformation is a growing industry. Freedom of speech, as Lincoln said in 1860, does not include the "right to mislead others, who have less access to history, and less leisure to study it." But the rights of capital over the media now trump those of honesty and fair public dialogue and will continue to do so until the public reasserts its legitimate control over the public commons, including the airwaves.

Telling the truth means summoning people to a higher vision than that of the affluent consumer society. Consider the well-studied but little-noted gap between the stagnant or falling trend line of happiness in the last half century and that of rising GNP. That gap ought to have reinforced the ancient message that, beyond some point, more is not better. If we fail to see a vision of a livable decent future beyond the consumer society, we will never summon the courage, imagination, or wit to do the obvious things to create something better than what is in prospect. So, what does a carbon neutral society and increasingly sustainable society look like? My list consists of communities with

- front porches;
- public parks;
- local businesses;
- windmills and solar collectors;
- local farms and better food;
- better woodlots and forests;
- local employment;
- more bike trails;
- summer baseball leagues;
- community theaters;
- better poetry;
- neighborhood book clubs;
- bowling leagues;
- better schools;
- vibrant and robust downtowns;
- with sidewalk cafes;
- great pubs serving microbrews;
- more kids playing outdoors;
- fewer freeways, shopping malls, sprawl, television;
- no more wars for oil or anything else.

Nirvana? Hardly! Humans have a remarkable capacity to screw up good things. But it is still possible to create a future that is a great deal better than what is in prospect. Ironically, what we must do to avert the worst effects of climate change are mostly the same things we would do to build sustainable communities, improve environmental quality, build prosperous economies, and improve the prospects for our children.

Finally, I am an educator and earn my keep in the quaint belief that if people only knew more, they would act better. Some of what they need to know is new, but most of it is old, very old. On my list of things people ought to know in order to discern the truth are a few technical things like (1) the laws of thermodynamics that tell us that economic growth only increases the pace of disorder, the transition from low entropy to high entropy; (2) the basic sciences of biology and ecology—for example, how the world works as a physical system; and (3) the fundamentals of carrying capacity, which apply equally to yeast cells in a wine vat, lemmings, and humans. But they ought to know, too, about human fallibility, gullibility, and the inescapable problem of ignorance. So I propose that schools, colleges, and universities require their students to read Marlowe's *Dr. Faustus*, Mary Shelley's *Frankenstein*, Melville's *Moby Dick*, and the book of Ecclesiastes. I would hope that they would be taught how to distinguish those things that we can do from those that we should not do. And they should be taught the many disciplines of applied hope that include the skills necessary to grow food, build shelter, manage woodlots, make energy from sunlight and wind, develop local enterprises, cook a good meal, use tools skillfully, repair and reuse, and talk sensibly at a public meeting.

Hope, authentic hope, can be found only in our capacity to discern the truth about our situation and ourselves and summon the fortitude to act accordingly. We have it on high authority that the Truth will set us free from illusion, greed, and ill will and perhaps, with a bit of luck, from self-imposed destruction.

Chapter 35

At the End of Our Tether? The Rationality of Nonviolence

(2008)

Somebody must begin it.
William Penn

Perhaps humankind will do the right thing, as Winston Churchill once said of Americans, but only after it has exhausted all other possibilities. In human relations, we've tried brute force, and that is the story of empires rising and falling and the lamentable catalogue of folly that we call history. In 1648 the creators of the Westphalian system of sovereign nation-states improved things slightly by creating a few rules to govern interstate anarchy in Europe. The architects of the post–World War II world improved things a bit more with the creation of international institutions such as the World Bank, the International Monetary Fund, and the United Nations. But war and militarization have a stronger hold on human affairs than ever, and sooner or later,

The title is adapted from H. G. Wells (1946, 1). Wells wrote, "This world is at the end of its tether. The end of everything we call life is close at hand and cannot be evaded." This article was originally published in 2008.

violence—whether by states, by terrorist groups, or simply by demented individuals—will devour the human prospect.

In the last few centuries we applied the same mindset to nature. We've bullied, bulldozed, and re-engineered her down to the gene, and that got us into more trouble and perplexities than anyone can comprehend. It is now proposed that we manage nature even more intensely—but the same goal with smarter methods will only delay the inevitable. Either way, we are rapidly creating what climate scientist James Hansen calls a "different planet" and one we are not going to like. We can quibble about the timing of disaster, but, given our present course, there is no good argument about its inevitability.

Whether to nature or human affairs, we continue to apply brute force with more powerful and sophisticated technology and expect different results—a definition, according to some, of insanity. True or not, it is a prescription for the destruction of nature and civilization that is woven into our politics, economies, and culture. The attempt to master nature and to control destiny through force has not worked and will not work, because the world, whether that of nature or that of nations, as Jonathan Schell puts it, is "unconquerable" (Schell 2003). The reasons are to be found in the mismatch between the human intellect and the complexity of nonlinear systems, and no amount of research, thought, or computation can fill that void of ignorance, which is only to acknowledge the limits of human foresight and the inevitability of surprises, unforeseen and unforeseeable results, unintended consequences, paradox, irony, and counterintuitive outcomes. But the limits of human intelligence do not prevent us from discerning something about self-induced messes.

So what kind of messes have we made for ourselves? Some are problems that are, by definition, solvable with enough rationality, money, and effort. The problem of powering the world by current sunlight, for example, is solvable given enough effort and money. But some situations are dilemmas, which by definition are not solvable by any rational means—although with enough foresight and wisdom they can be avoided or resolved at a higher level. British economist E. F. Schumacher once described the difference between "convergent" and "divergent" problems in much the same terms. In the former, logic tends to converge on a specific answer, while the latter "are refractory to mere logic and discursive reason" and require something akin to a change of heart and perspective (Schumacher 1977, 128). Donella Meadows, in a frequently cited article on the alchemy of change, concluded that of all possible ways to change social systems, the highest leverage comes, not with policies, taxes, numbers, and the usual menu of rational choices, but with change in how we think (Meadows 1997). The crucial issues we face are not so much problems as they are dilemmas. They cannot be solved by the application of more technology and smartness, but they can be transcended by a change of mindset.

Two dilemmas stand astride our age. The first has to do with age-old addiction to force in human affairs. We don't know exactly how or when violence became the method of choice, or the precise point at which it became wholly counterproductive (Schmookler 1984). But no tribe or nation that did not prepare for war could survive for long once its neighbors did. And since it makes no sense to have a good army if you don't use it from time to time, preparation for war tended to make its occurrence more likely. If it was ever rational, however, the bloody carnage of the past 100 years should have convinced even the dullest among us that violence within and between societies is now self-defeating and colossally stupid. Violence and threats have always tended to create more of the same—a deadly dance of action and reaction. The development of nuclear and biological weapons and the even more heinous weapons now in development have changed everything—everything but our way of thinking, as Einstein once noted. In an age of terrorism, the scale of potential destruction and the proliferation of small weapons of mass destruction mean that there is no sure means of security, safety, or deterrence anywhere for anyone. The conclusion is inescapable: from now on—whatever the issues—there can be no winners in any violent conflict, only losers. Nonetheless, the world now spends $1.2 trillion each year on weapons and militarism and is, unsurprisingly, less secure than ever. The United States alone spends 46 percent of the total, or $17,000 per second, more than the next 22 nations combined. It maintains over 737 military bases worldwide, but it is presently losing two wars while threatening to start a third. Economist Joseph Stiglitz estimates that the total cost of the Iraqi misadventure alone will be $2 trillion. Beyond the economic cost, it will surely leave a legacy of yet more terrorism, violence, and ruin in all of its many guises.

The word *realism* has always been a loaded word. In world politics it is contrasted with *idealism*, believed by realists to be the epitome of wooly-headedness. In realist theory, the power realities of interstate politics required military strength and the aggressive protection of the national interest defined as power. Realists were the architects of empires, world wars, cold wars, arms races, mutual assured destruction, the Vietnam War, and now the fiasco in Iraq. But one of the preeminent realists of the post–World War II era, Hans Morgenthau, was more of an idealist than commonly appreciated. He once proposed that governments give control of nuclear weapons to "an agency whose powers are commensurate with the worldwide destructive potentials of those weapons" (Joffe 2007). George Kennan, another post–World War II realist, similarly proposed international measures to prevent both nuclear war and ecological decline—ideas that are anathema to influential neoconservative realists now.

The second dilemma is the insolvability of long-term economic growth in a finite biosphere. As ecological economists like Herman Daly have said for decades, the economy is a subsystem of the biosphere, not an independent system. The "bottom line," therefore, is set by the laws of entropy and ecology, not by economic theory. The effort to make the economy sustainable by making it smarter and greener is all to the good, but altogether inadequate. It is incrementalism when we need systemic change that begins by changing the goals of the system. Economic growth can and should be smarter, and corporations ought to reduce their environmental impacts, and with a bit of effort and imagination it is possible for most of them to do so. Could we, however, organize all of the complexities of an endlessly growing global economy to fit within the limits of the biosphere in a mostly badly governed world in which greed, corruption, corporate competition, and consumerism dominate? As you read these words, the answer is being written in the disappearing forests of Sumatra, in the mountains being flattened in Appalachia, in the 1000 MW per week of new coal plants being built in China, in the billion dollars of advertisements spent each year to stoke the fires of Western-style consumption, in glitzy shopping malls, in the fantasy world of Dubai, in the temporizing of governments virtually everywhere, and by the corporate pursuit of short-term profit. Progress toward a truly green economy is incremental, not transformational, change—and a great deal of it is of the smoke-and-mirrors sort. If we had hundreds of years to make the necessary changes, we might muddle our way to a sustainable economy, but we don't have that much time. If we intend to preserve civilization, the inescapable conclusion is that we need a more fundamental economic transformation, and that means three things that presently appear to be utterly impossible: (1) a transition from economic growth (creation of more stuff) to development which genuinely improves the quality of life for everyone, first in wealthy nations and eventually everywhere; (2) the transformation of the consumer economy into one oriented first and foremost to needs not wants; and hardest of all, (3) summoning the compassion and wisdom to fairly distribute wealth, opportunity, and risk. The fact that these three seem wholly inconceivable to most of us indicates the scale of the challenge ahead and the necessity of a different manner of thinking.

Both dilemmas are intertwined at every point. To maintain economic growth, the powerful must have access to the oil and resources of poor third world nations whether they like it or not. Global trade, often to the disadvantage of poor nations, requires the use of military forces to patrol the seas, enforce inequities, strike quickly, and maintain pliant governments willing to plunder their own people and lands. The result is animosity that fuels global terrorism and ethnic violence. The power of envy and the desperate search for "a better life" requires the "haves" to build higher fences to keep the poor at bay. Profit and the fear of possible insurrection and worldwide turmoil drive the search for more advanced Star Wars kind of technology—robot armies, space platforms, and constant electronic surveillance. But, as Gandhi said repeatedly, our wealth and weapons make us cowards, and our fears condone the injustices that underpin our way of life and fuel the hostility that will some day bring it down.

In sum, (1) the time to heal our conflict with Earth and those between nations and ethnic groups is short; (2) both are dilemmas, not merely problems; (3) neither can be resolved by applying more of the kind of thinking that created them; (4) the connection between the two is our addiction to violence; and (5) neither can be solved without solving the other.

We are at the end of our tether and no amount of conventional rationality or smartness is nearly rational enough or smart enough. We need deeper, transformational change. The remorseless working out of big numbers, whether climate change, the loss of biological diversity, or the combination of hatred and the proliferation of heinous weaponry, is wreaking havoc on our pretensions of control. This is not the time for illusions or evasion; it is time for transformation.

Self-described realists will argue that, however necessary, humans are not up to change at this scale and pace—muddling along is the best that we can do. And for those inclined to wager, that is certainly the smart bet. But if that is all that can be said, we have no good reason for hope and might best prepare for our denouement. On the other hand, transformational change is not only necessary, but it may be possible as well. Do we have good reasons to transform the growth economy and transcend the use of force in world politics? Is the public ready for transformation? Is this an opportune time (a "teachable moment") to do so? Do we have better nonviolent alternatives?

There is a great deal of evidence to suggest a more hopeful view of possibilities than most "realists" are inclined to see. A recent BBC poll of attitudes in 21 countries, for example, shows that a majority, including a majority of Americans, are willing to make significant sacrifices to avoid rapid climate change—even though no "leader" has thought to ask them to do so. Can we craft a fair and ecologically sustainable economy that also sustains us spiritually? The present economy has failed miserably on all three counts. As Richard Layard puts it, "we are as a society no happier than fifty years ago. Yet every group in society is richer" (Layard 2005, 223). Beyond some minimal level, in other words, economic growth advances neither happiness nor well-being. But the outlines of a nonviolent economy are beginning to emerge in the rapid deployment of solar and wind technology, in a growing anticonsumer movement, in the slow food and slow money movements, and in fields like biomimicry and industrial ecology. In world affairs, the manifest failure of neoconservative realism in the Middle East and elsewhere may have created that teachable moment when we come to our senses and overthrow that outworn and dangerous paradigm for something far more realistic—security for everyone. And at least since Gandhi, we have known that there are better means and ends for the conduct of politics.

The transformative idea of nonviolence can no longer be dismissed as an Eastern oddity, a historical aberration, or the height of naïveté. At the end of our tether it is rather the core of a more realistic and practical global realism. There is no decent future for humankind without transformation of both our manner of relations and our collective relationship with the Earth. Gandhi stands as the preeminent modern theorist and practitioner of the art of nonviolence. His life and thought were grounded in the practice of ahimsa, a Sanskrit word that means unconditional love. To denote the practice of ahimsa, Gandhi coined the word *satyagraha*, which combines the Sanskrit word *sat*, meaning truth, with *graha*, meaning "holding firm to" (Schell 2003, 119). Gandhi honed the philosophy of nonviolence into an effective tool of change in India as Martin Luther King Jr. later did in the United States, but we've never known what to do with persons like Gandhi and King. On one hand we occasionally pay them lip service in public speeches and name holidays in their honor, but on the other hand we ignore what they had to say about how we live and how we conduct the public business. The time has come to pay closer attention to what they said and did and fathom what that means for us now.

The beginning of a more realistic realism is in the recognition that violence of any sort is a sure path to ruin on all levels and that the practice of nonviolence is a viable alternative—indeed our only alternative to collective suicide. But that implies changing a great deal that we presently take for granted, beginning with the belief in an unmovable and implacably evil enemy. Richard Gregg, an associate of Gandhi, for example, said that the goal of the practitioner of nonviolence

> is not to injure, or to crush and humiliate his opponent, or to "break his will" . . . [but] to convert the opponent, to change his understanding and his sense of values so that he will join wholeheartedly [to] seek a settlement truly amicable and truly satisfying to both sides. (Gregg 1935, 51)

As with war, the practice of nonviolence requires training, discipline, self-denial, strategy, courage, stamina, and heroism. Its aim is not to defeat but to convert and thereby resolve the particulars of conflict at a higher level. For Gandhi it required its practitioners, first, to transcend animosity and hatred to reach a higher level of being in "self-restraint, unselfishness, patience, gentleness" (Fischer 1962, 326). The aim is not to win conflict but to change the mind-set that leads to conflict and ultimately form a "broad human movement which is seeking not merely the end of war but [the end of] our equally non-pacifist civilization." In Gandhi's words, "true ahimsa should mean a complete freedom from ill will and anger and hate and an overwhelming love for all" (Fischer 1962, 207).

Gandhi applied the same logic to the industrial world of his day, regarding it as a "curse . . . depend[ing] entirely on [the] capacity to exploit" (Fischer 1962, 287). Its future, he thought, was "dark," not only because it engendered conflict between peoples, but because it cultivated "an infinite multiplicity of wants . . . [arising from] want of a living faith in a future state, and therefore also in Divinity" (Fischer 1962, 289).

The philosophy, strategy, and tactics of nonviolence have been updated to our own time and situation by many scholars, including Anders Boserup and Andrew Mack (1975), Richard Falk and Saul Mendlovitz (World Order Models Project), Michael Shuman and Hal Harvey (1993), Gene Sharp (1973, 2005), and the Dalai Lama (1999). Clearly we do not lack examples, precedents, alternatives, and better ideas than those now regnant. It is time—long past time—to take the next steps in rethinking and remodeling our economy and foreign policies to fit a higher view of the human potential. The first steps will be the hardest of all because the impediment is not intellectual but something else that lies deeper in our psyche. Over the millennia violence became an addiction of sorts. Our heroes are mostly violent men. Our national holidays mostly celebrate violence in our past. Most of our proudest scientific achievements have to do with the violent domination of nature. There is something in us that seems to need dependably loathsome adversaries even if, sometimes, they have to be conjured. And to that end we built massive institutions to plan and fight wars, giant corporations to supply the equipment for war, and a compliant media to sell us war as a patriotic necessity. In the process we made economies and societies dependent on arms makers and merchants of death and changed how we think and how we talk. We often speak violently and think in metaphors of combat and violence, so we "kill time" or "make a killing" in the market or wage futile wars on drugs, poverty, and terrorism. Worse, our children are being schooled to think violently by electronic games, television, and movies. We have made no comparable effort to build institutions for the study and propagation of peace and conflict resolution or to cultivate the daily habits of peace. We have barely begun to imagine the possibility of a nonviolent economy in which no one profits from war or violence in any form. And so it is surprising that we are continually surprised when our collective obsession with violence manifests yet again in violence down the street or in some distant place.

The transformation to a nonviolent world will require courageous champions at all levels—public officials, teachers, communicators, philanthropists, artists, statespersons, philosophers, and corporate executives. But it will most likely be driven by ordinary people who realize that we are all at the end of our tether and it is time to do something a great deal smarter and more decent. And "somebody must begin it."

Sources

Abell, R., et al. 2000. *Freshwater Ecoregions of North America: A Conservation Assessment*. Washington, DC: Island Press.

Abram, D. 1996. *The Spell of the Sensuous*. New York: Pantheon.

Alexander, C. 2001–2004. *The Nature of Order.* 4 vol. Berkeley: The Center for Environmental Structure.

Alexander, C., et al. 1977. *A Pattern Language*. New York: Oxford University Press.

Allen, R. 1980. *How to Save the World.* Totowa, NJ: Barnes and Noble.

Andrews, R. N. L. 1999. *Managing the Environment, Managing Ourselves*. New Haven: Yale University Press.

Appalachian Land Ownership Task Force. 1983. *Who Owns Appalachia?* Lexington: University of Kentucky Press.

Applebome, P. 1997. "Children Score Low in Adults' Esteem." *New York Times*, June 26.

Archer, D. 2009. *The Long Thaw*. Princeton: Princeton University Press.

Banuri, T., and A. Marglin. 1993. *Who Will Save the Forests?* London: Zed Books.

Bateson, G. 1979. *Mind and Nature*. New York: Dutton.

Bateson, G. 1975. *Steps to an Ecology of Mind*. New York: Ballantine Books.

Batra, R. 1993. *The Myth of Free Trade*. New York: Scribners.

Beard, C. A. 1913. *An Economic Interpretation of the Constitution of the United States*. New York: Free Press, 1935.

Becker, E. 1973. *The Denial of Death*. New York: Free Press.

Benyus, J. 1997. *Biomimicry: Innovation Inspired by Nature*. New York: William Morrow.

Berry, W. 1989. "The Futility of Global Thinking." *Harper's*, September.

Berry, W. 1987. *Home Economics*. San Francisco: North Point Press.

Berry, W. 1983. *Standing by Words*. San Francisco: North Point Press.

Berry, W. 1981. *The Gift of Good Land*. San Francisco: North Point Press.

Berry, W. 1974. *The Memory of Old Jack*. New York: Harcourt Brace Jovanovich.

Bloom, A. 1987. *The Closing of the American Mind*. New York: Simon and Schuster.

Bok, D. 1990. *Universities and the Future of America*. Durham: Duke University Press.

Bortoft, H. 1996. *The Wholeness of Nature*. New York: Lindisfarne Press.

Boserup, A., and A. Mack. 1975. *War without Weapons*. New York: Schocken.

Bowden, C. 1985. *Killing the Hidden Waters*. Austin: University of Texas Press.

Bronner, E. 1998 "College Freshmen Aiming for High Marks in Income." *New York Times*, January 12.

Brown, L. 1980. *Building a Sustainable Society*. New York: Norton.
Brown, L., et al. 1990. "Picturing a Sustainable Society." In *State of the World: 1990*, ed. L. Brown et al. New York: Norton.
Burtt, E. A. 1954. *The Metaphysical Foundations of Modern Science*. New York: Anchor Books.
Calaprice, A. 2005. *The New Quotable Einstein*. Princeton: Princeton University Press.
Caldwell, L. K. 1998. *The National Environmental Policy Act*. Bloomington: Indiana University Press.
Capra, F. 2002. *The Hidden Connections*. New York: Doubleday.
Capra, F. 1996. *The Web of Life*. New York: Anchor Books.
Carley, M., and I. Christie. 2000. *Managing Sustainable Development*. London: Earthscan.
Carson, R. 1984. *The Sense of Wonder*. New York: Harper.
Carson, R. 1962. *Silent Spring*. Boston: Houghton Mifflin.
Carter, J. 2002. "The Troubling New Face of America." *Washington Post*, September 5.
Catton, W. 1980. *Overshoot*. Urbana: University of Illinois Press.
Caudill, H. 1963. *Night Comes to the Cumberlands*. Boston: Little, Brown.
Chargaff, E. 1980. "Knowledge without Wisdom." *Harper's*, May, 41–48.
Cobb, E. 1977. *The Ecology of Imagination in Childhood*. Dallas: Spring Publications, 1993.
Costanza, R. 1987. "Social Traps and Environmental Policy." *BioScience* 37 (6): 407–12.
Crosby, A. 1997. *The Measure of Reality*. Cambridge: Cambridge University Press.
Dahl, R. 2002. *How Democratic Is the American Constitution?* New Haven: Yale University Press.
Dalai Lama. 1999. *Ethics for the New Millennium*. New York: Riverhead Books.
Daly, H. 1996. *Beyond Growth*. Boston: Beacon Press.
Daly, H. 1993. "The Perils of Free Trade." *Scientific American*, November, 50–57.
Daly, H. 1991. *Steady-State Economics*. Washington, DC: Island Press.
Daly, H. 1988. "Sustainable Development." Paper presented in Milan, Italy, March 1988.
Daly, H., ed. 1980. *Economics, Ecology, Ethics*. San Francisco: Freeman.
Daly, H., and J. Cobb. 1989. *For the Common Good*. Boston: Beacon Press.
Damasio, A. 1994. *Decartes' Error*. New York: Grossett.
Davis, D. B. 1984. *Slavery and Human Progress*. New York: Oxford University Press.
Day, C. 2002. *Spirit and Place*. Oxford: Architectural Press.
de Ségur, P.-P. 1980. *Napoleon's Russian Campaign*. Boston: Houghton Mifflin.
de Unamuno, M. 1977. *The Tragic Sense of Life in Men and Nations*. Princeton: Princeton University Press.
de Zengotita, T. 2002. "The Numbing of the American Mind." *Harpers*, April.

Dewey, J. 1990. *The School and Society*. Chicago: University of Chicago Press.
Dewey, J. 1981. "The School and Social Progress." In *The Philosophy of John Dewey*, ed. J. McDermott. Chicago: University of Chicago Press.
Dewey, J. 1954. *The Public and Its Problems*. Chicago: Swallow Press.
Diamond, S. 1981. *In Search of the Primitive*. New Brunswick, NJ: Transaction Books.
Ehrenfeld, D. 1979. *The Arrogance of Humanism*. New York: Oxford University Press.
Eiseley, L. 1979. *The Star Thrower*. New York: Harcourt Brace Jovanovich.
Eisler, R. 1987. *The Chalice and the Blade*. New York: Harper and Row.
Eliot, T. S. 1971. *T. S. Eliot: The Complete Poems and Plays*. New York: Harcourt, Brace and World.
Emerson, R. W. 1972. *Selections from Ralph Waldo Emerson*. Ed. S. Whicher. Boston: Houghton Mifflin.
Emmons, R. 2008. *Thanks*. New York: Houghton Mifflin.
Esdaile, C. 2007. *Napoleon's Wars*. New York: Viking Press.
Finley, M. I. 1980. *Ancient Slavery and Modern Ideology*. New York: Viking Press.
Fischer, L., ed. 1962. *The Essential Gandhi*. New York: Random House, 2002.
Fisher, T. 2001. "Revisiting the Discipline of Architecture." In *The Discipline of Architecture*, ed. A. Piotrowski and J. Robinson. Minneapolis: University of Minnesota Press.
Frank, R. 1999. *Luxury Fever*. Princeton: Princeton University Press.
Franke, R., and B. Chasin. 1991. *Kerala: Radical Reform as Development in an Indian State*. San Francisco: Institute for Food and Development Policy.
Frankl, V. 2004. *Man's Search for Meaning*. London: Rider.
Friedman, T. 2007. "What Was That All About?" *New York Times*, December 18, A33.
Friedmann, J. 1987. *Planning in the Public Domain*. Princeton: Princeton University Press.
Frisch, K. 1974. *Animal Architecture*. New York: Harcourt Brace Jovanovich.
Fromm, E. 1981. *To Have or to Be*. New York: Bantam Books.
Fromm, E. 1956. *The Art of Loving*. New York: Harper.
Galbraith, J. K. 2001. *The Essential Galbraith*. New York: Mariner.
Gallagher, W. 1993. *The Power of Place*. New York: Poseidon Press.
Gay, P. 1977. *The Enlightenment: An Interpretation*. New York: Norton.
Gelbspan, R. 1998. *The Heat Is On*. 2nd ed. Reading, MA: Perseus Books.
Georgescu-Roegen, N. 1971. *The Entropy Law and the Economic Process*. Cambridge: Harvard University Press.
Gleick, J. 1987. *Chaos: Making a New Science*. New York: Viking Press.
Goethe, J. W. 1952. *Goethe's Botanical Writings*. Honolulu: University of Hawaii Press.
Goodwin, B. 1994. *How the Leopard Changed Its Spots*. New York: Simon and Schuster.
Goodwyn, L. 1978. *The Populist Movement*. New York: Oxford University Press.

Gould, S. J. 1991. "Enchanted Evening." *Natural History*, September, 14.
Gowdy, J., and C. McDaniel. 1995. "One World, One Experiment." *Ecological Economics* 15:181–92.
Gray, J. G. 1984. *Re-Thinking American Education*. Middletown: Wesleyan University Press.
Gregg, R. 1935. *The Power of Nonviolence*. New York: Schocken, 1971.
Hall, P., and C. Ward. 1998. *Sociable Cities*. New York: Wiley.
Hansen, J. 2009. *Storms of My Grandchildren*. New York: Bloomsbury.
Hansen, J., et al. 2008. "Target Atmospheric CO_2: Where Should Humanity Aim?" *The Open Atmospheric Science Journal* 2:217–31.
Hardin, G. 1993. *Living within Limits: Ecology, Economics, and Population*. New York: Oxford University Press.
Hardin, G. 1986. *Filters against Folly*. New York: Viking Press.
Hardin, G. 1972. *Exploring New Ethics for Survival*. Baltimore: Penguin Books.
Havel, V. 2002. "A Farewell to Politics." *New York Review of Books*, October, 24.
Havel, V. 1992. *Summer Meditations*. New York: Knopf.
Hawken, P., et al. 1999. *Natural Capitalism*. Boston: Little, Brown.
Hawken, P. 1994. *The Ecology of Commerce*. New York: Harper.
Heilbroner, R. 1976. *Business Civilization in Decline*. New York: Norton.
Heinz Center. 2002. *The State of the Nation's Ecosystems*. New York: Cambridge University Press.
Heschel, A. J. 1951. *Man Is Not Alone: A Philosophy of Religion*. New York: Farrar, Strauss, Giroux.
Hirsch, F. 1976. *The Social Limits to Growth*. Cambridge: Harvard University Press.
Holdren, J. 2008. "One Last Chance to Lead." *Scientific American 2008 Earth 3.0 Supplement*, 20–21.
Homer-Dixon, T. 2000. *The Ingenuity Gap*. New York: Knopf.
Houghton, J. 1997. *Global Warming: The Complete Briefing*. 2nd ed. New York: Cambridge University Press, 2009.
Houghton, R. A., and G. Woodwell. 1989. "Global Climatic Change." *Scientific American*, April, 6–44.
Howard, A. 1979. *An Agricultural Testament*. Emmaus, PA: Rodale Press.
Illich, I. 1981. *Shadow Work*. Boston: Marion Boyars.
Illich, I. 1974. *Energy and Equity*. New York: Harper and Row.
Immanuel, K. 2005. "Increasing Destructiveness of Tropical Cyclones over the Past 30 Years." *Nature*, August 4, 686–88.
Intergovernmental Panel on Climate Change. 1991. *Climate Change*. New York: Oxford University Press.
Jackson, W. 1980. *New Roots for Agriculture*. Lincoln: University of Nebraska Press.
Jackson, W., and J. Piper. 1989. "The Necessary Marriage between Ecology and Agriculture." *Ecology* 70 (6): 1591–93.

James, W. 1987. "The Ph.D. Octopus." In *William James: Writings 1902–1920*, ed. B. Kuklick. New York: Library of America. (Original work published 1903.)

Janis, I. 1972. *Victims of Groupthink*. Boston: Houghton Mifflin.

Joffe, P. 2007. "The Dwindling Margin for Error: The Realist Perspective on Global Governance and Global Warming." *Rutgers Journal of Law and Public Policy* 5 (1): 89–176.

Jones, H. M. 1973. *The Age of Energy*. New York: Viking Press.

Jung, C. 1965. *Memories, Dreams, Reflections*. New York: Vintage Books.

Kammen, M. 1987. *A Machine That Would Go of Itself*. New York: Knopf.

Kaptchuk, T. 2000. *The Web That Has No Weaver*. New York: McGraw-Hill.

Keller, E. 1983. *A Feeling for the Organism*. New York: Freeman.

Kellert, S. 1996. *The Value of Life*. Washington, DC: Island Press.

Kellert, S., and E. O. Wilson, ed. 1993. *The Biophilia Hypothesis*. Washington, DC: Island Press.

Kellert, S., et al., ed. 2008. *Biophilic Design*. New York: Wiley.

Kelly, M. 2001. *The Divine Right of Capital*. San Francisco: Berrett-Koehler.

Kelman, S. 1988. "Why Public Ideas Matter." In *The Power of Public Ideas*, R. Reich, ed. Cambridge: Harvard University Press.

Kemmis, D. 1990. *Community and the Politics of Place*. Norman: University of Oklahoma Press.

Kohak, E. 1984. *The Embers and the Stars*. Chicago: University of Chicago Press.

Kohr, L. 1980. "Slow Is Beautiful." *CoEvolution Quarterly*, Summer.

Kohr, L. 1957. *The Breakdown of Nations*. New York: Dutton, 1978.

Kurlansky, M. 2006. *Nonviolence*. New York: Modern Library.

Lansing, S. 1991. *Priests and Programmers*. Princeton: Princeton University Press.

Latour, B. 1993. *We Have Never Been Modern*. Cambridge: Harvard University Press.

Lauber, V. 1978. "Ecology, Politics, and Liberal Democracy." *Government and Opposition* 13 (Spring): 2.

Layard, R. 2005. *Happiness: Lessons from a New Science*. New York: Penguin Books.

Leakey, R., and R. Lewin. 1995. *The Sixth Extinction*. New York: Doubleday.

Lear, J. 2006. *Radical Hope*. Cambridge: Harvard University Press.

Ledewitz, B. 1998. "Establishing a Federal Constitutional Right to a Healthy Environment in US and in Our Posterity." *Mississippi Law Journal* 68:2.

Leopold, A. 1991. *The River of the Mother of God and Other Essays by Aldo Leopold*. Ed. S. Flader and J. B. Callicott. Madison: University of Wisconsin Press.

Leopold, A. 1966. *A Sand County Almanac*. New York: Ballantine Books.

Leopold, A. 1953. *Round River*. New York: Oxford University Press, 1987.

Levins, R. 1998. "Looking at the Whole: Toward a Social Ecology of Health," Kansas Health Foundation, Robert H. Ebert Lecture.

Levy, M., and M. Salvadori. 1992. *Why Buildings Fall Down*. New York: Norton.
Lewis, C. S. 1947. *The Abolition of Man*. New York: Macmillan.
Lieven, A. 2002. "The Push for War." *London Review of Books* 4:19.
Linklater, A. 2002. *Measuring America*. New York: Walker.
Livingston, J. 1982. *The Fallacy of Wildlife Conservation*. Toronto: McClelland and Stewart.
Logsdon, G. 1994. *At Nature's Pace*. New York: Pantheon.
Lopez, B. 1989a. "American Geographies." *Orion Nature Quarterly* 8 (4) (September): 52–61.
Lopez, B. 1989b. *Crossing Open Ground*. New York: Vintage.
Lovelock, J. 2006. *The Revenge of Gaia*. London: Penguin Books.
Lovins, A. 1990. "The Role of Energy Efficiency." In *Global Warming*, ed. J. Leggett. New York: Oxford University Press.
Lovins, A., and H. Lovins. 1982. *Brittle Power*. Andover, MA: Brick House.
Lovins, A., et al. 2002. *Small Is Profitable*. Old Snowmass, CO: Rocky Mountain Institute.
Lyle, J. 1994. *Regenerative Design for Sustainable Development*. New York: Wiley.
Lynd, S. 1982. *The Fight against Shutdowns*. San Pedro, CA: Singlejack Books.
Margulis, L. 1998. *The Symbiotic Planet*. London: Phoenix.
Martin, C. 1992. *In the Spirit of the Earth*. Baltimore: Johns Hopkins University Press.
Marx, L. 1987. "Does Improved Technology Mean Progress?" *Technology Review* 90:1.
Maslow, A. 1971. *The Farther Reaches of Human Nature*. New York: Viking Press.
Maslow, A. 1966. *The Psychology of Science*. Chicago: Gateway Books.
Matthews, R. 1995. *If Men Were Angels*. Lawrence: University of Kansas Press.
McDonough, W., and M. Braungart. 2002. *Cradle to Cradle*. Washington, DC: North Point Press.
McDonough, W., and M. Braungart. 1998. "The Next Industrial Revolution." *Atlantic Monthly* 282 (4): 82–92.
McHarg, I. 1969. *Design with Nature*. Garden City, NY: Doubleday.
McKibben, B. 2010. *Eaarth*. New York: Times Books.
McKibben, B. 2003. *Enough*. New York: Times Books.
McNeill, J. R. 2000. *Something New under the Sun*. New York: Norton.
Meadows, D. 2001. "Economic Laws Clash with Planet's." *EarthLight* 15 (Spring).
Meadows, D. 1997. "Places to Intervene in a System." *Whole Earth* (Winter): 78–84.
Meadows, D., et al. 1972. *The Limits to Growth*. New York: Universe Books.
Meine, C. 1988. *Aldo Leopold: His Life and Work*. Madison: University of Wisconsin.
Merton, T. 1985. *Love and Living*. New York: Harcourt Brace Jovanovich.

Midgley, M. 2006. *The Essential Mary Midgley*. London: Routledge.
Midgley, M. 1990. "Why Smartness Is Not Enough." In *Rethinking the Curriculum*, ed. M. Clark and S. Wawrytko. Westport, CT: Greenwood Press.
Miles, J. 2001. "Global Requiem: The Apocalyptic Moment in Religion, Science, and Art." *CrossCurrents*, January 16. www.crosscurrents.org.
Miller, R. 1989. "Editorial." *Holistic Education Review*, Spring.
Miller, S. 1998. "A Note on the Banality of Evil." *Wilson Quarterly* 22 (4): 54–59.
Miller, W. L. 1998. *Arguing about Slavery*. New York: Vintage Books.
Moravic, H. 1988. *Mind Children*. Cambridge: Harvard University Press.
Morello, L., and S. Goodman. 2009. "House Bill Shifts Focus to Climate Change Adaptation." *New York Times*, April 1.
Morris, D. 1990. "Free Trade: The Great Destroyer." *The Ecologist* 20 (September / October): 190–95.
Mumford, L. 1970. *The Myth of the Machine: The Pentagon of Power*. New York: Harcourt Brace Jovanovich.
Mumford, L. 1966. "Utopia, the City, and the Machine." In *Utopias and Utopian Thought*, ed. Frank Manuel. Boston: Beacon Press.
Mumford, L. 1946. *Values for Survival*. New York: Harcourt, Brace.
Mumford, L. 1938. *The Culture of Cities*. New York: Harcourt Brace Jovanovich.
Mumford, L. 1934. *Technics and Civilization*. New York: Harcourt, Brace, and World.
Myers, N. 1998. *Perverse Subsidies*. Winnipeg: International Institute for Sustainable Development.
Nabhan, G., and S. Trimble. 1994. *The Geography of Childhood*. Boston: Beacon Press.
National Academy of Sciences. 1991. *Policy Implications of Greenhouse Warming*. Washington, DC: National Academies Press.
Newman, J. H. 1982. *The Idea of a University*. Notre Dame: Notre Dame University Press.
Nordhaus, W. 1990a. "Global Warming: Slowing the Greenhouse Express." In *Setting National Priorities*, ed. H. Aaron. Washington, DC: Brookings Institution.
Nordhaus, W. 1990b. "Greenhouse Economics: Count before You Leap." *Economist*, July 7.
Norgaard, R. 1987. "Risk and Its Management." Paper given to the Society of Economic Anthropology, April 1987.
Oakes, J. 1998. *The Ruling Race: A History of American Slaveholders*. New York: Norton.
Odum, H. T., and E. C. Odum. 2001. *A Prosperous Way Down*. Boulder: University Press of Colorado.
Ophuls, W. 1977. *Ecology and the Politics of Scarcity*. San Francisco: Freeman.
Orwell, G. 1981. *A Collection of Essays*. New York: Harcourt Brace Jovanovich.
Orwell, G. 1958. *The Road to Wigan Pier*. New York: Harcourt Brace Jovanovich.

Pascal, B. 1941. *Pensées*. New York: Modern Library.
Pelikan, J. 1992. *The Idea of the University: A Reexamination*. New Haven: Yale University Press.
Perrow, C. 1984. *Normal Accidents*. New York: Basic Books.
Pevsner, N. 1990. *An Outline of European Architecture*. London: Penguin Books.
Polanyi, M. 1958. *Personal Knowledge: Towards a Post-Critical Philosophy*. New York: Harper Torchbook, 1964.
Porter, T. 1995. *Trust in Numbers*. Princeton: Princeton University Press.
Postman, N. 1988. *Conscientious Objections*. New York: Knopf.
Postman, N. 1982. *The Disappearance of Childhood*. New York: Dell.
Power, T. M. 1988. *The Economic Pursuit of Quality*. Armonk, NY: Sharpe.
Price, C. 1993. *Time, Discounting, and Value*. London: Routledge.
Raffensperger, C., and J. Tickner. 1999. *Protecting Public Health and the Environment*. Washington, DC: Island Press.
Rafferty, M. 1980. *The Ozarks: Land and Life*. Norman: University of Oklahoma Press.
Redclift, M. 1987. *Sustainable Development*. New York: Methuen.
Reece, E. 2006. *Lost Mountain*. New York: Riverhead Books.
Reich, R. 1991. *The Work of Nations*. New York: Knopf.
Repetto, R. 2008. "The Climate Crisis and the Adaptation Myth." New Haven: Yale School of Forestry and Environmental Studies, Working Paper Number 13.
Ricketts, T., et al. 1999. *Terrestrial Ecoregions of North America: A Conservation Assessment*. Washington, DC: Island Press.
Rocky Mountain Institute. 1998. *Green Development*. New York: Wiley.
Rogers, C. 1961. *On Becoming a Person*. Boston: Houghton Mifflin.
Rogers, R. 1997. *Cities for a Small Planet*. London: Faber and Faber.
Rogers, R., and A. Power. 2000. *Cities for a Small Country*. London: Faber and Faber.
Roosevelt, T. 1913. *An Autobiography*. New York: Macmillan.
Roszak, T. 1992. *The Voice of the Earth*. New York: Simon and Schuster.
Rowe, S. 1990. *Home Place: Essays on Ecology*. Edmonton: NeWest.
Rudofsky, B. 1964. *Architecture without Architects*. Albuquerque: University of New Mexico Press.
Ruskin, J. 1989. *The Seven Lamps of Architecture*. New York: Dover.
Ruskin, J. 1968. *Unto This Last*. New York: Dutton.
Sachs, W., ed. 1992. *The Development Dictionary*. London: Zed Books.
Sachs, W. 1989. "A Critique of Ecology." *New Perspectives Quarterly* Spring: 16–19.
Sacks, O. 1993. "To See and Not See." *New Yorker*, May 10, 59–73.
Sahlins, M. 1972. *Stone Age Economics*. Chicago: Aldine.
Saul, J. R. 1993. *Voltaire's Bastards: The Dictatorship of Reason in the West*. New York: Vintage Books.
Schell, J. 2003. *The Unconquerable World*. New York: Metropolitan Books.

Schell, J. 2000. *The Fate of the Earth*. Palo Alto: Stanford University Press.
Schmookler, A. B. 1984. *The Parable of the Tribes*. Berkeley: University of California Press.
Schumacher, E. F. 1977. *A Guide for the Perplexed*. New York: Harper and Row.
Schumacher, E. F. 1973. *Small Is Beautiful: Economics as if People Mattered*. New York: Harper Torchbooks.
Schumpeter, J. 1962. *Capitalism, Socialism, and Democracy*. New York: Harper Torchbooks.
Schwartz, B. 1986. *The Battle for Human Nature*. New York: Norton.
Schwenk, T. 1989. *Sensitive Chaos*. London: Rudolf Steiner Press, 1996.
Scientific American. 1989. "Managing Planet Earth." *Scientific American Special Issue*, September.
Scott, J. C. 1998. *Seeing Like a State*. New Haven: Yale University Press.
Sharp, G. 2005. *Waging Nonviolent Struggle*. Boston: Porter Sargent.
Sharp, G. 1973. *The Politics of Nonviolent Action*. Boston: Porter Sargent.
Shepard, P. 1996. *Traces of an Omnivore*. Washington, DC: Island Press.
Shepard, P. 1982. *Nature and Madness*. San Francisco: Sierra Club Books.
Shepard, P. 1977. "Place in American Culture." *North American Review* Fall:22–32.
Shepard, P. 1973. *The Tender Carnivore and the Sacred Game*. New York: Scribners.
Sherman, T. 1996. *A Place on the Glacial Till*. New York: Oxford University Press.
Shuman, M., and H. Harvey. 1993. *Security without War*. Boulder: Westview.
Simmons, I. G. 1989. *Changing the Face of the Earth*. Oxford: Blackwell.
Simon, J. 1980. *The Ultimate Resource*. Princeton: Princeton University Press.
Sinsheimer, R. 1978. "The Presumptions of Science." *Daedalus* 107 (2): 23–35.
Smith, P. 1990. *Killing the Spirit*. New York: Viking Press.
Smith, P. 1984. *Dissenting Opinions*. San Francisco: North Point Press.
Snyder, G. 1990. *The Practice of the Wild*. San Francisco: North Point Press.
Snyder, G. 1974. *Turtle Island*. New York: New Directions.
Solomon, S., et al. 2009. "Irreversible Climate Change Due to Carbon Dioxide Emissions." *Proceedings of the National Academy of Sciences* 106 (6): 1704–9.
Speth, J. G. 1989. "A Luddite Recants." *Amicus Journal* (Spring).
Speth, J. G. 2008. *The Bridge at the End of the World*. New Haven: Yale University Press.
Steindl-Rast, D. 1984. *Gratefulness, the Heart of Prayer*. Mahwah, NJ: Paulist Press.
Steiner, G. 2001. *Grammars of Creation*. London: Faber and Faber.
Stern, N. 2009. *The Global Deal*. New York: PublicAffairs.
Stern, N. 2007. *The Economics of Climate Change*. New York: Cambridge University Press.
Stone, C. 1974. *Should Trees Have Standing?* Los Altos, CA: William Kaufman.
Sturt, G. 1984. *The Wheelwright's Shop*. New York: Cambridge University Press.

Sunstein, C. 2001. *Designing Democracy*. New York: Oxford University Press.
Sunstein, C. 1993. *The Partial Constitution*. Cambridge: Harvard University Press.
Suzuki, D. 1997. *The Sacred Balance*. Amherst, NY: Prometheus Books.
Tainter, J. 1988. *The Collapse of Complex Societies*. Cambridge: Cambridge University Press.
Tasch, W. 2008. *Slow Money*. White River Junction, VT: Chelsea Green.
Teune, H. 1988. *Growth*. Newbury Park, CA: Sage.
Thompson, D. 1992. *On Growth and Form*. New York: Dover.
Thompson, W. I. 1987. "The Cultural Implications of the New Biology." In *GAIA: A Way of Knowing*, ed. W. I. Thompson. Great Barrington, MA: Lindisfarne Press.
Thornton, J. 2000. *Pandora's Poison*. Cambridge: MIT Press.
Tidwell, M. 2003. *Bayou Farewell*. New York: Vintage Books.
Todd, J. 1991. "Ecological Engineering, Living Machines and the Visionary Landscape." In *Ecological Engineering for Wastewater Treatment*, ed. C. Etnier and B. Guterstam. Gothenburg, Sweden: Boksgaden.
Todd, J. 1990. "Living Machines." *Annals of Earth* 8:1.
Todd, J., and N. Todd. 1984. *Bioshelters, Ocean Arks, City Farming: Ecology as the Basis for Design*. San Francisco: Sierra Club Books.
Toulmin, S. 2001. *Return to Reason*. Cambridge: Harvard University Press.
Toulmin, S. 1990. *Cosmopolis*. Chicago: University of Chicago Press.
Tsui, E. 1999. *Evolutionary Architecture*. New York: Wiley.
Tuan, Y. F. 1977. *Space and Place: Their Perspective of Experience*. Minneapolis: University of Minnesota Press.
Tuchman, B. 1984. *The March of Folly*. New York: Knopf.
Tucker, M. E. 2003. *Worldly Wonder*. Chicago: Open Court.
Turnbull, C. 1983. *The Human Cycle*. New York: Simon and Schuster.
Turner, F. 1980. *Beyond Geography: The Western Spirit against the Wilderness*. New York: Viking Press.
Turner, J. 1996. *The Abstract Wild*. Tucson: University of Arizona Press.
Union of Concerned Scientists. 1992. "Warning to Humanity." Boston: Union of Concerned Scientists.
U.S. Geological Survey. 1998. *Status and Trends of the Nation's Biological Resources*. 2 vol. Washington, DC: Superintendent of Documents, U.S. Government Printing Office.
Van der Ryn, S., and S. Cowan. 1996. *Ecological Design*. Washington, DC: Island Press.
Victor, D., et al. 2009. "The Geoengineering Option." *Foreign Affairs* 88 (2): 64–76.
Vitruvius. 1960. *The Ten Books of Architecture*. New York: Dover.
Wackernagel, M., and W. Rees. 1996. *Our Ecological Footprint*. Philadelphia: New Society Publishers.
Wann, W. 1990. *Biologic*. Boulder, CO: Johnson Books.

Weart, S. 2003. *The Discovery of Global Warming*. Cambridge: Harvard University Press.
Weaver, R. M. 1948. *Ideas Have Consequences*. Chicago: University of Chicago Press, 1984.
Webb, W. P. 1975. *The Great Frontier*. Austin: University of Texas Press.
Wells, H. G. 1946. *Mind at the End of Its Tether*. New York: Didier.
Wheeler, W. M. 1962. In *The Practical Cogitator*, 3rd ed., ed. C. Curtis and F. Greenslet, 226–29. Boston: Houghton Mifflin.
White, L. 1967. "The Historical Roots of Our Ecologic Crisis." *Science* (March): 155.
Whitehead, A. N. 1974. *Science and Philosophy*. New York: Philosophical Library.
Whitehead, A. N. 1967. *The Aims of Education*. New York: Basic Books.
Whybrow, P. 2005. *American Mania: When More Is Not Enough*. New York: Norton.
Whyte, W., and K. Whyte. 1988. *Making Mondragon*. Ithaca: Cornell University.
Wiesel, E. 1990. Remarks before the Global Forum, Moscow, January.
Wilkinson, R. 1973. *Poverty and Progress*. New York: Praeger.
Willis, D. 1995. *The Sand Dollar and the Slide Rule*. Reading, MA: Addison-Wesley.
Wilson, E. O. 2002. *The Future of Life*. New York: Knopf.
Wilson, E. O. 1992. *The Diversity of Life*. Cambridge: Harvard University Press.
Wilson, E. O. 1984. *Biophilia*. Cambridge: Harvard University Press.
Wilson, P. 1988. *The Domestication of the Human Species*. New Haven: Yale University Press.
Wittfogel, K. 1956. *Oriental Despotism*. New Haven: Yale University Press.
World Commission on Environment and Development. 1987. *Our Common Future*. New York: Oxford University Press.
World Resources Institute. 1985. *Tropical Forests: A Call for Action*. Washington, DC: World Resources Institute.
Worster, D. 1987. "The Vulnerable Earth." *Environmental Review* 11 (Summer): 2.
Worster, D. 1985. *Rivers of Empire*. New York: Pantheon.
Wright, F. L. 1993. "An Organic Architecture." In *Frank Lloyd Wright Collected Writings*, ed. B. Pheiffer, vol. 3, 299–334. New York: Rizzoli International Publications.
Zakaria, F. 2003. *The Future of Freedom*. New York: Norton.
Zamoyski, A. 2004. *Moscow 1812*. New York: Harper Perennial.

Permissions

Many of these essays were previously published in books and periodicals and have been mildly altered for this collection. We express grateful acknowledgment for permission to reprint the following selections:

No copyright claim is made for "Gratitude," a variation on a commencement speech made by the author, which, by his wishes, remains in the public domain.

"Loving Children: A Design Problem" was previously published as "Loving Children: The Political Economy of Design" in David W. Orr, *The Nature of Design: Ecology, Culture, and Human Intervention* (Oxford, UK: Oxford University Press, 2004). Reprinted by permission of Oxford University Press, Inc.

"Place as Teacher" was previously published as "Origins of the Ideas" and the following chapters were reprinted from David W. Orr, *Design on the Edge: The Making of a High-Performance Building* (Cambridge, MA: MIT Press, 2006): "The Origins of Ecological Design" and "The Design Revolution: Notes for Practitioners." Reprinted by permission of the MIT Press.

"The Liberal Arts, the Campus, and the Biosphere" was previously published as "The Liberal Arts, the Campus, and the Biosphere: An Alternative to Bloom's Vision of Education" and the following chapters were reprinted from David W. Orr, *Ecological Literacy: Education and the Transition to a Postmodern World* (Albany: State University of New York Press, 1992): "The Problem of Sustainability," "Two Meanings of Sustainability," "Ecological Literacy," and "Place and Pedagogy." Reprinted by permission of SUNY Press.

"All Sustainability Is Local: New Wilmington, Pennsylvania" was previously published as "A World That Takes Its Environment Seriously" and the following chapters were reprinted, with some modifications, from David W. Orr, *Earth in Mind: On Education, Environment, and the Human Prospect*, 10th anniversary edition (1994; updated, Washington, DC: Island Press, 2004): "The Problem of Education," "Some Thoughts on Intelligence," "Reflections on Water and Oil," and "What Is Education For?" Reprinted by permission of Island Press.

We gratefully acknowledge permission from Wiley-Blackwell to reprint in this book, in somewhat altered form, 18 articles by David W. Orr originally published in the journal *Conservation Biology* between the years 1988 and 2008.

About David W. Orr

David W. Orr is the Paul Sears Distinguished Professor of Environmental Studies and Politics and Special Assistant to the President of Oberlin College. He is the recipient of six honorary degrees and other awards, including the Millennium Leadership Award from Global Green, the Bioneers Award, the National Wildlife Federation Leadership Award, and a Lyndhurst Prize acknowledging "persons of exceptional moral character, vision, and energy." He has been a scholar in residence at five universities and has lectured at hundreds of colleges and universities throughout the U.S. and Europe.

His career as a scholar, teacher, writer, speaker, and entrepreneur spans fields as diverse as environment and politics, environmental education, campus greening, green building, ecological design, and climate change. He is the author of seven books and coeditor of three others. His first book, *Ecological Literacy* (SUNY, 1992), was described as a "true classic" by Garrett Hardin. A second book, *Earth in Mind* (Island Press, 1994/2004), is praised by people as diverse as biologist E. O. Wilson and writer, poet, and farmer Wendell Berry. Both books are widely read and used in hundreds of colleges and universities.

In 1987 he organized studies of energy, water, and materials use on several college campuses that helped to launch the green campus movement. In 1989 Orr organized the first ever conference on the effects of impending climate change on the banking industry. Cosponsored by then governor Bill Clinton, the conference featured prominent bankers throughout the mid South and leading climate scientists, including Stephen Schneider and George Woodwell.

In 1996 he organized the effort to design the first substantially green building on a U.S. college campus. The Adam Joseph Lewis Center was later named by the U.S. Department of Energy as "One of Thirty Milestone Buildings in the 20th Century" and by the *New York Times* as the most interesting of a new generation of college and university buildings. The Lewis Center purifies all of its wastewater and is the first college building in the U.S. powered entirely by sunlight. But most important, it became a laboratory in sustainability that is training some of the nation's brightest and most dedicated students for careers in solving environmental problems. The story of that building is told in two books, *The Nature of Design* (Oxford, 2002), which Fritjof Capra called "brilliant," and a second, *Design on the Edge* (MIT, 2006), which architect Sim van der Ryn describes as "powerful and inspiring."

Orr's political writings appear in *The Last Refuge: Patriotism, Politics, and the Environment in an Age of Terror* (Island Press, 2004) and articles such as "The Imminent Demise of the Republican Party" (www.commondreams.org) written in January of 2005.

In an influential article in the *Chronicle of Higher Education* in 2000, Orr proposed the goal of carbon neutrality for colleges and universities and subsequently organized and funded an effort to define a carbon neutral plan for his own campus at Oberlin. Seven years later, hundreds of colleges and universities, including Oberlin, had made that pledge.

He was instrumental in initiating and funding a 2-year $1.2 million collaborative project to define a 100-days climate action plan for the Obama administration (www.climateactionproject.com). He helped, as well, to launch a project with prominent legal scholars across the U.S. to define the legal rights of posterity in cases where the actions of the present generation might deprive posterity of "life, liberty, and property." He is also active in efforts to stop mountaintop removal in Appalachia and develop a new economy based on ecological restoration and wind energy. He is presently engaged in a decade-long effort to develop an integrated model of resilient, post-carbon development in his hometown of Oberlin. All of this while working as a high-level adviser to four grand, grandchildren. He is the author of *Down to the Wire: Confronting Climate Collapse* (Oxford University Press, 2009).

Index

NOTE: page numbers followed by "n" refer to footnotes.